D0983712

Sir Charles Lyell's Scientific Journals on the Species Question

Yale Studies in the History of
Science and Medicine, 5

Sir Charles Lyell, Bart., 1797–1875

Sir Charles Lyell's Scientific Journals on the Species Question

Edited by
Leonard G. Wilson

New Haven and London, Yale University Press, 1970

Library of Congress catalog card number: 77–99848
Standard book number: 300–01231–4

Designed by Helen Frisk Buzyna,
set in Baskerville type,
and printed in the United States of America by
The Colonial Press Inc.,
Clinton, Massachusetts.

Distributed in Great Britain, Europe, Asia, and
Africa by Yale University Press Ltd., London; in
Canada by McGill University Press, Montreal; and
in Latin America by Centro Interamericano de Libros
Académicos, Mexico City.

To the family of

Charles Antony Lyell V.C.

2d Baron Lyell of Kinnordy
and great, great nephew of Sir Charles Lyell

Killed in action in Tunisia
27 April 1943

Preface

In 1961, while studying the correspondence and notebooks of Sir Charles Lyell at Kinnordy House, Kirriemuir, Scotland, through the courtesy and hospitality of Lord Lyell of Kinnordy, I discovered this series of seven notebooks in which Sir Charles Lyell had entered his notes and reflections on the question of the transmutation of species. These notebooks, or "scientific journals" as Lyell called them, were of obvious historical interest, and I decided immediately that they deserved publication. However, I was not able to have them transcribed and edited immediately, and, when the task of transcription and editing was begun, it proved to be prolonged and laborious.

These journals record the private thoughts of one of the central figures in the revolution of scientific thought brought about by the publication of Charles Darwin's *Origin of Species*. Darwin's theory of the transmutation of species by natural selection transformed the science of biology and provoked a religious crisis by destroying at once the credibility of both the Biblical account of the creation of man and the argument for the existence of God from the presence of design in nature. The latter, the underlying argument of natural theology, had permeated both scientific and religious thought in Victorian times.

The journals show that, as early as 1855, Lyell had come independently to consider seriously the scientific possibility of the transmutation of species, although then, and until much later, he remained convinced that the weight of evidence was opposed to transmutation. They also record Lyell's conversation with Darwin on April 16, 1856, when for the first time Darwin described to Lyell his theory of natural selection. The journals include copies of portions of a number of Lyell's letters to Darwin and others. The copies of his letters to Darwin are particularly important, because apart from a very few letters, I have not been able to locate Lyell's letters to Darwin and suspect that this immense correspondence, so important for the history of science, has not survived. Thus, these few copies of Lyell's letters to Darwin become the more important for their rarity. In his references to books, articles, and lectures throughout the journals, Lyell reveals the broad and diverse aspects of geology, palaeontology, natural history, and anthropology that he brought to bear on the species question. Lastly, these journals record the depth of Lyell's reassessment of his own scientific and religious views. His acceptance of the evolution of man by natural selection was reluctant and painful, yet he had always realized that, if the theory as a whole were true, it must apply to man and must explain the origin of the human mind and soul.

In preparing these journals for publication, I have treated them as historical documents whose publication, once undertaken, should be definitive. Each journal is published in its entirety, even when portions may seem so trivial and irrelevant as to cry out to be omitted. The difficulty with omission would be that the reader would remain uncertain of what had been omitted and of the criteria of judgment that had been exerted in making omissions. Any exact scholar would then still feel obliged to consult the original documents.

I have punctuated the text and corrected obvious spelling slips, but have retained archaic spellings and abbreviations. The edition tries to remain as faithful as possible to the manuscript notes without including errors that the reader might attribute as readily to the printer as to Lyell. For the sake of clarity of presentation, however, I have taken certain liberties with archaic superscripts, capitalization, and abbreviations in the headings. In addition, where Lyell jotted down dates after headings, these have been placed above the headings for the sake of clarity and consistency. The square bracketed [ ] numbers in the text refer to the consecutive manuscript pages of the journals. In some instances Lyell left pages blank for no clear reason. In the first journal he had left blank pages at the beginning and used these to complete the index for that journal because he did not have sufficient room at the end. As a result, the page numbers for the index of Journal I change abruptly, from pages at the end of the notebook to pages at the beginning.

I am grateful to Lord Lyell of Kinnordy for permission to

have these journals microfilmed and published. I am also grateful to his mother, the Lady Lyell, for much courteous and kindly hospitality during my several visits to Kinnordy House to work on these and other Lyell papers.

Mrs. Elizabeth Musgrave, my secretary while I was at Yale University School of Medicine, transcribed the largest portion of the journals, but significant portions were also transcribed by Mrs. Beatrice Fiorino at Yale and by Mrs. Leontine M. Hans, my research associate at the University of Minnesota. Mrs. Hans also read the whole transcript together with me to check it, line by line, against the original notebooks, and searched out many of Lyell's references identified in the footnotes. These were often remarkably obscure and difficult to find. In the searching of references we received unfailing help from the staff of the Bio-Medical Library of the University of Minnesota, most especially from Mrs. Mary Mueller, who attended to our numerous and frequently perplexing interlibrary loan requests. The final manuscript was typed by my secretary, Mrs. Helen Mammen. I am deeply grateful to all these persons. Without their long-continued, painstaking, and conscientious help this edition could not have been prepared.

The research on which this work is based has been supported by two grants from the National Science Foundation to Yale University, NSF Grant No. G–16458 and NSF Grant No. GS–1277, and by a grant from the National Science Foundation to the University of Minnesota, NSF Grant No. GS–1819. These grants paid a portion of my travel expenses in going to Scotland, the cost of microfilming the notebooks and of preparing prints from the microfilms, part of the cost of preparing the transcript, and other, lesser expenses.

I am deeply grateful to Yale University, where I taught and worked from 1960 until 1967, and to the University of Minnesota, where I came in July 1967 and where the work of editing and annotating these journals and the writing of the introduction has been done. Numerous friends, associates, and students at both these universities have been a steady source of encouragement, help, and support.

L. G. W.

Minneapolis, Minnesota
1968

Contents

Manuscript Collections Referred to

Darwin mss., Cambridge. These are the personal papers and books of Charles Darwin, including many letters to him, at the Cambridge University Library, Cambridge, England.

Darwin-Lyell mss., APS library, Philadelphia. These papers include most of the letters written by Charles Darwin to Sir Charles Lyell plus many letters to Lyell from other scientists, particularly geologists.

Hooker mss. The papers of Sir William Jackson Hooker and Sir Joseph Dalton Hooker in the library of the Royal Botanic Garden, Kew, England.

Huxley mss. The papers of Thomas Henry Huxley, including letters received by him, at the Lyon Playfair Library, Imperial College of Science and Technology, London, S.W.7, England.

Kinnordy mss. Letters, journals and notebooks of Sir Charles Lyell and other members of the Lyell family belonging to the Rt. Hon. Lord Lyell of Kinnordy, Kinnordy House, Kirriemuir, Angus, Scotland.

Lyell mss., Edinburgh. A large body of correspondence to Sir Charles Lyell at Edinburgh University Library, Old College, South Bridge, Edinburgh 8, Scotland.

Mantell mss. The papers of Gideon Algernon Mantell (1790–1852), surgeon of Lewes, Sussex, England, including correspondence received by him, at the Alexander Turnbull Library, Wellington, New Zealand.

Introduction

These notebooks record the development of Sir Charles Lyell's ideas concerning the origin of species from 1855 to 1861, years that were for him a period of profound, prolonged, and painful reassessment of the possibility of the transmutation of species.

Ever since 1832 when, in the second volume of his *Principles of Geology*,[1] he had discussed Lamarck's theory with devastating criticism, Lyell had been committed to the view that species were real and stable entities that might be driven to extinction but could not be altered. His own work on the succession of Tertiary formations had acquainted Lyell with thousands of species of Tertiary fossil shells, and he saw that in the long course of geological time through the Tertiary period, successive fossil faunas did not change abruptly, but gradually. From one period to another, some of the species became extinct, to be replaced by new ones, but a larger proportion of species survived. Over a series of periods, however, the relentless extinction of old species and their replacement by new ones gradually produced an almost complete change of fauna.

Lyell had interpreted the extinction of species as an inevitable consequence of two factors: first, of the changes in physical conditions, both in particular localities and over the whole earth's surface, brought about by the steady action of geological processes, and second, of the accompanying fluctuations in the populations of other species on which the life of a particular species depended or by which it was menaced. Lyell was convinced that species were real entities, in part by the fact that in order to live successfully a species had to be adapted to a particular mode of life, a particular set of physical conditions, and a particular set of relationships with surrounding species. Its elaborate and exact adaptations to enable it to live in one

1. Charles Lyell, *Principles of Geology*, 3 vols., London, Murray, 1830–33.

enormously complex set of conditions necessarily made it unfit for life in any other set of conditions. Furthermore, the very exactness of its adaptations meant that no one character of a species could be changed arbitrarily without throwing the interdependent complex of adaptations out of balance. Lyell, therefore, thought that before a species could modify itself so as to be adapted to new conditions, it would be displaced by other species already fitted to the new conditions and thereby would be rendered extinct.

Lyell's discussion of the relationship of species to their environments was deeply influenced by his reading of Lamarck's *Philosophie Zoologique* (1809) in 1827, just as his discussion of the reality and fixity of species was developed in reaction to Lamarck's view of their artificiality and transformability. Lamarck argued that the structures and habits of plants and animals are produced by their adaptation to the environment. In a sense Lyell agreed, because he held that both structure and peculiarities of habit existed in order to bring about the adaptation of a species to its environment. Yet he differed fundamentally from Lamarck in that he thought that these adaptations were a manifestation of design in nature. They represented a particular detail in the overall plan of creation—a plan that comprehended the immensities of both space and time and included in its provisions the entire succession of geological epochs and changes.

Lyell's view of the continued and uniform succession of species through geological time was not evolutionary. His scientific attitude of mind was oriented fundamentally to the eighteenth century and to the world view suggested by Isaac Newton's *Principia.* In this view, the world was a vast ordered scheme, its phenomena determined everywhere and at all times by natural laws. These laws prevailed throughout the world because they had been established in the beginning by God who had created the world.

The Newtonian world view was basically ahistorical in that it considered the natural order of the world to continue unchanged once it had been established by its creator. The planets had continued to revolve about the sun from the moment of creation, held in their elliptical orbits by the unchanging laws of gravitation and inertia. Newton's system did not allow for a succession of changes on the earth's surface nor was he aware that the earth might have had a history prior to the present appearance of things.

For Charles Lyell, considering in his mind's eye the natural order of the world in the 1820s, the Newtonian scheme was no longer adequate. Newton's view of the world had been timeless; Lyell had to envision a natural order that would allow for a long succession of changes and a process of historical development on earth. This necessity was imposed on him by the great development in the science of geology, which had begun in the late seventeenth century, had continued vigorously throughout the eighteenth century, and was in full flood in the early nineteenth century. The development of geology had shown two things: first, that the stratified rocks of the earth's surface, which had been laid down horizontally under water, were not of one age but represented a series of formations deposited at different times; second, that successive formations must often have been separated by wide periods of time and great events, because they contained the remains of plants and animals so different as to be referred to different creations, and because the strata originally horizontal beneath the water were now not only elevated to form dry land but were often found in an inclined or even vertical position. The elevation and displacement of the strata on such a grand scale suggested catastrophic disturbances and shakings of the earth's surface.

In 1795 in his *Theory of the Earth,* James Hutton had argued for a long succession of gradual and uniform changes in the history of the earth.[2] Hutton had been deeply impressed by the series of changes represented by elevated rock strata which had been formed from sediments accumulated in the sea over a long period of time. The sediments must have originated from the gradual wearing down of some preexisting land, and every rock stratum was evidence for the existence of the land from which its sediments had come. Hutton had been struck particularly by unconformities between strata such as that he had found at the Siccar Point near St. Abb's Head, Berwick-

2. James Hutton, *Theory of the Earth with proofs and illustrations,* 2 vols., Edinburgh, Cadell and Davies, 1795.

shire. There, slightly inclined beds of Old Red Sandstone rest on the edges of vertical strata of an older (Silurian) schist, which Hutton referred to as Primary, and the junction is revealed, and the edges of the schist laid bare, at low tide. The vertical strata of the schist are clearly ripple marked, a detail that confirms the fact that they were laid down under water in a horizontal position. There they must have been accumulated from the detritus of preexisting land over a very long period of time, because the series of strata is very thick. They must gradually have been hardened to form solid rock and then elevated from the sea bottom to form land and not only elevated but raised from their horizontal position to their present vertical one. In this position, they were subjected to erosion, which wore down the edges of the strata to a uniform, smooth surface. Then, the vertical schist strata again sank beneath the sea, and the sediments that were to form the Old Red Sandstone strata gradually accumulated over them and hardened into rock. A second elevation raised the Old Red Sandstone strata to form the present land of Berwickshire, and over a long period of time the waves of the North Sea wore away the sandstone to reveal its junction with the schist at the base of the cliffs of the Siccar Point. John Playfair, Hutton's friend and biographer, recorded his astonishment that Hutton, by his genius, could demonstrate such a long series of changes at this one spot.

Another of Hutton's observations that had influenced his thinking was the discovery of granite intrusions in the bed of the Tilt, a small stream in Glen Tilt in the Highlands of Scotland. Because it was found in the core of mountains and was a crystalline rock, granite had been thought to represent the original and primitive earth surface on which the stratified rocks had later been deposited. If that were true, then granite was older than any of the stratified rocks and represented the beginning of geological history. Hutton, however, was able to prove that the Glen Tilt granite had forced its way into the surrounding stratified schist in a molten condition, because in the vicinity of the granite dykes the white schist strata were darkened and altered in a way that could only be brought about by enormous heat. The granite was, therefore, not older than the stratified rock but younger, and the oldest known rocks were consequently stratified. But each stratified rock presupposed the presence of previously existing land and the long preexistence, too, of a steady process of the wearing down of land and the deposition of sediments in the sea. It was this vista of successive erosion and deposition of sediments, their consolidation into rock, and subsequent elevation to form land that Hutton saw stretching endlessly into the remote past and that he expressed in 1785 by his statement that in the earth "we find no vestige of a beginning—no prospect of an end." [3]

Hutton was not on the whole very familiar with the fossils of the rock strata he studied, and he was aware of their importance only insofar as he knew that certain strata possessed characteristic marine shells that were conclusive evidence that the strata had been laid down beneath the sea. He did not know that, during the past history of the earth, there had appeared a series of populations of animal species each of which had, in succession, become extinct.

In the late eighteenth century, a number of remarkable fossil animals were discovered in both Europe and America. In 1796, Georges Cuvier showed that the fossil Mammoth of Siberia was an elephant, but belonged to a species quite distinct from either of the two living species—the Asian and the African elephants. In 1806, he also showed that the animal whose bones had been found so extensively at Big Bone Lick in Kentucky and elsewhere in America was not an elephant but belonged to a distinct though closely related genus, which he named the Mastodon. In North America, the large Megalonyx had also been discovered by Thomas Jefferson in Virginia, and Cuvier showed its relationship both to the gigantic extinct Megatherium of South America and to the much smaller living species of South American tree sloths.

The discovery of the Mammoth, the Mastodon, the Megalonyx, and Megatherium revealed the existence of a group of gigantic and wide-ranging extinct animals that seemed to have lived at a period so recent that naturalists at first thought they still might be alive in some remote district. In addition to these

3. James Hutton, "Theory of the Earth," *Trans. Roy. Soc. Edinburgh,* 1788, *1*, 209–304.

animals, Cuvier found in the gravel and alluvial deposits of the Paris basin a number of other fossil animals, the fossil skeletons of a daman (or *Hyrax*), rhinoceros, hippopotamus, tapir, elephant, and a species of Mastodon much smaller than that of North America. Taken together, these fossils represented a rich and varied fauna that had lived in Europe just before the modern period.

All the fossils that Cuvier described from the gravel and alluvium belonged to the family *Pachydermata,* and, although they were all of species distinct from any living ones, he could, with the exception of Mastodon, assign them to living genera. This was not the case with a number of the fossil bones that he obtained from the gypsum quarries of Montmartre. Among these, he found pachyderms unrepresented by any living genus, and he had to establish two new genera, Palaeotherium and Anoplotherium, for them. Together with these fossil pachyderms, he found the bones of a number of fossil carnivores, an opossum (a genus now confined to America), and various birds, reptiles, and fishes. Again he had revealed a fossil fauna of astonishing richness that was nonetheless completely different as to species, and in several instances as to genus, from the fossil fauna of the gravel and alluvium.

Cuvier was also aware of the numerous fossil crocodiles and other reptiles found in the chalk quarries of the mountain of St. Pierre at Maestricht, Belgium. These represented another distinct and still older fauna that was described by Faujas de Saint Fond in 1803. The discovery of fossil crocodiles in the chalk beds at Maestricht and elsewhere had been particularly important because Napoleon's expedition to Egypt in 1798 had permitted French scientists, and particularly Étienne Geoffroy Saint-Hilaire, to make a careful study of the Nile crocodile and later to compare the fossil crocodile skeletons with it. This comparison showed that the fossil crocodiles were of species quite distinct from the living ones, but it also raised the question of the relationship of the living to the fossil crocodiles. Étienne Geoffroy Saint-Hilaire, who was sympathetic to the ideas of Lamarck, concluded that the fossil crocodile species were ancestral to the living ones. He later extended this idea to cover the general relationship between fossil populations of species and their modern living counterparts.

In 1812, Cuvier collected his various memoirs on different fossil animals, published during the preceding years, and issued them together as his *Recherches sur les ossemens fossiles*.[4] He included an introductory essay, "Discours sur les révolutions de la surface du globe," in which he sought to show that the succession of distinct fauna that had existed during the past history of the earth (revealed largely by his own researches) was to be accounted for by a series of great revolutions, or catastrophic disturbances, of the surface of the earth that destroyed each fauna in turn. In the Tertiary strata of the Paris basin, which included the gypsum beds of Montmarte, there was an alternation of freshwater and marine sediments, each accompanied by an abrupt change in fossil life, a fact that suggested to Cuvier that these catastrophes may have taken the form of an invasion of the sea over the land.

Also in 1812, there was discovered at Lyme Regis in Dorsetshire, in beds of the Blue Lias, one of the series of Secondary formations lying beneath the Chalk in England, the skeleton of a large reptile. It was at first taken to be a crocodile, but after several additional skeletons had been found and carefully compared, the Rev. William Daniel Conybeare of Oxford decided in 1820 that this animal was essentially a lizard with paddle-like limbs adapted for life in the sea, and he named it Ichthyosaurus. In 1821 Gideon Mantell, a surgeon of Lewes in Sussex, discovered the remains of a new and gigantic fossil reptile in the Wealden (Jurassic), a formation lying just below the Chalk. The astonishing character of this fossil animal was that it had been a land animal of enormous size, and its teeth indicated that it had lived by eating plants. In 1824, on account of the resemblances of the fossil skeleton to that of the small modern iguana of South America, Mantell named the fossil animal Iguanodon.

In 1823, an almost perfect skeleton of yet another large fossil reptile was found in the Blue Lias at Lyme Regis. Fragmentary remains of this animal had been found earlier, and because it

4. Georges Cuvier, *Recherches sur les ossemens fossiles,* 4 vols., Paris, 1812.

appeared to be intermediate between the Ichthyosaurus and the crocodile, the Rev. Mr. Conybeare, who described it, named it Plesiosaurus. The fossil was especially remarkable for the great length of its neck, which was equal to the combined length of the body and tail. Lyell, then a young law student in London with an enthusiasm for geology, was particularly impressed by Plesiosaurus. In February 1824, he wrote to Gideon Mantell:

> The new animal is a very perfect skeleton, and a prodigy, for it has forty cervical vertebrae, whereas existing quadrupeds range from seven to nine, reptiles from three to nine. Aves reach no higher than twenty, the swan being the maximum. What a leap have we here, and how many links in the chain will geology have to supply.[5]

This passage suggests that in 1824 Lyell, in harmony with the thought of his time, took for granted the existence of the "scale of nature" or "chain of being." He also assumed that fossil discoveries were helping to fill in missing portions of the chain. What is also evident is that he was finding that fossils sometimes added surprisingly to the length and complexity of the chain. In 1824, the Rev. William Buckland published his description of Megalosaurus, another enormous fossil reptile whose remains were found in the Oolite at Stonesfield, northwest of Oxford, and in the second edition of his *Recherches sur les ossemens fossiles* (1821–24) Georges Cuvier had described the Pterodactyls, or fossil flying reptiles, of the Secondary period—both unexpected and astonishing links on the great chain of being.

In 1824, as one of the secretaries of the Geological Society of London, Lyell helped to edit the first volume of the new series of its *Transactions* in which the descriptions of Plesiosaurus and Megalosaurus were published. In 1826, he reviewed in a long article for the *Quarterly Review* the contents of this volume, and he took the occasion both to consider the state of

5. Charles Lyell to Gideon Mantell, February 17, 1824, Mantell mss. In K. M. Lyell, *Life, Letters and Journals of Sir Charles Lyell,* 2 vols., London, Murray, 1881, *1,* 151.

geology at the time and to survey the exciting array of then recent fossil discoveries. He was also in a sense reviewing the third edition of Cuvier's *Ossemens Fossiles,* which had appeared in 1825, because he had frequently to refer to it for information about new fossils.[6]

The effect of the flood of fossil discoveries was, Lyell saw, to give a lengthened perspective of the history of the world and a far broader view of the plan of creation. Of the fossils he wrote:

> None of these fossil plants or animals appear referable to species now in being, with the exception of a few imbedded in the most recent strata; yet they all belong to genera, families, or orders established for the classification of living organic productions. They even supply links in the chain, without which our knowledge of the existing systems would be comparatively imperfect. It is therefore clear to demonstration, that all, at whatever distance of time created, are parts of one connected plan. They have all proceeded from the same Author, and bear indelibly impressed upon them the marks of having been designed by One Mind.[7]

The study of geology thus immensely enlarged the scope of natural theology. Instead of one fixed and stable plan of creation, manifested in the present and having existed unchanged since its first appearance, the present appearance of nature became for Lyell merely a fragment of the overall plan that required the full immensity of time for its complete unfolding. He saw a "connection of the course of things which come within our view, with the past, the present and the future," which seemed to him best expressed by Bishop Butler in his *Analogy of Religion.*[8] Lyell quoted Butler's statement that

> We are placed in the middle of a scheme, *not a fixed but a progressive one,* every way incomprehensible—incomprehen-

6. Georges Cuvier, *Recherches sur les ossemens fossiles ou l'on rétablit les caractères de plusieurs animaux dont les revolutions du globe ont détruit les espèces,* 3d ed., 5 vols. in 7, Paris and Amsterdam, Dufour and d'Ocagne, 1825.

7. Charles Lyell, "Art. IX.—*Transactions of the Geological Society of London.* Vol. i, 2d. Series. London, 1824," *Quart. Rev.,* 1826, *34,* 507–40, p. 538.

8. Ibid., p. 539.

> sible in a manner equally with respect to what has been, what now is and what shall be hereafter.[9]

In the *Analogy of Religion,* Butler argued that we can have no direct knowledge of the spiritual world, which lies entirely beyond our experience, except by analogy with the world which we know. For Lyell, this concept of analogy seemed to apply directly to the understanding of the history of the earth. We could only learn of conditions on the earth's surface during past ages by analogy with conditions existing at present. He had already found in his early investigations of rocks, and he was to find again and again, evidence that suggested many analogies between past and present conditions on the earth's surface. He saw analogies between the ripple marks of a modern beach and ancient ripple-marked sandstones, between the assemblage of the green alga Chara, freshwater shells, and Caddis-fly larval cases in a freshwater loch in Scotland and the same assemblage of organisms found fossil in a great depth of ancient freshwater marl in Auvergne, and so on. These analogies helped to confirm Lyell's belief that he stood in the midst of a natural order that, although the same causes might produce different effects at different times, had existed uniformly throughout time.

Lyell and the Concept of Progressive Development

In his 1826 article in the *Quarterly Review,* Lyell was prepared to admit that the fossils so far discovered did indicate some measure of progressive change through time. He wrote:

> An opinion was entertained soon after the commencement of the study of organic remains, that in ascending from the lowest to the more recent strata, a gradual and progressive scale could be traced from the simplest forms of organization to those more complicated, ending at length in the class of animals most related to man. And such is still the general inference to be deduced from observed facts, though some

9. Ibid., Lyell's italics.

recent exceptions to this rule are too well authenticated to justify an implicit reliance on such generalizations.[10]

The principal exception he probably had in mind was the discovery in 1814 of the jaw of a small mammal in the Stonesfield Slate, a series of beds lying low down in the Oolite formation of Oxfordshire. Two jaws had been obtained in 1814 by William John Broderip; he had given one to the Rev. William Buckland and unfortunately misplaced the other one. In 1818, Buckland showed the fossil to Georges Cuvier at Oxford during a visit that Cuvier made to England. On rapid examination, Cuvier identified the jaw as that of a Didelphys, or opossum, that is, of a marsupial mammal, and in 1824, Buckland referred to this identification in his description of Megalosaurus found in the same beds.[11] The discovery of this mammal was startling to all previous conceptions, because the Oolite was a very old formation lying well down in the series of Secondary formations in which, so far, reptiles had been the highest fossil vertebrates discovered. No other fossil mammals had been discovered earlier than the Tertiary formations of the Paris basin.

The Stonesfield mammal was such an anomaly in what seemed to be the emerging pattern of fossil succession that the French geologist Constant Prévost, who visited England during the summer of 1824, believed there must have been an error either in the identification of the fossil or in the interpretation of the geological age of the deposits in which it was found. He questioned whether the Stonesfield Slate actually lay beneath the Oolite.[12] However, there could be little doubt about the stratigraphic position of the slate, because access to it was gained by shafts sunk for sixty feet through the overlying Oolite and because experienced English geologists considered the Stonesfield Slate a formation they could recognize

10. Ibid., p. 513.

11. William Buckland, "Notice on the Megalosaurus or Great Fossil Lizard of Stonesfield," *Trans. Geol. Soc. London,* 1824, ser. 2, *1,* 390–96.

12. Constant Prévost, "Observations sur les schistes calcaires oolitiques de Stonesfield en Angleterre, dans lesquels ont été trouvés plusieurs ossemens fossiles de Mammifères," *Ann. Sci. Nat.,* 1825, *4,* 389–417.

as one of the Oolite slates whose outcrops extended diagonally across the whole breadth of England.

After Prévost had expressed his doubts concerning the Stonesfield mammal, Buckland sent his specimen to Cuvier who confirmed that it was a marsupial related to Didelphys, but he now considered it sufficiently distinct to be placed in a genus of its own. Then in 1828, William Broderip found the other Stonesfield specimen, which had been misplaced among his own belongings since 1814, and wrote a description of it.[13] On close examination Broderip found that his specimen was not only a species different from Buckland's but that it was so different it ought to be placed in a different genus. This discovery seems to have had a considerable influence on Lyell's thinking, for not only was there a fossil mammal in the Stonesfield Slate, but there were two distinct genera of fossil mammals.[14] And, but for what could only be called chance discovery, geologists would not have known anything about them.

In the summer of 1827, Lyell read Lamarck's *Philosophie zoologique* during his travels as a young barrister through Devonshire and Somersetshire on the western circuit of the Court of Assizes. This great book seems to have influenced him profoundly in two respects. It convinced him that if the theory of the progressive development of animals and plants through geological time were true, then this development must have culminated in the appearance of man, who would, therefore, be related by descent to the other animals. Lyell rebelled against this conclusion, perhaps because it was so opposed to the whole framework of ideas within which he had been educated and because he thought that the rational and intellectual powers of man distinguished him decisively from the animals. Second, his reading of Lamarck made Lyell more deeply aware of the many-sided relationship of a species to its environment and the significance that this relationship had in determining the structure and peculiarities of function and habit of each species.

13. William John Broderip, "Observations on the jaw of a fossil mammiferous animal found in the Stonesfield slate," *Zool. J.*, 1828, *3*, 408–12.

14. Charles Lyell, *Principles*, 1830, *1*, 150.

In May 1828, Lyell traveled with Roderick Murchison and Mrs. Murchison to the Auvergne district of France, where they spent the next several weeks in studying both the extinct volcanoes and volcanic formations of that country and the accumulations of freshwater sediments that had been formed in ancient lakes there. Near the end of July, they went south to Montpellier and, after a few weeks rest at Nice, proceeded along the coast into Italy. In September, Lyell parted from the Murchisons at Padua and went southward through Italy. In mid-November, he sailed from Naples to Messina and from there set out on a geological exploration of Sicily that, after a visit to the Val del Bove, the ascent of Etna, and a long circuit to Syracuse and along the southern coast of the island, ended at Palermo on January 1, 1829. He returned northward through Italy, Switzerland, and France and on February 26, was back in London.

This long expedition of nearly ten months was decisive in the development of Lyell's ideas. It provided him with a multitude of analogies between past and present conditions and strengthened his belief in the uniformity of the physical world through time. As well as analogies between volcanic action, volcanic rocks, and the effects of earthquakes on the uplift of strata in the past and the present, the analogies he saw included those between the biological environments of the past and the present. The community of fossil species that had lived in the ancient freshwater lakes of Auvergne was similar, both in the kinds of species represented and the kinds of associations they formed, to the community of living species in the modern Scottish lake. The assemblage of fossils in the marls of Auvergne, therefore, suggested the same kind of environment of clear, tranquil, freshwater as a modern lake. The community of shells and corals living in the Mediterranean had also lived there in the remote past, before the island of Sicily had been raised above the sea. Conditions in the Mediterranean at that period must, therefore, have been very similar to those existing at the present time.

The fact that the groups of fossils found in rocks each represented a particular biological environment had a special significance for Lyell when, in writing the *Principles,* he began early in 1830 to consider the problem of changes of climate. In his *Quarterly Review* article of 1826, he had summarized the evidence that the earth's climate had been much warmer during past geological periods than it was at the present time.[15] Perhaps, too, there had been a greater uniformity of temperature between the equator and the poles than existed in the modern period. The predominance of univalve shells and large corals in Secondary marine formations represented a biological community similar to that in modern tropical seas. The plants of the Coal formations, which were then thought to be Palms and Ferns, were of a size and luxuriance that today would appear only under the conditions of a tropical climate. The gigantic reptiles, which lived throughout the Secondary period, also suggested warm conditions, because today the larger reptiles are all confined to warm climates.

The explanation commonly given before 1830 for warmer world climates during the geological past was that the earth originally had been formed as a hot molten mass that had gradually cooled. After it had cooled sufficiently to form a hard crust and the conditions were established that would support life, there was still a very long period during which sufficient heat was radiated from the interior of the earth to maintain, together with the heat received from the sun, the equivalent of a tropical climate over the entire surface of the earth. This explanation of the warmer climates of the past in northern latitudes was set forth by Sir Humphry Davy in a small volume, published posthumously, his *Consolations in travel,* which appeared at the very time when Lyell was developing his theory of climate.[16]

For Lyell, such an explanation as that given by Davy was of little value, because it postulated the existence of a cause governing conditions on the surface of the earth during the past, which had no parallel in the present and, therefore, could not be tested by analogy. Lyell sought instead to discover the physical factors that determined climate in different regions of the earth's surface and that determined the average degree

15. Charles Lyell, "Art. IX . . . ," p. 525.

16. Sir Humphry Davy, *Consolations in travel, or the last days of a philosopher,* London, Murray, 1830, 281 pp.

of heat or cold of the climate of the world taken as a whole. He was especially indebted to the work of Alexander von Humboldt, who had collected together a wealth of data on the distribution of mean annual temperatures over the surface of the earth and had plotted isothermal lines. Lyell also used information on winds, temperatures, and ocean currents that had been gathered at the hydrographic office of the British admiralty.

As a result of his studies, Lyell concluded that the climate of any local region depended on its relationship to surrounding areas of land and sea, to mountain ranges, winds, and ocean currents. Furthermore, the degree of heat or cold of the climate of the earth as a whole also depended on the pattern of distribution of areas of land and sea. If the proportion of areas of land to areas of sea were high in northern latitudes, for instance, the climate would be cold because snow and ice would accumulate on the land thereby cooling the surrounding area. On the other hand, a high proportion of land to sea in tropical regions would exert a warming influence on world climate, because the heat of the sun falling on the land would be radiated into the atmosphere raising its temperature and creating warm winds that would tend to warm surrounding areas. Areas of sea in high latitudes tended to exert a moderating influence on climate by tempering both the heat of summer and the cold of winter, but in tropical regions they would exert less warming influence than would areas of land.

The consequence of Lyell's theory that the pattern of distribution of areas of land and sea controlled climate was that geological processes, which in the long course of time changed the distribution of land and sea, would also change climate. When the climate of northern Europe had been much warmer during the Secondary period, Lyell thought that what were now great land masses in Europe had then been merely archipelagoes of islands in an area of predominating sea.

The dependence of climate on geography and of change of climate on changes in geography had still further consequences for Lyell's understanding of the development of organic life. Climate so largely determines the conditions under which animals and plants must live that in effect it determines what they shall be. Lamarck had shown how profound was the adjustment and how detailed the adaptation of each species to the special conditions of its environment. He saw, in fact, a relationship so close that he believed that the environment itself evoked the adaptations. Lyell now saw the plant and animal species of any geological period as an interacting complex related to climate and geography and determined in its general character by climate and geography.

From this standpoint, the large reptiles of the Secondary period need not represent a particular stage in the progressive development of animal life. Instead, they might represent the animals best adapted to live under the conditions of warm climate, luxuriant vegetation, swamps, river deltas, and low scattered islands that seemed to have prevailed during much of that period. As geography and climate changed, however, new sets of conditions would require new species of plants and animals adapted to the new environment.

With this understanding of the determinate relationship of species to their environments in each geological period, it would be difficult to see how any continued and progressive development of animal or plant life through geological time could occur. Lyell seems to have come to doubt whether, in fact, it had occurred. There was no discernible thread of progress in geological or geographical changes which, on the contrary, seemed to fall into recurrent patterns. Lyell's conviction that the history of life on earth did not represent a single line of development, but had been, instead, a series of manifold adjustments to fluctuating environmental conditions—adjustments made by replacing species adapted to the old conditions with species adapted to the new—was one of his most persistent and deep-seated beliefs. Even after Charles Darwin published the *Origin of Species* and Lyell had come to accept part, at least, of Darwin's theory, Lyell wrote to him on August 28, 1860 about why reptiles had not evolved independently into mammals on oceanic islands that had been isolated throughout a long period of geologic history.

> Perhaps you will say that this would be a greater puzzle to the out-&-out progressionists who may be converted by you

> to transmutation . . . for you do not believe in that constant advance, as an ever working tendency, & allow for stationary & retrograde movements.[17]

Thus, for Lyell in 1860, one of the attractive features of Darwin's theory was that it did not require him to accept a continuous progressive development of the living world, but a much more complex and many-sided development in relation to environmental changes.

Lyell's consideration of the life of fossil species was influenced by the fact that he was an accomplished amateur nattturalist. He, therefore, wished to understand the conditions that had governed fossil species, by analogy with those that determine the life of existing species. In the second volume of the *Principles,* published in 1832, he studied the relationship of each species to its physical environment, the interactions between species and the struggle for existence. He showed that each species is adapted to, and consequently dependent on, a particular set of physical conditions and the presence of a particular assemblage of other species. On the other hand, because geological changes are slowly but inexorably destroying environments and altering geography and climate, they are thereby steadily removing the conditions essential for the life of a species. Similarly, the migrations and fluctuations in numbers of other species steadily threaten the life of an individual species. Because each species is, and necessarily must be, adapted to a particular set of conditions both physical and biological, the destruction of these conditions will drive it into extinction.

Between 1832, when Lyell developed his views of the succession of species in relation to geological history, and 1850, the range of his knowledge of palaeontology and the scope of application for his ideas increased enormously, but his basic theoretical viewpoint changed little. The doctrine of progressive development, on the other hand, enjoyed increasing popularity. This popularity was based partly on the rapid growth of knowledge of the palaeontological record. In 1838, Roderick Murchison published his *Silurian System* in which he

17. Charles Lyell, Scientific Journal No. VI (p. 468).

described the fossils of a long series of primary rocks in southern Wales and Shropshire.[18] Murchison showed that these rocks, which previously had been assumed to contain few or no fossils, in fact possessed a very rich fossil fauna that included numerous representatives of every major class of invertebrates but no vertebrates, not even fishes. Murchison's results were confirmed almost simultaneously by the work of the New York State Geological Survey. In New York State, there was a magnificent series of Silurian rocks of great depth, extending undisturbed for hundreds of miles and displaying in luxuriant richness the fossil life of palaeozoic seas. On the one hand, the New York strata demonstrated the existence of tranquil and uniform conditions throughout long periods of time at an early stage in the history of the earth; on the other, they tended to show that, in these ancient seas, there had been a rich development of forms of invertebrate life in the absence of any vertebrates whatever.

In the 1830s, Louis Agassiz began his studies of fossil fishes and showed how they began to appear in the uppermost Silurian strata with only a few representatives and afterward reached a rich and varied development in the Devonian period. Similarly, Agassiz showed later how such invertebrate groups as the Echinoderms and Mollusca appeared first in the fossil record with only a few representatives, whereas in later geological formations, the number of fossil genera and species in the group increased enormously. Agassiz gave to the first-appearing representatives of a group the name "prophetic types," their being prophetic of the full development of the group at later geological periods.

The doctrine of progressive development was based, therefore, on the fossil evidence for the apparently successive appearance of plant and animal groups through time—the invertebrates in the Silurian, the bony fishes in the Devonian, the reptiles after the Carboniferous, and the birds and mammals in the Tertiary period. It was a doctrine popular in Great Britain in the 1840s, especially among those geologists who wished to reconcile geology with religion. They held that the

18. Roderick I. Murchison, *The Silurian System,* 2 pts., London, Murray, 1839, 768 pp.

fossil record showed the progressive development of the plan for the living creation in the mind of the creator. It was to them as if the parade of remarkable and often astonishing fossils that had come to light were so many concrete illustrations of the creation described so succinctly in Genesis. The fossils of successive geological formations represented a series of successive creations, each a stage in the foreordained plan of creation. The concept of progressive development was directly opposed to Lyell's doctrine of the uniformity of the physical world throughout geological history, because it introduced a kind of event, namely, creation, that had occurred at intervals during the geological past, but did not occur at the present time. Moreover, creation was not the kind of event to be accounted for by the processes of the physical world, either now or during the past. It required the intervention of God, the first cause, in the physical world. The great charm of the theory of progressive development, therefore, was that it retained a powerful element of the marvelous and unaccountable in the history of the world. The more gigantic, bizarre, or unusual the fossils discovered, the mightier and more resourceful was the creative power that had brought them into being.

In Britain, this view of the meaning of the fossil record was advocated by Hugh Miller in his *Foot-Prints of the Creator* (1847),[19] by Roderick Murchison in *Siluria* (1854), and by Adam Sedgwick, who argued for it in a long preliminary dissertation to the fifth edition of his *Discourse on the Studies of the University of Cambridge* (1850).[20] Progressive development was particularly attractive to Miller and Sedgwick because it not only preserved for them the concept of the world as having been created by God, but made man, who appeared last, the capstone of creation. Yet, it never failed to astonish Charles Lyell that such men as Miller and Sedgwick did not see where they were going, because there was an alternative interpretation of the evidence of progress in the fossil record. It might equally well be used in support of Lamarck's theory of the

19. Hugh Miller, *The Foot-Prints of the Creator; or the Asterolepis of Stromness*, Edinburgh, 1847, 313 pp.

20. Adam Sedgwick, *Discourse on the Studies of the University of Cambridge*, 5th ed. with additions and a preliminary dissertation, London, Parker, 1850, 322 pp.

transmutation of species and their common relationship by descent, a theory that would make man cousin to the orangutan. In 1844, this very application had been made in an anonymous work, *Vestiges of the Natural History of Creation,* actually written by Robert Chambers, an Edinburgh publisher.[21] Chambers attributed the progressive development apparently visible in the fossil record to "development" conceived of as a fundamental natural law. He believed that development was an inherent and necessary characteristic of the natural order and that the successive appearance of new forms of life throughout geological history was strictly the consequence of a pervasive natural law. Chambers' book was attacked immediately and violently by a multitude of critics; conspicuous among them was Adam Sedgwick.[22]

For himself, Lyell remained deeply opposed to the doctrine of progressive development. The idea of successive creations, each constituting a decisive break in the natural order of events, would destroy his concept of a natural order existing uniformly through time. Creation, by having no analogy in the present, lay beyond the means of study or even of intelligent discussion. Yet Lyell himself continued to use the word to describe the first appearance of new species. However, he thought of new species as coming in steadily throughout the course of geological time and as being produced in some way unknown, but as a consequence of the ordinary processes of nature. He could not accept a necessary and continued development through time because of the ever-present necessity that species should be adapted to the prevailing conditions of climate and geography, conditions that were cyclical and recurrent rather than progressive.

In addition to his fundamental bias, Lyell thought the geologists who advocated progressive development wildly reckless in the generalizations they drew from fossil evidence. They made little allowance for the extreme incompleteness of the fossil record and for the fragmentary picture it gave, therefore, of the life of ancient geological periods. Each rock, Lyell con-

21. Robert Chambers, *Vestiges of the Natural History of Creation,* London, Churchill, 1844, 390 pp.

22. Adam Sedgwick, "Natural History of creation," *Edinburgh Rev.,* 1845, *82,* 1–85.

sidered, represented a sediment laid down in a particular locality and, therefore, reflected a particular biological environment. It could be expected to contain only the remains of the plants and animals that had lived there. Even of these, not all would have structures capable of being preserved or of leaving their imprint behind. Thus, in considering the formation of coal, Lyell had come to the conclusion that coal had been formed in immense low-lying coastal swamps similar to the Great Dismal Swamp of Virginia, which he had visited. Thus, the fossil plants embedded in coal, being swamp plants, would give little information about the upland plants of the Carboniferous period, any more than the plants of the Great Dismal Swamp would today be a reliable guide to what grew on the upper slopes of the Appalachians.

Similarly, the incompleteness of the fossil record was indicated by the occurrence of isolated species in geological formations far removed from those containing related species. An outstanding example was the Stonesfield fossil marsupial. If marsupials were present in England early in the Jurassic period, there were probably other marsupials living at the same time distributed elsewhere in the world, because the modern representatives of this group occur in America and Australia. Yet none but the Stonesfield fossils had been found, and only the jawbones of these. Geologists had discovered as yet only the merest fragment of the animal forms that had lived during the Jurassic period, or at any other period of the past for that matter.

Lyell was confident that additional fossil mammals would be discovered in Secondary rocks, and in 1847 Professor Wilhelm Plieninger of Stuttgart found, in a bed of Triassic age at Würtemberg, two fossil molar teeth which, he decided, must have belonged to a small insectivorous mammal. He called it Microlestes. Here was a mammal older even than those at Stonesfield. This animal could not have been the lone representative of its species or its group, but two small molar teeth were all the remains that had been found of the fossil mammals of the whole world during the Triassic period.

In 1851, Lyell took the occasion of his anniversary address as President of the Geological Society of London to subject to the severest criticism the idea that the fossil record necessarily revealed a pattern of progressive development. He argued that the fossil record was by its very nature too fragmentary to permit sweeping generalizations based on the absence of particular classes of living forms. By virtue of the very fact that the fossils in a particular bed represented only the community of plants and animals at one locality, they could not be taken to represent the entire life of the earth's surface at the time they were deposited. Even in the area of a particular sediment, only a fraction of the plants and animals would be preserved as fossils and only a fraction of that fraction would be discovered by geologists. Lyell emphasized and reemphasized the extreme incompleteness of the fossil record. The oldest strata known happened all to be marine formations, but areas of land could have existed during the Silurian period, for instance, yet we would have no knowledge of them. The oldest land flora then known, in the Carboniferous, Lyell said, possessed conifers and even, according to some botanists, palms, although neither of these was by any means the most primitive land plants. From the Cretaceous through the Tertiary period, all the classes of land plants were represented and four or five almost complete changes of species had occurred, yet there had been no significant advance in plant organization or complexity.[23]

Perhaps the most striking fact, which Lyell presented, was the very pale reflection that sediments in modern seas gave of the contemporary living world. In 1850, Edward Forbes and Robert MacAndrew had made a series of dredgings off the coast of Great Britain. They had found a relative abundance of the remains of marine invertebrates and fish bones in the deposits, but no remains of Cetacea or land mammals. "If reliance could be placed on negative evidence," Lyell said, "we might deduce from such facts, that no cetacea existed in the sea, and no reptiles, birds or quadrupeds on the neighbouring land." [24] He also mentioned that the few elytra of beetles, which he himself had been able to find in modern freshwater deposits in Britain,

23. Charles Lyell, *Address delivered at the Anniversary Meeting of the Geological Society of London, on the 21st of February, 1851*, London, Taylor, 60 pp.
24. Ibid., p. 38.

gave a very incomplete picture of the more than 11,000 species of insects now living on the island.

These observations showed that the apparent absence of the higher groups of plants and animals in the older strata might be at least in part illusory, for they could well have been present on the earth and have left no trace. Reptiles had been thought to be absent from all periods older than the Lias, but, recently, evidence of their presence had been found in both Permian and the oldest Carboniferous strata. Lyell multiplies his examples, but his whole argument is to show that the fossil record did not sustain the idea of a progressive development of living forms through time.

Reexamination of the Species Question

In his 1851 address, Lyell had publicly delineated his position on the question of progressive development with the accomplished force of a skilled barrister. Yet, privately, he may have had some doubts. In December 1853, he and Lady Lyell sailed to the Canary Islands with her brother-in-law and sister, Charles and Frances Bunbury, to spend several months studying the volcanic geology of these islands and of the adjacent island of Madeira. However, in Madeira and the Canaries, Lyell also encountered a peculiar geographical distribution of species similar to that which Charles Darwin had found in the Galapagos Islands. It was his first direct experience of the natural history of oceanic islands. He seems initially to have been struck by the unusual plants of Madeira. He noted:

> At Santa Cruz some magnificent Laurels like evergreen oaks. Of these there were many native trees at Funchal till they cut them down & planted English oaks which are shabby in comparison. The pale green Euphorbia piscatoria with some few flowers on it, adorned the lava rocks between Brazen Head & Porto Nova Camisso.[25]

On January 2, 1854, he left Lady Lyell and the Bunburys on Madeira and sailed with Captain Keppel on the British man-of-war *St. Jean d'Acre* to the island of Porto Santo to

25. Charles Lyell, Notebook 186, p. 61, Kinnordy mss.

spend a week making geological observations and collections. On his return, he made further extensive field excursions around Madeira and near Santo Jorge found a bed of lignite containing the fossilized leaves of many plants, under basalt, at an elevation of about 1,000 feet above sea level. Charles Bunbury identified a number of the fossil leaves and found among them a mixture of ferns with Dicotyledons, unusual among Tertiary formations but similar to the forest flora of present-day Madeira. This discovery showed that Madeira had been covered with vegetation similar in general character to its present one since the time when it was only half its present size.

This one discovery thus showed that Madeira probably had existed continuously as a land surface since the Miocene period and, Lyell suspected, probably always was one of a group of islands in the open ocean. Everything Lyell saw of the volcanic geology of Madeira tended to convince him that it had been built up very gradually, over a long period of time, from the accumulated outpourings of many volcanic eruptions. This view was directly opposed to Leopold von Buch's theory of craters of elevation, according to which the inclined beds of lava and basalt had originally been poured out in a nearly horizontal position under the sea and later upraised to their present inclined position.

Of the native animal species that lived on Madeira, the insects, and particularly the beetles, were remarkable for the large number of species peculiar to the island group and still more remarkable for the number of species peculiar to individual islands.

Even more remarkable than the beetles were the land shells of these islands. In 1833, the Rev. Richard Thomas Lowe had described some seventy-one species of land shells from the Madeiran group of which forty-four were new species. He had found very few of these species also living in the Canary Islands, whereas within the Madeiran group only two species were common to both Madeira and Porto Santo, though the latter were only thirty miles apart.[26] When he arrived in Madeira in De-

26. Richard T. Lowe, "Primitiae Faunae et Florae Maderae et Portus Sancti; sive Apecies quaedam novae vel hactenus minus rite cognitae Animaluim et Plantarum in his Insulis degentium breviter descriptae," *Cambridge Phil. Soc. Trans.*, 1833, *4*, 1–70.

cember 1853, Lyell seems already to have been alert to the importance of the land shells, because Lady Lyell immediately began to make a collection of the living species. On January 6, 1854, she wrote to her sister-in-law Marianne Lyell, "I have taught Antonia [her maid] to kill snails & clean out the shells & she is very expert." [27]

On Saturday, February 18, 1854, the Lyells, Bunburys, and Georg Hartung sailed on the steamer *Severn* from Funchal, Madeira, to Santa Cruz, Teneriffe, where they stayed at the Lazzaretto. Lyell was immediately struck by the presence of camels on Teneriffe and the hills dotted with a large succulent plant, Euphorbia Canariensis. Four days later, February 22, the Lyells and Georg Hartung sailed to the Grand Canary Island, where Lyell and Hartung made many geological excursions on horseback. Lyell was astonished at the landscape. He noted:

> I never was in a country where the vegetation was so exclusively & unEuropean & so peculiar
>
> The Palms were only seen here & there, but the cactus-like Euphorbias & other species & the plants before ment.d very remarkable & large. . . . Madeira veget.n European in comparison.[28]

Lyell had also been struck by the absence of any native mammals except bats on the islands. He later wrote in the *Principles*:

> When we have travelled over large and fertile islands, thirty miles or more in diameter, such as the Grand Canary and Teneriffe, and have seen how many domestic animals, such as camels, horses, asses, dogs, sheep and pigs, they now support, we cannot but feel amazed that not even the smaller wild animals, such as squirrels, field mice and weasels, should be met with in a wild state.[29]

The absence of mammals might be accounted for simply by the distance of these islands from the mainland, but that very

27. M. E. Lyell to Marianne Lyell, January 6, 1854, Kinnordy mss.
28. Charles Lyell, Notebook 193, p. 131, Kinnordy mss.
29. Charles Lyell, *Principles,* 11th ed., 1872, *2,* 415.

explanation implied that mammals had not been created for every area of the world in which they might live successfully.

Lyell remained in the Canary Islands until the middle of March 1854, and during this time he spent some two weeks on Palma. After his return to London in April, he settled into his usual round of scientific activities, and during the next year, he was occupied chiefly in the preparation of a fifth edition of his *Manual of Elementary Geology,* which he completed in February 1855. However, at the same time, he was trying to get his collections from Madeira and the Canaries identified. He told Charles Bunbury:

> Of my 4 species of Bryozoa from the Grand Canary one is recent & three unknown so says the first rate authority Mr. Busk. One lunulite, one Retepora, one Eschora & one Flustra. I imagine the age may be Miocene or falunian but this is a mere guess as yet.[30]

He also mentions a visit with Joseph Hooker to Charles Darwin's house in November 1854, during which they evidently had a discussion of the species question.

When the *Manual* had been completed in February 1855, Sir Charles and Lady Lyell went to Berlin for several weeks to visit relatives and stopped at Paris on their return. The summer was taken up with scientific meetings and a long visit to Scotland in August and September. It was thus not until November 1855, when Georg Hartung came to London from Germany, that Lyell and Hartung began to go over together the results of their geological explorations in Madeira and the Canaries a year and a half earlier. Their renewed studies again confronted Lyell with the extraordinary features of the animal and plant species of these islands. On Sunday, November 17, he wrote to his sister Fanny:

> Mr. Hartung is working away very steadily with me on our joint papers on the Madeira and Canary Islands. Like any point in Nature on which you fix your attention, you soon find that all others seem to connect themselves with it. The

30. Charles Lyell to Charles J. F. Bunbury, November 13, 1854, Kinnordy mss.

> discovery that the living insects in the different islands of the Madeira group, such as Porto Santo & the three Desertas, are in a great degree, as well as the land shells & I believe many of the plants, distinct, or not of the same species, & found no where else in the world & Geology may help to explain this & the fact that a certain portion of the animals & plants do agree with the Canaries & with Africa & southern Europe. . . . It seems to me that many species have been created, as it were expressly for each island since they were disconnected & isolated in the sea. But I can show that the origin of the islands, which are of volcanic formation, dates back to a time when the surrounding sea was inhabited by a third or 4th only of the species now existing & all the rest (species of fossil shells, corals etc.) have died out. But I must not run on as it w.[d] take me too long to point out how all these bear on one & the same theory—of the mode of the first coming in of species.[31]

A week later, on November 26, 1855, Lyell read an article by A. R. Wallace published the previous September in the *Annals and Magazine of Natural History,* "On the Law which has regulated the Introduction of New Species." In this paper, Wallace had assembled evidence from the palaeontological record and from the geographical distribution of plants and animals to show that "Every species has come into existence coincident in both space and time with a pre-existing closely allied species." This seems to have struck Lyell so forcibly that he entered some notes on it in the first of the series of seven notebooks that he was to devote to the species question and that are published here.

The Contents of the Notebooks

During the winter months of 1856, Lyell continued to study the taxonomy and geographical distribution of his Madeiran and Canary Island species. Georg Hartung had a recurrence of the tuberculosis that had originally forced him to go to Madeira, and this breakdown in his health obliged him to suspend

31. Charles Lyell to Frances Lyell, November 17, 1855, Kinnordy mss.

his scientific work. In the meantime, Lyell became acquainted with Thomas Vernon Wollaston, then thirty-four-years old, a nephew of William Hyde Wollaston, the chemist. Like Hartung, Wollaston had spent a number of years in Madeira for the sake of his health, and in 1854, he had published a work on Madeiran insects.[32] Of Wollaston's help, Lyell wrote to Charles Bunbury on February 19, 1856:

> I have learnt much conchologically & entomologically from him. The beetles are wonderful in their distribution, each island & almost every rock having its own, so many peculiar to the Madeiras, several hundred & apterous altho' of genera not wingless in Europe or elsewhere. . . .
>
> Almost every land shell different from every one living in Porto S.°, & the fossil helices of Madeira in like manner distinct from the Portosantan whether belong[ing] to the living or extinct shells.[33]

Through February and March 1856, Lyell continued to work on Madeiran land shells and insects. He was anxious to determine whether the Madeiran and Canary Islands had always existed as separate islands in the ocean or whether they were the surviving remnants of a former continental land area, which had extended continuously from Europe and Africa. The idea of a lost continent of Atlantis, which had sunk beneath the ocean in recent geological times, had been suggested in 1846 by Edward Forbes to account for the presence in southwestern Ireland of a flora not found closer than northern Spain. However, everything that Lyell saw, both of the geology and the species of the Madeiras and Canaries, tended to convince him that these had existed for a very long time as separate oceanic islands and had been such at least since the Miocene period.

The first scientific journal contains notes on the general question of species, the idiosyncrasies of Madeiran and Canary species, and copies of several of Lyell's letters to T. V. Wollaston. The large number of species peculiar to individual islands sug-

32. T. Vernon Wollaston, *Insecta Maderensia; being an account of the Insects of the Islands of the Madeiran Group,* London, 1854, 634 pp.

33. Charles Lyell to Charles J. F. Bunbury, February 19, 1856, Kinnordy mss.

gested that each island had long been isolated. On March 28, he wrote to Leonard Horner:

> The Madeiras are like the Galapagos, every island & rock inhabited by distinct species. What a wonderful contrast with the British Isles (above a hundred in number) where the same fauna prevails everywhere, or if not strictly so, has at least in its distribution no respect to the barriers offered by channels of salt water.[34]

Questions of distribution of species led him next to the question of how species might be carried across areas of ocean to colonize islands. Perhaps it was for this reason that he decided to visit Charles Darwin at Down House in the village of Down, Kent, about fourteen miles from London. Lyell knew that for many years Darwin had been convinced of the transmutation of species and had, since the completion of his work on barnacles a year and a half earlier, made the question of the origins and migrations of species his principal concern. Darwin had taken up the breeding of pigeons and had been testing the power of seeds to survive immersion in salt water to see how species might vary and spread.

Sir Charles and Lady Lyell were at Down from Sunday April 13, 1856, until the afternoon of Wednesday, April 16. As was his custom, Lyell took books and papers with him so that he could devote part of each day to work and would neither consume too much of Darwin's time in social conversation nor lose too much of his own.

On Sunday, April 13, Lyell and Darwin evidently talked of the geology of volcanic islands, the migration of plants and shells, and other topics.[35] Their conversation was clearly directed toward the problem of the original colonization of oceanic islands with species of plants and animals. Lyell wrote in his notebook:

> Man may in an advanced state have flourished for ages in some part of the Old World. He belongs to an Old World

34. Charles Lyell to Leonard Horner, March 28, 1856, Kinnordy mss.
35. Charles Lyell, Scientific Journal No. I (pp. 52–53).

form of Anthropomorphous animals as distinguished from the New World Platyrhine animals or quadrumana.

Darwin thinks that Agassiz's embryology has something in it, or that the order of development in individuals & of similar types in time may be connected.[36]

On Monday, April 14, they discussed the volcanic geology of Madeira, and on Tuesday, Lyell wrote a letter embodying the results of their discussion of volcanic geology to Georg Hartung who was then in the Azores.

On the morning of Wednesday, April 16, Darwin and Lyell discussed the species question and Darwin explained his theory of natural selection fully to Lyell. Lyell entered an outline of the main features of the theory in his scientific journal.[37]

April 16 1856

With Darwin: On the Formation of Species by Natural Selection—(Origin Qy?)

Genera differ in the variability of the species, but all extensive genera have species in them which have a tendency to vary. When the condit.[s] alter, those individuals, which vary so as to adapt them to the new circums.[s], flourish & survive while the others are cut off.

The varieties extirpated are even more persecuted & annihilated by organic than inorganic causes. The struggle for existence ag.[t] other species is more serious than ag.[t] changes of climate & physical geography. The extinction of species has been always going on. The number of species which migrated to the Madeiras was not great in proport. to those now there, for a few types may have been the origination of many allied species.

The young pigeons are more of the normal type than the old of each variety. Embryology, therefore, leads to the opinion that you get nearer the type in going nearer to the foetal

36. Charles Lyell, Notebook 213, pp. 30–31, Kinnordy mss.
37. Charles Lyell, Scientific Journal No. I (pp. 54–55).

> archetype & in like manner in Time we may get back nearer to the archetype of each genus & family & class.
>
> The reason why Mr. Wallace['s] introduction of species, most allied to those immediately preceding in Time, or that new species was in each geol.[1] period akin to species of the period immediately antecedent, seems explained by the Natural Selection Theory.

His reference to Wallace at the end of this passage suggests that, since his reading of Wallace's paper on the previous November 25, Lyell had felt an urgent need to explain the close relationship pointed out by Wallace between the species living in a given locality and those that had preceded them there at an earlier geological period. Dr. H. Lewis McKinney has suggested that this was probably the first occasion on which Darwin described his theory of natural selection to Lyell.[38] This inference is rendered very probable by the fact that there are no references to natural selection in Lyell's notebooks before this date, whereas they occur frequently thereafter. Neither are there any earlier references to natural selection in the surviving letters from Darwin to Lyell. However, there are few such references before 1859, and Darwin was probably unwilling to discuss his theory in correspondence before the publication of the Darwin-Wallace papers in 1858. Certainly, Lyell had known for a long time, in fact since 1837, that Darwin doubted the fixity of species, and he had been aware that Darwin had also long been gathering materials for a book on the question of the origin of species. Since the completion of his work on barnacles in 1854, the collection of information relating to the species question had been Darwin's principal concern.

One of the strongest indications that this Wednesday morning, April 16, 1856, was the first occasion on which Darwin explained his theory to Lyell is that Lyell immediately and very strongly urged him to publish it. Darwin demurred on the grounds that the theory could not be published without a massive amount of detailed and interconnected evidence that he had gathered and was still gathering and attempting to organ-

38. H. Lewis McKinney, "Alfred Russel Wallace and the Discovery of Natural Selection," *J. Hist. Med.*, 1966, *21*, 333–57, p. 350.

ize. Lyell nevertheless urged that Darwin publish at least a short essay to present the main outlines of his theory, leaving till later the publication of his full array of evidence. This would assure his priority. On May 3, Darwin wrote to Lyell:

> With respect to your suggestion of a sketch of my views, I hardly know what to think, but will reflect on it, but it goes against my prejudices. To give a fair sketch would be absolutely impossible, for every proposition requires such an array of facts. If I were to do anything, it could only refer to the main agency of change—selection—and perhaps point out a very few of the leading features, which countenance such a view, and some few of the main difficulties. But I do not know what to think; I rather hate the idea of writing for priority, yet I certainly should be vexed if any one were to publish my doctrines before me. Anyhow, I thank you heartily for your sympathy.[39]

On Wednesday April 30, Lyell wrote to Charles Bunbury a letter in which, after discussing the botany of Madeira and the Canary Islands, he wrote:

> When Huxley, Hooker, and Wollaston were at Darwin's last week, they (all four of them) ran a tilt against species farther I believe than they are deliberately prepared to go. Wollaston least unorthodox. I cannot easily see how they can go so far, and not embrace the whole Lamarckian doctrine.[40]

This passage, taken by itself, gives the impression that Lyell was present at the meeting at Darwin's when species were discussed, but in fact, he was not and seems to have learned of the substance of the discussion from Wollaston. This fact is established by a recently discovered letter written by Lyell to Darwin the following day, May 1, 1856. After a discussion of

39. Charles Darwin to Charles Lyell, May 3, 1856, in Charles Darwin, *Life and Letters,* Francis Darwin, ed., 3 vols., London, Murray, 1887, *2,* 67–68.

40. Charles Lyell to Charles Bunbury, April 30, 1856, Kinnordy mss. Published in K. M. Lyell, *Life, Letters and Journals of Sir Charles Lyell,* 1881, *2,* 211–13.

the distribution and modes of dispersion of species of land shells, Lyell wrote: "I hear that when you & Hooker & Huxley & Wollaston got together you made light of all species & grew more & more unorthodox." [41] After further remarks about Oswald Heer's ideas on the geographical distribution of plants, Lyell added: "I wish you would publish some small fragment of your data, *pigeons* if you please & so out with the theory & let it take date & be cited & understood."[42]

The meeting to which Lyell refers seems to have taken place at Down sometime between Sunday, April 20, and Sunday, April 27. The scientific interests of the different men assembled suggest that Darwin had deliberately invited them to discuss species, because each had in a different way been concerned with the species question. Thomas Henry Huxley had criticized the application of theory of progressive development to questions of comparative anatomy and embryology; Joseph Hooker had devoted years to the study of the geographical distribution of plants; and Thomas Vernon Wollaston had studied the species distribution of both insects and land shells in Madeira and the Canary Islands. Darwin was perhaps seeking to test in advance the reception his theory might be given by scientists.

After a further conversation with Lyell in London, Darwin also wrote to Joseph Hooker on May 9:

> I had a good talk with Lyell about my species work, and he urges me strongly to publish something. I am fixed against any periodical or Journal, as I positively will *not* expose myself to an Editor or a Council allowing a publication for which they might be abused. If I publish anything it must be a *very thin* and little view giving a sketch of my views and difficulties; but it is really dreadfully unphilosophical to give a resumé, without exact references, of an unpublished work. But Lyell seems to think I might do this, at the suggestion of friends, and on the ground, which I might state, that I had been at work for eighteen years, and yet could not publish for several years, and especially as I could point out difficul-

41. Charles Lyell to Charles Darwin, May 1, 1856, Darwin mss. This letter was drawn to my attention by Professor Robert Stauffer, who obtained a xerox copy for me from the Cambridge University Library, England.

42. Ibid.

ties which seemed to me to require especial investigation. Now what think you? [43]

Evidently, after receiving a reply from Hooker, Darwin wrote to him again on May 11.[44] Hooker seems to have advised against the publication of a preliminary essay on the grounds that it might detract from the novelty and value of his later book. Yet Darwin was still undecided. On June 10 he wrote to his cousin William Darwin Fox:

Sir C. Lyell was staying here lately, & I told him somewhat of my views on species, & he was sufficiently struck to suggest, (& has since written so strongly to urge me) to me to publish a sort of Preliminary Essay. This I have begun to do, but my work will be horridly imperfect & with many mistakes so that I groan & tremble when I think of it.[45]

On June 29, 1856, Lyell seems to have written again to Darwin urging him to publish, for he copied a portion of his letter into a notebook.

Y.[r] anecdote of my saying that I ought in consistency to have gone for transmut.[n]—that I have uniformly taken the other side in all edit.[s] but have shown much inclin.[n] to appreciate the simulation of permanent varieties, of the character of species—that I have urged you to publish & set forth all that can be s.[d] ag.[st] me—that in no book has the gradual dying out & coming in of spec.[s] been more insisted upon, nor the necessity of allowing for our ignorance & not assuming breaks in the chain because of no sequence & of admitting lost links owing to small one observed or observable—that finally you hope your book will convince in whole or in part—. To this I c.[d] reply in a new Ed. of Manual or P. of G. wh. w.[d] act in setting the case well before the public—also that in com-

43. Charles Darwin to Joseph D. Hooker, May 9, 1856, in Lyell, *Life and Letters, 2,* 68–69.
44. Ibid., *2,* 69–71.
45. Charles Darwin to William Darwin Fox. Postmarked June 10, 1856. This letter is the property of Mr. Christopher Pearce of Canterbury, Kent, England, and a xerox copy was obtained for me by Mr. Richard French, Rhodes Scholar at Magdalen College, Oxford. Published with permission of Sir George Darwin.

munication with C. L. he has f.[d] an approximat.[n] in some points; on many that I shall be a fair judge. No unnecessary intervention of unknown or hypothetical agency.[46]

During the summer of 1856, Darwin seems to have tried to organize his ideas on the geographical distribution of species in relation to geological history. He felt very strongly that the geographical distribution of plants and animals must be explained in terms of existing geography and ordinary means of dissemination. On these points, Darwin was far more in agreement with Lyell than, from his published letters, he seems to have realized. Darwin seems to have thought that Lyell accepted Edward Forbes's theory of a former immense extension of the European continent into the Atlantic. However, Lyell, from the evidence of his own letters and notebooks, believed that the Madeiras and Canaries had been separate islands at least since the Miocene, and he cited Matthew Fontayne Maury's depth chart of the Atlantic to show the improbability of a former land connection between these islands and Europe. At the same time, in order to explain certain phenomena now known to result from glaciation, Lyell was willing to consider relative changes of the levels of land and sea in the British Isles in recent geological periods on a scale that modern geologists not only do not accept, but would consider fantastic.

By October 1856, Darwin had come to the conclusion that he could not publish a short sketch of his theory, but must begin to write a full account as well as he could from his accumulated information. He, therefore, began to write the manuscript of the large book on which he was at work in June 1858 when he received Alfred Russel Wallace's paper, containing Wallace's discovery of the theory of natural selection.

Further evidence that April 16, 1856, was the first occasion on which Darwin informed Lyell fully about his theory of natural selection is the profound effect this conversation seems to have had on Lyell's thinking. Prior to April 16, 1856, Lyell's notes on the species question are exploratory and tentative. After that day, although he remains reserved and tentative in

46. Charles Lyell to Charles Darwin, June 29, 1856, in Charles Lyell, Notebook 213, pp. 101–02, Kinnordy mss.

his conclusions, Lyell explores new directions. Without voicing his assumptions, he nonetheless asks if species have undergone transmutation, and, if this transmutation has been brought about by the continued action of natural selection on varieties, what then will the implications be? How will such a theory influence geology, natural history, and man's concept of himself?

In his second scientific journal, which Lyell began on May 1, 1856, he discussed the theory of progressive development and the reasons for its attractiveness to so many writers. He also discusses what is meant by a species. Several times, he refers to the significance of the new theory for the origin of man and the question, which it raises, of how the reasoning powers of man may have evolved out of the irrational. He draws an analogy between the slow growth of forest trees as opposed to their cutting down, which requires only a short time, and the slow evolution of species as opposed to their rapid extinction. In one passage in which he speculates on the meaning of the appearance of new species, Lyell reveals how profoundly the acceptance of the transmutation of species would touch on his religious feelings and his whole view of the world. He wrote:

> If in deciphering records relating to many millions, perhaps millions of millions of past ages, we discover much that is irreconcilable with all the popular creeds which exist now and all that have ever existed, it is no sign of our being false interpreters for it will not shake what has been common to the greater number of faiths in all ages & among all races, a belief in the Unity of the system, the intelligence, order & benevolence of the Deity. It will not alter our hopes of a future state—it cannot lessen our idea of the dignity of our race to gain such victories over Time.[47]

Similarly in an entry written in February 1857 in his third scientific journal, Lyell again tried to express his sense of the universality of the species question.

> The ordinary naturalist is not sufficiently aware that when dogmatizing on what species are, he is grappling with the

47. Charles Lyell, Scientific Journal No. II (p. 121).

> whole question of the organic world & its connection with the time past & with Man.[48]

On July 25, 1856, Lyell set out on a long geological tour through Germany, Austria, and Switzerland and did not return to London till the end of October. The initial entries in the third scientific journal were made near the beginning of this journey. After his return, there are additional entries dated from November 28, 1856 to February 9, 1857, but then there is a long gap of a year and a half with no further entries until July 11, 1858, which was after the reading of the Darwin-Wallace papers before the Linnaean Society on July 1, 1858. This gap seems to have been occasioned in the first instance by Lyell's preoccupation with a series of exciting fossil discoveries made in the Purbeck strata in December 1856.

When Lyell published the fifth edition of his *Manual* in 1855, only six species of fossil mammals were known from before the Tertiary period. These included five species of Marsupials: the *Microlestes* of the Trias, the *Spalacotherium* of the Purbeck beds discovered in 1854, the two genera of Stonesfield marsupials, *Amphitherium,* represented by two species, and *Phascolotherium,* and one species of placental mammal *Stereognathus ooliticus* also found at Stonesfield in September 1854. Lyell had been especially pleased by the two discoveries of *Spalacotherium* and *Stereognathus* in 1854, because they confirmed his belief in the extreme incompleteness of the fossil record and justified his opinion that geologists could not assert the absence of groups of higher plants and animals during earlier geological periods simply on a basis of negative evidence.

Early in December 1856, Mr. Samuel H. Beckles, an amateur geologist, consulted with Lyell in London on the desirability of exploring further the Middle Purbeck bed at Durlestone Bay near Swanage on the coast of Dorsetshire where Mr. W. R. Brodie had discovered *Spalacotherium* two years earlier. Lyell evidently encouraged Beckles to go ahead and excavate this ancient bed of soft marl. Two months later, he could write:

> As the fruit of his second day's excavations (Dec. 11th) Mr.

48. Charles Lyell, Scientific Journal No. III (p. 164).

> Beckles sent me the lower jaw of a mammal of a new genus, a discovery soon followed by others in rapid succession, so that at the end of three weeks there were disinterred from an area not exceeding 40 feet in length by 10 feet in width, the remains of five or six new species belonging to three or four distinct genera, varying in size from that of a mole to that of a hedgehog, besides the entire skeleton of a crocodile, the shell or carapace of a freshwater tortoise, and some smaller reptiles.[49]

As the fossils were unearthed, Beckles sent them directly to Lyell. In addition, Mr. Brodie sent a number of specimens he had collected since his discovery of *Spalacotherium* two years earlier. Dr. Hugh Falconer examined and interpreted the fossils for Lyell as they arrived, and some of the specimens were referred to Professor Richard Owen. If one wonders how so many fossils could be discovered so quickly where so few had been discovered earlier, it should be understood that Mr. Beckles had had removed, at his own expense, some 3,000 tons of stone in order to lay bare the portion of the dirt bed they were examining. These fossils, therefore, represented the fruits of a considerable capital investment in scientific research.

By the middle of March 1857, some fourteen specimens of mammals, belonging to eight or nine genera had been obtained from the Middle Purbeck. Lyell wrote:

> As all of them have been obtained from an area less than 500 square yards in extent, and from a single stratum not more than a few inches thick, we may safely conclude that the whole lived together in the same region, and in all likelihood they constituted a mere fraction of the mammalia which inhabited the lands drained by one river and its tributaries. They afford the first positive proof as yet obtained of the coexistence of a varied fauna of the highest class of vertebrata with that ample development of reptile life which marks all the periods from the Trias to the Lower Cretaceous inclusive and with a gymnospermous flora, or that state of the vegetable

49. Charles Lyell, *Supplement to the Fifth Edition of a Manual of Elementary Geology*, 2d ed., London, Murray, 1857, 40 pp., p. 18.

> kingdom when cycads and conifers predominated over all kinds of plants, except the ferns, so far at least as our present imperfect knowledge of fossil botany entitles us to speak.[50]

The Purbeck fossils were, therefore, a brilliant confirmation of Lyell's belief that ancient Secondary faunas and floras did involve an association of a rich variety of species belonging to different groups of animals, including higher groups such as the mammals, and that knowledge of the fossil record was as yet extremely fragmentary. Moreover, the characteristics of the association of species, Lyell thought, probably reflected the climatic conditions of the geological period in which they had lived. Apart from the opossums of North and South America, all the living species of Marsupials were confined to Australia. However, it was entirely possible that at the time of the Purbeck the climate and conditions of Europe may have approximated those of modern Australia.

> The advocates, however, of the doctrine of progressive development will offer a different explanation of the phenomena. They will refer the large admixture of marsupials in the Stonesfield and Purbeck fauna to chronological rather than to climatal conditions,—to the age of the planet rather than to the state of a portion of its dry land.[51]

But, Lyell pointed out:

> There flourished in the Pliocene period throughout Europe, Asia and America, so far as we yet know, a placental fauna, consisting of species now for the most part extinct, which was coeval with the extinct Pliocene marsupials of Australia. Such facts although far too limited to enable us to generalize with confidence, seem rather to imply that at certain periods of the past, as in our own days, the predominance of certain families of terrestrial mammalia has had more to do with conditions of space than of time, or in other words has been more governed by geographical circumstances than by a law

50. Ibid., pp. 26–27.
51. Ibid., p. 29.

> of successive development of higher and higher grades of organization, in proportion as the planet grew older.[52]

The brilliant fossil discoveries of 1857, then, seem to have caused Lyell to retreat from his tentative acceptance of the possibility of transmutation of 1856 to a reaffirmation of his earlier position, and the gap in the entries in his scientific journals on the species question seems not to have been accidental.

During the summer of 1857, still another distraction took Lyell's attention away from the species question. During a geological tour in Switzerland, he became convinced, from a study of the effects of glaciers in the Alps, of the general truth of the glacial theory. While at St. Gallen in mid-August, Lyell decided to go to Italy to revisit Vesuvius and Etna, where he had not been since 1828. His purpose was to obtain evidence from the structure of these volcanic mountains that they had been built up gradually, by many repeated eruptions, and that they had not been upheaved in a catastrophic manner. Lyell wished to destroy once and for all Leopold von Buch and Élie de Beaumont's theory of craters of elevation. His study, made with Georg Hartung, of the geology of Madeira and the Canary Islands had shown that this theory was extremely improbable, and Lyell thought that if he revisited Etna, which had been introduced in evidence by Élie de Beaumont, he could remove the last foundation for the theory.

In October 1857, Lyell spent less than two weeks in his renewed exploration of Etna, but he was able to gather a sufficient amount of striking information to present a long paper, "On the Structure of Lavas which have consolidated on steep slopes," to the Royal Society of London the following June.

Sir Charles and Lady Lyell did not return from Italy to London until just before Christmas 1857, and during the winter months of 1858, Lyell was occupied in reading works on volcanic geology for his paper on lavas. Little more than a week after the delivery of this lecture on June 10, 1858, Lyell received from Charles Darwin the letter written on June 18 with which Darwin transmitted Wallace's paper "On the Tendency

52. Ibid.

of Varieties to depart indefinitely from the original Type." Darwin wrote apparently with reference to their conversation at Down, two years earlier, on April 16, 1856: "Your words have come true with a vengeance—that I should be forestalled. You said this when I explained to you here very briefly my views of 'Natural Selection' depending on the struggle for existence." [53]

With his interest again aroused in the species question Lyell renewed his entries in his Scientific Journal No. III. He made extensive notes on Louis Agassiz's "Essay on Classification," which formed the introduction to the latter's *Contributions to the Natural History of the United States of America,* published in 1857.[54]

In September, however, Lyell went again to Italy and Sicily to renew his study of Vesuvius and Etna. This time he traveled alone and spent nearly five weeks in an exhaustive exploration of Etna; he did not return to London until late in October. His first entry on species[55] after his return is dated October 26, 1858.

In his fourth scientific journal, which he began on February 8, 1859, Lyell is clearly preoccupied with questions relating to the origin of man and the antiquity of the human race on the earth. On March 30, he wrote to Leonard Horner:

> We have been breakfasting at L.[d] Stanhope's & I told across the table the story of Dr. Falconer having just found stone hatchets & flint arrowheads with remains of extinct mammals in caves near Palermo & this leading to antiquity of Man . . . Lord Shaftesbury[56] was fully aware of the antiquity of the earth, but was a good deal put out by the idea of man being so ancient. Macaulay[57] came in well to the rescue showing how ignorant we were of chronology & how the septuagint & common version differed.[58]

53. Charles Darwin to Charles Lyell, June 18, 1858, Darwin-Lyell mss.
54. Louis Agassiz, *Contributions to the Natural History of the United States of America,* Boston, Little, Brown, 1857, 452 pp.
55. Scientific Journal No. III (p. 195).
56. Antony Ashley Cooper, seventh Earl of Shaftesbury (1801–85), philanthropist.
57. Thomas Babington Macaulay (1800–59) was successively a barrister, member of Parliament, an administrator of India, and a historian of England.
58. Charles Lyell to Leonard Horner, March 30, 1859, Kinnordy mss.

This interest was heightened further in May 1859, when Joseph Prestwich announced that he had found a flint implement, clearly the product of human workmanship, in situ in gravel at a depth of seventeen feet below the surface at St. Acheul near Abbeville in France. This discovery confirmed the numerous discoveries of flint implements at St. Acheul by the French antiquary Boucher de Perthes and showed that the presence of man in Europe extended back at least into the later Pliocene period and that man had then been contemporary with many species of animals now extinct.

Until April 1859, Lyell had been occupied with writing and correcting for the press his long paper on the structure of lavas. As a result of his second visit to Etna in the autumn of 1858, he made a number of long additions to the version delivered before the Royal Society on June 10, 1858. With this paper out of the way, Lyell's attention returned to the question of the origin of species and of man. On June 17, he wrote a long letter on the subject to Thomas Henry Huxley in which he said in part:

> If we found all the leading Classes, Orders, Families & Genera, or could reasonably hope to find them, or could fairly infer that they did exist in the oldest Periods, then we might by development get the species, or I could conceive the Genera, in the course of millions of ages. But once admit the probable want of Placental Mammalia in the Lower Silurian & we require such an event as the first appearance of that type at some subsequent Period, an event which might compare with the first coming in of any other new type—ending with Man & it becomes difficult to know where to stop. . . .
>
> If the Lamarckians are right we shall in time discover extinct fossil varieties of Men intermediate between some of the quadramana & Man? If we cannot come to any conclusion in regard to Man, the bringing to light of the mode of coming in of antecedent types is hopeless. In going back from the recent to the older periods we meet the whole difficulty at the first step in our retrospect by not finding any

> representative of Man in the Miocene period, or of any 'implement-making' brute of the Order Primates.[59]

In July 1859, Lyell went to Abbeville and Amiens to examine the localities where flint implements had been found. On August 6, he wrote to his sister-in-law Mrs. Henry Lyell:

> We have had splendid weather, not too hot for me at least, & the Amiens & Abbeville case, so far as implying that Man was coeval with the Mammoth, appears to be made out. It is only a pity we have no human skulls to show how near they come to the negro, caucasian or australian or other races. I obtained 65 recently found flint hatchets all dug up in 10 weeks since Prestwich was at Amiens & Abbeville & more than 30 others had been met with in the same short interval at Amiens alone.[60]

From Amiens, Sir Charles and Lady Lyell went to Paris and thence by diligence to Le Puy-en-Velay in the old volcanic region of southern France to investigate a report of the discovery of fossil human bones. At Le Puy, George Poulett Scrope joined Lyell who was having diggings made on the Montagne de Denise, but the results proved inconclusive.

On his return to England in the beginning of September, Lyell went immediately to Scotland to spend a few days with his brother and sisters at Shielhill, a house on their family estate, before attending the meeting of the British Association at Aberdeen. He brought with him to Scotland the proof sheets of Charles Darwin's new work on *The Origin of Species,* which the publisher John Murray had sent him at Darwin's request. Despite the many distractions of the British Association meeting, including the visit of a delegation from the association to Queen Victoria and Prince Albert at Balmoral Castle, Lyell read the *Origin of Species* while in Scotland and was much excited by it. On October 3, he wrote to Darwin:

59. Charles Lyell to Thomas H. Huxley, June 17, 1859, Huxley mss.
60. Charles Lyell to K. M. Lyell, August 6, 1859, Kinnordy mss.

My Dear Darwin,

I have just finished your volume, and right glad I am that I did my best with Hooker to persuade you to publish it without waiting for a time which probably could never have arrived, though you lived to the age of a hundred, when you had prepared all your facts on which you ground so many grand generalisations.

It is a splendid case of close reasoning and long sustained argument throughout so many pages, the condensation immense, too great perhaps for the uninitiated, but an effective and important preliminary statement, which will admit, even before your detailed proofs appear, of some occasional useful exemplifications, such as your pigeons and cirripedes, of which you make such excellent use.[61]

Lyell's entries in his fourth scientific journal during the autumn of 1859 deal primarily with the related questions of Darwin's theory and the antiquity of man. He was also in active correspondence with Darwin and in mid-November appears to have told Darwin of his own extensive reading on the species question since 1855.[62] In November, Lyell also read Joseph Hooker's "Introductory Essay" to the flora of Tasmania in which Hooker gave his support to Darwin's theory.[63]

After the publication of the *Origin of Species* on November 24, Darwin received a number of letters from friends or acquaintances who had read the book and sent him their opinion, usually critical, of it. He forwarded some of these letters to Lyell who copied portions of the criticisms into his fifth scientific journal, begun on December 6. This notebook is largely taken up with the discussion and controversy provoked by Darwin's book.

With so much discussed and written on Darwin's theory, the question arises as to how much all of this had influenced Lyell's own thinking. His opinion is perhaps suggested by an entry

61. Charles Lyell to Charles Darwin, October 3, 1859, in K. M. Lyell, *Life, Letters and Journals*, 1881, *2*, 325–26.

62. This is inferred from Darwin's reply. See Charles Darwin to Charles Lyell, November 23, 1859, Darwin-Lyell mss.

63. Charles Lyell to Joseph D. Hooker, November 13, 1859, Hooker mss.

near the beginning of his sixth scientific journal, begun on May 3, 1860. Of the theory of miraculous creation, he wrote:

> Mr. Darwin has written a work which will constitute an era in geology & natural history to show that the rival hypothesis of unlimited variability is the more probable of the two, & that the descendants of common parents may become in the course of ages so unlike each other as to be entitled to rank as distinct species, from each other or from some of their progenitors.[64]

The word "probable" in the statement "the rival hypothesis of unlimited variability is the more probable" indicates the nature of Lyell's position. He considered Darwin's theory a hypothesis that the vast amount of evidence assembled in the *Origin of Species* made probable. Lyell had a capacity for suspended judgment in the face of seemingly overwhelming evidence, which during this period was a source of repeated astonishment and exasperation to Darwin. He could never be sure of Lyell's opinion. Lyell was willing to consider the most diverse kinds of evidence, to trace out the skeins of complex reasoning, to suggest further consequences of the theory, and to point out evidence that seemed to be decisive in favor of Darwin's view, yet he refused to draw final conclusions. In Lyell's cautious judgment, the consequences of Darwin's theory were so comprehensive and staggering that it required an equally comprehensive study of its implications for every aspect of the world of living nature. Moreover, Lyell was sensitive to questions inherent in Darwin's theory that might have escaped a less critical mind. At one point he asks:

"What is this Variety-making power? That is the question." [65]

On August 28, 1860, Lyell again wrote to Darwin on the absence of mammalia in oceanic islands. Because of his interest in the Stonesfield and Purbeck mammals, Lyell had long been fascinated by the problem of the first origin of mammals, and whether this had been a single or multiple origin. In September, he also corresponded with Darwin about the possible

64. Charles Lyell, Scientific Journal No. VI (p. 407).
65. Ibid. (p. 410).

multiple origin of dogs from several different wild canine species, such as the wolf and the jackal. He included copies of several of his letters to Darwin, or portions of them, in his seventh scientific journal, begun on September 25, 1860.

Through 1860, Lyell continued to gather and examine evidence concerning the antiquity of man. In April 1861, he revisited Abbeville in France and a week or two later, on an expedition with Joseph Prestwich and John Evans to a site near Bedford where flint hatchets similar to those at Amiens had been found, Lyell took a chill that marked the onset of an illness that confined him to bed for the next three weeks.[66] During 1860 and 1861, he was also gathering material for a new edition of the *Manual of Geology,* and he planned to incorporate in it a chapter or two on the question of the antiquity of man.[67] After the delay incurred by his illness, this work kept him in London through the summer of 1861. On August 28, he wrote to Leonard Horner:

> I enjoy my mornings' work much & the references one has here in books & people cannot be had at the sea side. Falconer is in town. I never saw him in such good health. I have seen Huxley, Sharpey, Carpenter, Wilkinson & others in this deserted city.[68]

Lyell's work was slowed in late August and September by a visit to England of the Swiss palaeobotanist Oswald Heer who came to examine a number of Miocene fossil plants that William Pengelly had obtained from beds of lignite at Bovey Tracey, Cornwall. Lyell also read very attentively Heer and Gaudin's monograph on Tertiary flora. With all the additional information on palaeobotany, the antiquity of man, and other palaeontological subjects that he was accumulating for his new edition of the *Manual* or *Elements,* for he had determined to revert to the original title of the work, Lyell found that he must expand it to two volumes. This prospect disturbed him.

66. Charles Lyell to W. S. Symonds, May 12, 1961, copy, Kinnordy mss.
67. Charles Lyell to Georg Hartung, May 16, 1861, copy, Kinnordy mss.
68. Charles Lyell to Leonard Horner, August 20, 1861, Kinnordy mss.

> As soon as I prepared the new chapters for the press it struck me more forcibly than ever that I was committing three blunders at once, 1st making the Elements too large and expensive—2dly putting in to them what belonged more to the Principles & 3dly forcing all the holders of my books, if they wanted to know my latest new views to buy over again most of what they had already in one at least of my works.[69]

In consequence, he decided to take out of the *Elements* his whole discussion of the antiquity of man and of the species question and to incorporate it in a new work that he decided to prepare as soon as possible. The result of this decision was that Lyell began in December 1861 to write the work that appeared late in 1862 as the *Antiquity of Man.* There was no longer any need for him to continue his scientific journals on species and on man because he was now able to incorporate his ideas directly into the manuscript of the *Antiquity of Man.* The entries in Scientific Journal No. VII, therefore, terminate in December 1861.

69. Charles Lyell to Leonard Horner, November 5, 1861, Kinnordy mss.

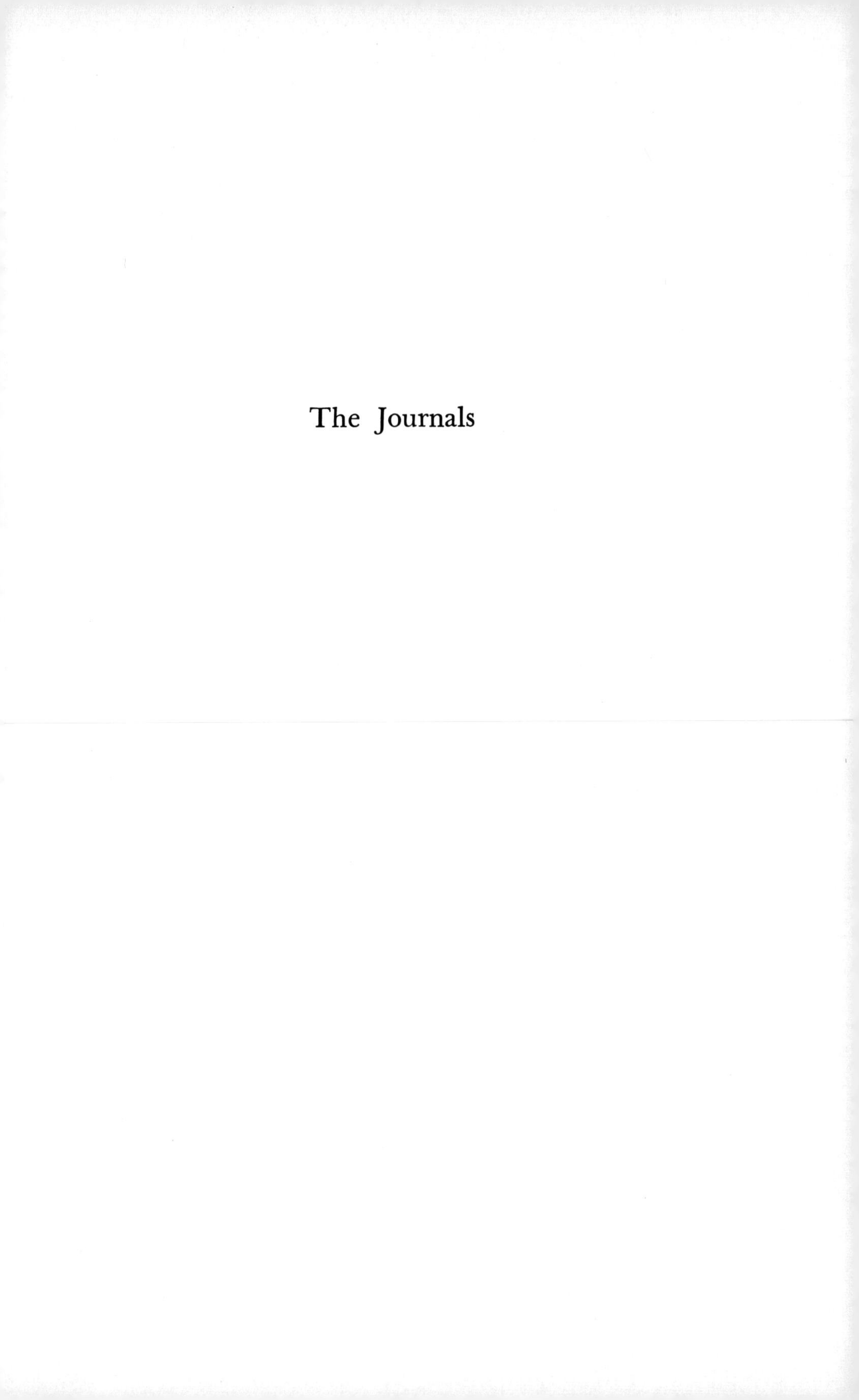

The Journals

Scientific Journal
No. I

November 28, 1855 [p. 1]

Wallace, Index Book 1, p. 31 [1]

Each species must probably be intended to exist for a given term, not an hour or day but for many generations—assume that it must continue for a period during wh. considerable geogr[l]. changes occur on the earth—ergo for thousands probably 100.[ds] of thousands of years—[2]

Let a being not omniscient have the permission to lay down the condit.[s] of a new being—wh. shall be called into existence provided it can endure.

1.st it may be a region where the range of the thermometer never exceeds 15 F. & most of the spec.[s] of the district may be incapable of [p. 2] surviving a range of 16 or 17 degrees.

2.[dly] All the attributes & instincts of coexisting species must be consid.[d] & their present and future power of increase or diminution—

3.[dly] As these last have been partly governed by a consideration of geograph. & climatal condit.[s] which have passed away & of species now extinct it will follow that the new species will partake of the characters which others of the same genus have had in reference to past condit.[s] of the animal, vegetable & inorganic worlds.

It will also bear on it the stamp of the future changes destined to affect all its future contemporaries. Hence there are [p. 3] innumerable reasons connected with the past & future as well as the present which will cause the new species to resemble those wh. exist or wh. lately existed.

4. A new aquatic species will require different attributes from an air-breathing if there be once in 50 or 500 years an intense frost, or drought, or sultriness, or electrical state of the atmosphere this must be allowed for. Epidemics & their un-

known causes physiological & other & this not only of the species but of those which are to supply them with food—or to keep down some hostile species.

5. Greater prolific power may guard against one contingency, a slight change of colour ag.[t] another. Suppose for example an insect [p. 4] destined to live thru the summer into the autumn be of a bright green & concealed by that colour while the leaves are green but not so in autumn—a slight admixture of yellow may cause it to be less readily found & devoured by birds in autumn & so carry it over its period of future danger which might exterminate it 100,000 years hence.

The Galapagos may be sufficiently analog.[s] in many respects to S. America to cause a common type of facies tho' diff.[g] in many [respects].[3]

Creation seems to require omnipotence therefore we cannot estimate it, but the above calcul.[s] stop far short of omniscience.

If there are many large lakes [p. 5] as in N. America they may for the first time coexist with a climate like that of N. America & with the antecedence of a glacial epoch & all the organic beings once fitted for that epoch & not yet extinct. If a continent had come into existence in the Southern Hemisphere of the same size & with the same geogr.[l] condit.[s] as N. America the antecedents being all different the species w.[d] be so too, & a newly introduced spec.[s] w.[d] not be the same.

The power of cattle & plants introd.[d] by man to succeed may seem oppos.[d] to this view, but if left to themselves nearly all w.[d] die out. They are comparable to spec.[s] carried occasionally by accidental causes unconnected with human interference wh. must have always occurred & have caused mere temporary [p. 6] disturbance.

Assuming therefore certain laws similar to those which require gills in a fish & lungs in a bird & certain constit.[n] to endure great cold & to enable a species to fight its way & stand its ground against others, a new species cannot at any given period be created arbitrarily. Assuming that Omnipotence rules by self-imposed laws, by which alone Man can comprehend the Universe.

Species must die out (see P. of G.)[4] & new ones come in or the Earth would be depopulated, but an ignorant naturalist might invent condit.^{s} capable of enduring or preserving a new species for a year only or a century [p. 7] or perhaps for the time required but at the expense of 2 or 3 other preexisting plants or animals whose term is not nearly run out, in which case the new creation instead of preventing the unpeopling of the earth would accelerate it.

The condit.^{s} of existence in climate [and] geogr.^{y} [and of the] plants & animals present, alone, to say nothing of [those present in] the past & future, are innumerable; even the known circum.^{s}, & the unknown probably far greater. It is scarcely conceivable that they can ever coincide twice in the history of the planet, [even] if it were to endure for millions of millions of years. Hence that the same species sh.^{d} reappear may well be regarded as impossible, or that a carbonif.^{s} spec.^{s} could thrive now, or if so that its reappearance sh.^{d} be consistent with the well-being [p. 8] of all the species wh. are now living, man included.

Rudimentary organs are a great mystery. They favour the Lamarckian hypothesis[5] tho' the arguments against such variability of species are too powerful to allow us to believe in such an hypothesis—as that the abortive legs of a snake-like reptile are the remains of a quadruped altered into a snake. It is a mystery like abortion & monstrosities & the monad of a man. The rudimentary organs without function imply at least the adherence to some system or law & it makes the assumption that a given term allotted to every species is probably one condition of creation, & the specific centres may [p. 9] be another.

It may be asked why sh.^{d} omnipotence tie itself down by restrictions—the answer is first that it does so—2.dly Man is enabled thereby to study & understand the mechanism of the org.^{c} & inorganic worlds. This may be a very small part of the reasons of fixed laws but one is enough. The St. Helena plants & insects may have lasted for the allotted term. Longevity in species may be as varied as in individuals & there may be some exceptions, but as there is an average in the individuals of a species so there may be a mean durat.^{n} in the spec.^{s} of a class or Order or Genus for the successive generations. No species may ever have been created which was not destined to run thru a certain mean [p. 10] line of generations so that its power of reproducing and perpetuating its spec.[s] shall not have been given to it in vain. Yet as some individ.[s] die soon after their birth or at the instant of it, so may species, & a volc.[c] island like St. Helena may be destroyed by the ocean or peopled by man & many spec.[s] may be permitted to be prematurely cut off. We cannot affirm this. How many other islands may have existed on wh. these species flourished—or a continent in mid-ocean now submerged & the spec.[s] found in St. Helena may have had their day, & their myriads of centuries, & the advent of Man was the signal for their extermination.

The rate of extermination is now [p. 11] so rapid as compared to the power of species to accommodate themselves to new circums.[s] that the Lamarckian Hypoth.[s] seems inadequate.

We may imagine periods when a very new species may be introduced without a prototype but this w.[d] be an exception to the general rule. The disturbance caused by Man very great & the deviation. [The recent appearance of Man is] the great argument for progression, so far as being a great innovation, but not so much as a great development—an Eocene Macacus against the theory of anthropomorphous species coming last.[6]

Of innumerable ways in which Omnipotence might fit a new species to all the present and future condit.[s] of its existence, there may be one which is preferable to all others, and if so this will cause the new species to be in all probability allied to preexisting and extinct [p. 12] or with many coexisting species of the same genus. What is called here necessity may merely mean that it pleases the Author of Nature not simply to ordain fitness, but the greatest fitness, how far consideration of mere beauty in form, of endless variety for variety's own sake, may enter into these laws is matter for speculation, but independently even of such condit.[s], there may be a propriety absolutely exceeding all others, & this may always be chosen & may cause many allied species.

The relation of the ruminant animals to Man may be the carrying out of the law of the mutual dependence of all species on each other, & not any peculiarity of the actual creation, nor any proof of its superior dignity consid.[d] apart from the Human [p. 13] Race.

The starting of each species from one point seems consistent with the simplicity of Nature. It is as easy to foresee whether a single pair will continue in the first generation as 10 first pairs for there will be this number in the 2.[d] or 3.[d] generation & then the problem of continuance for a million years will have to be solved. If I place a species each on two or more islands of a group at first, or a species capable of spreading in time to the same islands, the same forethought is required in an omniscient being, or one capable of grappling with so profound a problem. The idea of providing for the continuance of the species by a precaution, such as placing it simultaneously at first in a multitude of spots so that some one of these may succeed, is obviously to assimilate [p. 14] the creative power to a being of limited prescience. It may be that the occasion of spreading to another island of the group may not be destined to occur for a myriad of centuries & in this time the island will be submerged; if so, the species will not be placed there but in some other isl.[d], if a myriad be too short a period.

The creation therefore of species from single original stocks may imply a higher foreknowledge of future events.

But it is usually introduced (the many original birth-places) from ignorance of past geogr.[l] and geol.[l] changes which have facilitated transportation of species—by ice—for example & causes most unexpectedly discovered.

If the distribution of species in any one group of islands is best explained [p. 15] by the doctrine of specific centres, it seems reasonable to adopt it as the probable law. To dispense with such an hypothesis, as in the case of those of assumed equivocal generation, just because we are baffled in our attempts to explain the migration of species, is unphilosophical. The doctrine accounts for very many phenomena. We may *a priori* anticipate some mystery owing to our limited knowledge of the present, still more of past conditions. That more enigmas do not present themselves is perhaps more to be wondered at.

Suppose permission be given to introduce a spec.[s] into a new & large uninhab.[d] island provided it shall be one capable of enduring 10,000 years—& 5 guesses be allowed.

To the first it may be objected that after [p. 16] the first 10 centuries there will be a frost or a drought, to the 2.[d] as to the

3.^{d} guess, after 3000 yrs an increase of an inimical species & so on. All 5 may fail even though I have chosen some Europeo-Asiatic species with a very wide range & one which has existed for a long geol.[1] period & is still flourishing—To endure 10,000 yrs it w.^{d} probably require some S. American attributes if the isl.^{d} be situated like the Galapagos, common to many species of the hemisphere—some state of the electrical or other condit. of the atmosphere.

The Salvages accord.^{g} to Woodward & a rock called Cima off Baxio, P.^{o} Santo have each a peculiar land shell.[7]

Maury's[8] charts, says W. [Wollaston] show that in order that certain species sh.^{d} [p. 17] be common to the Azores & Europe they must have gone far round as they could never have crossed the Atlantic Abyss.

St. Helena—see Darwin's voyage for peculiar fauna of.[9]

December 2, 1855

E. Forbes, Fauna & Flora, p. 341 [10]

As there are certain plants common over England of the Germanic flora, wh. are not Irish & wh. tho' belonging to the same European Flora, are restricted to certain parts of Engl.^{d}, so if Engl.^{d} [were] submerged [over] all but some areas, species might be peculiar [to those areas], altho' not created there.

In like manner if we found in one of a group of Atlantic islands a peculiar species, it may have belonged to a once larger area of land.

December 2, 1855 [p. 18]

E. Forbes, *Ibid.*, p. 342

[The fact of] The reptiles not having reached Ireland, (those of the Germanic type) & certain animals like the mole & birds of short flight, leads to two conclusions—1.st the union of Ireland & England was not of indefinite length so as to give time for the spread W. (& north.^{d}) of these species—2.dly Ireland had not existed for very long as an independ.^{t} isle or part of an

Atlantis unsubmerged—had this been the case it w.[d] have contained the remnants of the reptilian fauna of the Atlantis.

The glacial submergence accounts [p. 19] for the small period during which Ireland was land within Mio- or Pliocene eras.

The glacial period was fatal to reptile life, but if England united to the continent after its separation from Ireland, then reptiles might migrate especially from the south into England after the climate had become milder, & when the extreme glacial cold was over—whereas Ireland's union with England was during the retreat of the cold.

The absence in Irish bogs and peat as well as in the Scotch [bogs] of elephant & rhinoc.[s] implies recent emergence. The Irish Elk may have migrated readily.

Works Relating to Distribution & Geological Changes of Existing Species—[pp. 20–21]

1. E. Forbes, Fauna & Flora[11]
2. Hooker's New Zealand—Introduction[12]
3. Trimmer, Joshua, succession map of geogr.[y] of Engl.[d] [13]
4. Hooker, Introduction (& Thomson) to Indian Flora[14]
5. De Candolle's introduction[15]
6. Glacial period map—Murchison's Europe[16]
7. Prestwich, forthcoming paper on tertiary & glacial drifts[17]
8. Austen's paper on the English Channel [18]
9. E. Forbes' map of Provinces of Mollusca[19]
10. " " " of Atlantis & of glacial extent of areas in Atlantic[20]
11. De la Beche's map of England & of 100 fathom line[21]
12. Hopkins, effects of Gulf Stream & calculations as to change of climate & degrees of F. altered—[22]

December 2, 1855 [p. 22]

E. Forbes, Flora & Fauna,[23] Continued

p. 344—17. The Sangatti cliff & the elephant bed of Brighton strengthen the grounds for the connexion of Engl.[d] & the continent at a very late period.

Besides the narrowness of the English Channel as contrasted with S.[t] George's channel tho' this does not apply to the Galloway approach of the 2 islands near Mull of Cantyre, but there the proximity was too far north for certain animals to get round, those of southern habits.

p. 346 bot.—(25-) Great Germanic plain destruction of—qy. glacial beds submerged?

Ib. (24)

Would not the Southern plants have been incapable of migrating into Kent when the cold prevailed—perhaps [p. 23] some of them w.[d] not.

December 3, 1855

Geographical Changes

Woodward[24] remarks that Sir J. Richardson[25] argued from the fish of shallow soundings in the N. Atlantic & from the littoral shells & the birds of short flight & the plants such as Calluna vulgaris common to Europe and Newfoundland & northern part of N. America, that there was a continuity of land in the pliocene period in the boreal & north temperate zone. If so, go back to miocene period & we may suppose [that there then existed] a connexion between the opposite lands of the Atlantic somewhat farther South & uniting [perhaps] Madeira & Azores? & N. America.

For the White Mountains were islands even in glacial period.

December 7 [1855] [p. 24]

Progressive Development

Rupert Jones[26] says that naked cuttlefish have been found fossil in the Devonian of the Hartz mountains in Germany—qy.? where reported?

Rise of Alps

9000 ft. in Pliocene or pleistocene? period, according to D. Sharpe.[27] This w.[d] agree with my floating ice and erratics on

Jura. If as much sinking in Pliocene period, then the Atlantic continent would be proved as a probable hypothesis—qy.? as to Prof. Ramsay's[28] reasons for suspecting it on glacial grounds wh. excluded his paper from the G. S. Journal?

The land has risen in Scotland 3000.[d] feet, in N. America New Engl.[d] [p. 25] [?] thousand since glacial epoch, but the whole pliocene period was many times longer in duration than the pleistocene, to say nothing of the Falunian, as an epoch in which also many existing plants and animals flourished.[29] Hence we are at liberty to assume that those parts of the Ocean, which are not deeper than the continents are high, may have been land within the era of species now living.

Hence it is always more probable that an island even like S[t]. Helena derived plants of a species of small migrating powers by the former union with a continent than that it had two birth-places, provided that the generality of facts favor the single origin of species.

December 8, 1855 [p. 26]

Mollusca

If we have not a naked cephalopod in the Silurian, so neither have we as yet an ascidian—why? because both w.[d] leave no solid relics behind them.

We ought never to wonder that we have not the very highest [form] unless we have the very simplest form of the same class or family.

If we have no snake why expect a crocodile?

If no ———, why a shark or a lepidosteus.

December 18, 1855

Rise of the Alps

Smith of Jordanhill[30] found glacial species in Museums of Switzerland, Aur and Berne (of shells.)

[p. 27] It might perhaps have been expected that the examination of the vicinity of the Congo would have thrown some light on the origin, if I may so express myself, of the Flora of S.t Helena. This however has not proved to be the case, for neither has a single indigenous species, nor have any of the principal genera characterizing the vegetation of that Island, been found either on the banks of the Congo, or on any other part of this coast of Africa. There appears to be some affinity between the vegetation [p. 28] of the banks of the Congo, and that of Madagascar, and the Isles of France and Bourbon. This affinity, however consists more in a certain degree of resemblance in several natural families and extensive or remarkable genera, than in identity of species, of which there seems to be very few in common.

R. Brown,[31] Appendix to Captain Tuckey's narrative of the Congo Expedition (p. 476) 1818.[32]

December 22, 1855 [p. 29]

Areas of Subsidence & Elevation

[1.st] Suppose the Pacific generally during the period of the growth of coral reefs to have been an area of subsidence.

2.dly that this subsidence has been going on since the beginning of the Miocene Period.[33]

3.dly That on the sinking area there have been many volcanos & chains of volcanos.

4.thly that where these last have been in action, there is a two-fold counter-acting power preventing the land from being submerged. 1.st the partial upheaval owing to injection & expansion. 2.dly the outpouring of melted matter at the surface.

[p. 30] It will follow from the above premises that the sunken area will be studded with islands of which a portion will have been saved partly by the original great height of the mountains, whether of granite or other rock or by the less amount of sinking, the rate of subsidence not being everywhere equal, & partly by volcanic action.

And as many volcanos will have become extinct in succession

since the beginning of the Miocene Period several of the archipelagos [p. 31] of atolls will represent archipelagos of extinct volcanos which w.[d] have been quite submerged, but for the coral which grew on them.

Hence if the ocean be 10,000 ft. or more deep, round an atoll, it may not imply a thickness of coral of many thousand feet, but merely a volcano which was built up while the area was sinking & would have been lost but for the growth of coral.

Hence also we need not suppose coral-building & reef-forming corals to have existed thro' several tertiary periods in the regions of atolls, but merely that they came into existence in the Pliocene or Pleistocene [p. 32] eras & prevented the disappearance of some island.[34]

The continent of South America may have undergone a great moment of upheaval while the Andes volcanos were accumulated & grew. They may have quadrupled the elevation of the Western extremity of the continent by injection & overflowing or outpouring.

To the Westward, all was sinking, save when Juan Fernandez (as Darwin reminds us) saved the land from being all lost in the Galapagos.[35]

Volcanic action on a very great scale in the globe may cause the great heaving & [p. 33] sinking to which oceans & continents are due, but volcanos may exert a subordinate superficial influence—exaggerating the elevat.[ns] in the one case & diminishing locally the disappearance of land in the other—in other words causing mountain-chains in land & groups of chains of islands in areas of subsidence, or in oceanic tracts.

If volcanos were naturally connected with upheaval on so grand a scale, as is required to produce the greater movements of lands, the extensive spaces above the level of the sea [p. 34] would be chiefly those where volcanos are most active, & instead of volcanos being so often near the sea, they w.[d] be chiefly inland?

Is there then any natural connection between the distribut.[n] of active & pliocene volc.[s] & the distribution of seas & continents.

Perhaps the Andes are more active because, on the border of an area of sinking, water of the sea being supplied. Possibly the insular posit.^{n} of many volc.^{s} is connected with their being in areas of sinking, in the middle of which they create partial elevation. [p. 35] If the Canary & Madeira islands had in the Miocene period sea beaches & were even then volcanic islands, the connection of such islands with the land goes back to a period anterior to certain Miocene deposits.

On the other hand are there not indications of sinking in the Almendrado? [36] I suspect there are, & this may have taken place in the Miocene Period & may have separated island from island, oscillation having caused marine fossil remains to be now above the sea-level.

Darwin says that the existence of the Canaries as islands in the Miocene period is against the theory of a pleistocene Atlantis.[37]

[p. 36] The S. Jorge forest, Madeira, shows that in the Pliocene Period this peculiar "Atlantic isl.^{d} flora", existed, & this is against a pliocene continent? A continental flora w.^{d} have penetrated; Darwin says that the Azores were connected with the Madeiras & Canaries, but his argument is not so much by identity of species as by analogy of genera, & this I sh.^{d} explain by similarity of condit.^{s} having ruled, in each island, when the species came into being.

If, at any period since the plants of S. Jorge existed, Madeira had been connected with the continent, the flora of the continent w.^{d} have mixed with the insular [p. 37] flora & this last with the continental flora. At the time of the Woodwardic eadicene, Madeira was insulated & so where the Las Palmas laminaria zone shells flourished, the G.^{d} Canary was insulated. Had they ever since been united with each other (Madeira and the Canaries) or with the Continent, the Flora of the several islands w.^{d} not have been so far peculiar to each. The depth of water between Madeira and Africa or Portugal [is] 9000.^{d} feet, but a number of non-volcanic eminences may have existed when the islands were forming? Suppose the Canaries to have oscillated in level. [See Fig. 1.]

January 8, 1856

[p. 38] Such oscillation may have caused them to have been as high as the Andes during a Miocene and older Pleiocene period after wh. they may have sunk to their present level, gradually during a newer Pleiocene epoch acquiring, in each island as it became isolated, a few peculiar species, & some species belonging to limited continental areas surviving in each group of islands & appearing to have been created for such islands, although in truth having once enjoyed a wider range before the submergence had proceeded so far.

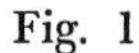

Fig. 1

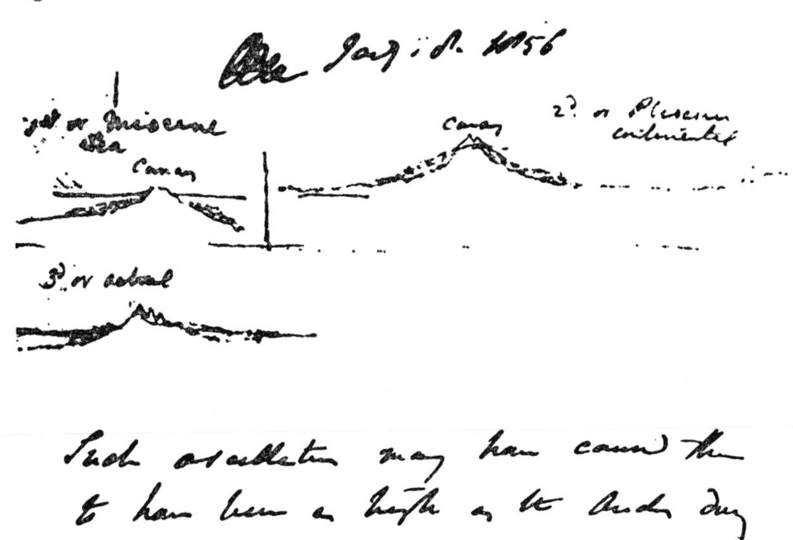

[p. 39] The interval which separated the production of the older & newer formations in Palma & Fuerteventura &, probably the oldest parts of Grand Canary & the more modern, may correspond to epochs of subsidence. First Meiocene eruptions in a Meiocene sea, & upheaval of Meiocene marine strata, then continued rise so as to unite G.d Canary & Azores & Europe. Then a long epoch of sinking during a suspension of volcanic action & the consequence, separation during the older Pliocene period of the islands. Renewal of volcanic eruption overflowing the Older Pliocene fossil plants of S. Jorge. Encroachment of the sea & format.n of valleys—in pleistocene & post-plio. period.[38] By this means we might reconcile volcanic eras with upheaval & suspension of volcanic [p. 40] action into subsidence.

January 31, 1856 [p. 41]

Copy of Letter in the Royal Society
to Sir Joseph Banks[39]

Baltimore County Maryland
January 10.th, 1794.

Sir/

I wish to send your Society information of certain remarks I made in a journey I took to Niagara in 1789, in as few words as possible.

It has been long since known that our part of the World has been convulsed by some cause heretofore, the strata on this side the Allagany [*sic*] lie in confusion. Beyond these mountains the same has not happened, the strata there lie regular, even and parallel to the horizon. [p. 42] It is curious enough in the bank of [Lake] Ontario beneath 8 or 10 strata of earth very distinct, to see a stratum of stone about a foot thick in the middle of the Bank, at the foot of it another stratum of stone 2 feet thick, this last in the bed of Ontario, cracked in pieces, and lying where Nature placed it. All the earth seems to have been carried away.

The fort of Niagara stands in an angle of [Lake] Ontario and the [Niagara] river, the country about it very level 8 miles up the river, there is a bank rising in the land 150 or 200 yards high (I guess) from the top of this bank the [p. 43] country upward is also level bounded by the Lakes Erie and Huron, at this place is the landing, 7 or 8 miles above the landing lies the cataract of Niagara which has wore away a limestone Rock from the bank aforesaid to its present place. There is evidence enough on the spot of the cataract's going continually upward, I forbear mentioning any, because if this matter be worth examining into, you may easily get better testimony from the spot than I can give you.

The Bank at Fort Detroit is, (I hear) 10 feet high, at the Cataract I guess 200 yards, then the water brakes over at the landing aforesaid. [p. 44] It must have been a sea about Detroit.

If it can be determined at what rate per annum the Cataract

wears away the rock, the age of the World or of the Deluge will be known; it is said on the spot to be half a rod a year.

I forbear troubling you further & remain Sir

Yr mst obedt humble servt

Ja.s Calder[40]

February 5, 1856 [p. 45]

The Alps were united with the Jura, or formed part of a great extent of land, or were of vast height, when the Rigi & other Miocene congloms [conglomerates] were forming along the margin of a slowly subsiding continental or large-insular tract.

1.st Period high Alpine land.
2.d " submergence 5. to 7000 feet below the Miocene Ocean.
3.d " emergence of Alps flanked with marine miocene strata, union of Alpine & Jurassic land.
4.th Another period of subsidence of 9000 feet below the glacial Ocean—D. Sharpe.[41]
5.th Period—Another of re-emergence again uniting the Jura & the Alps.

Madeira, the Canaries & the Europeo-African continent may have been [p. 46] joined to the Azores during the 1.st of these Periods & disunited during the 2.d when the *Clypeaster altus* beds were formed.[42]

During the 3.d Period an emergence common to the Alps & the Atlantic islands may have taken place. Clypeaster beds [would then have been] upraised [producing a] 2.d union of Atlantic isles with the continent.

In the 4.th Period a subsidence also common to both may have occurred—disuniting Alps, Jura, Africa & Madeira.

In the 5.th the Alps rose again but the Atlantic islands continued separate.

So that the changes in Physical Geography commonly assumed for the Alps are not so much greater than those which we introduce to explain [p. 47] the distribution of species in the Atlantic islands—for the last great upward movement of the Alpine region has no parallel in the African. [See Fig. 2.]

Fig. 2

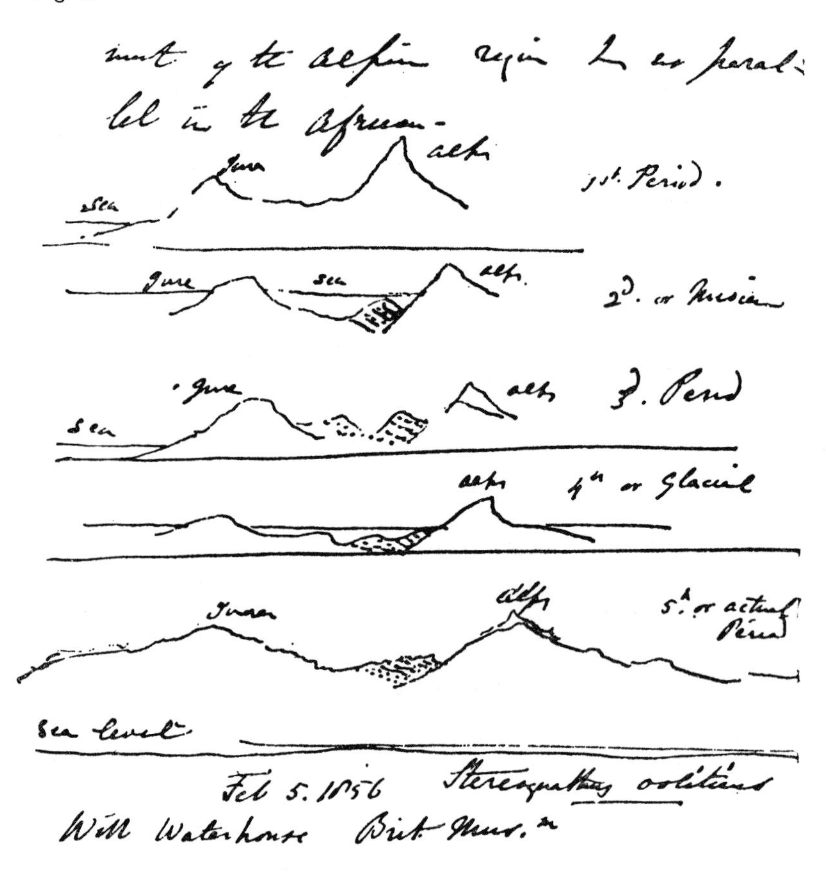

February 5, 1856

Stereognathus Ooliticus[43]

With Waterhouse[44] [at] Brit. Mus.[m]

The teeth of Stereognathus ooliticus had 3 lines of equal cusps unlike any living type of mammals & not like Migale or any shrew tho' the side view first sent to W. was like it. Stereogn.[s] oolit. was about size of guinea-pig both ends of the lower jaw broken off.

Catalogue of Madeira & Porto Santo Land Mollusca, Living & Fossil, by T. V. Wollaston, London, February 5, 1856 [p. 48]

x	means	fossil	in	Madeira proper
xx	"	"	"	Porto Santo
xxx	"	"	"	Madeira & P.° Santo
xxxx	"	"	"	wholly extinct (apparently)
[x]	"	"	"	Southern Deserta

Arion			
		1. ater L.	Mad.
Limax			
		2. gagatei Drap.	Mad.
		3. cinereus Müll.	Mad.
		4. flavus L.	Mad.
		5. agrestis L.	Mad.
		var. punctata	Mad.
		var. maculata	Mad.
Testacella			
		6. haliotidea, Drap.	Mad.
		7. Maugei, Fer.	Mad.

[p. 49]

Vitrina			
	x	8. Teneriffae, Quoy et Gaim	Mad.
		9. media, Lowe	Mad., P.° S.°
		10. Lamarckii, Per.	Mad.
Helix			
		11. membranacea, Lowe	Mad.
	x	12. furva, Lowe	Mad.
		var. α	Mad.
		var. β	Mad.
	x	13. Erubescens, Lowe	Mad.

		var. α	Mad.
		var. β	Mad.
		var. γ	Mad.
		var. δ	Des. Aust.
		var. ε	P.° S.°
xx	14.	*phlebophora,* Lowe	
		var. α (chlorata)	P.° S.°
		var. β (planata)	P.° S.°
		var. γ (decolorata)	P.° S.°
		var. δ (scrobiculata)	Ferro.
xxxx	15.	psammophora, Lowe	P.° S.°
xxxx	16.	chrysomela, Pf.	P.° S.°
xxxx	17.	fluctosa, Lowe	P.° S.°
x	18.	undata, Lowe	Mad.
	19.	vulcania, Lowe	Des. Bar., Des. Gr.
	20.	leonina, Lowe (insular var. of 19)	Des. Aust.

[p. 50]

xxxx	21.	Lowei, Fer.	
		var. α	P.° S.°
		var. β	P.° S.°
xx	22.	Portosanctana, Sow.	
		var. α	P.° S.°
		var. β	P.° S.°
		var. γ	P.° S.°
	23.	armillata, Lowe	Mad.
	24.	Michaudi, Desh.	P.° S.°
	25.	pisana. Müll	Mad., P.° S.°
xx	26.	Subplicata, Sow.	P.° S.°
xxxx	27.	Bowdi[t]chiana, Fer.	Mad., P.° S.°
	28.	punctulata, Sow.	
		var. α (setulosa)	P.° S.°
		var. β (avellana)	Des. Aust., P.° S.°
		var. γ (solida)	P.° S.°
(x)x	29.	*vulgata,* Lowe	

		var. α (trifasciata)	Mad., Des. Gr., Des. Aust.
		var. β (canicalensis)	Mad.
		var. γ (giramica)	Mad., Des. Bor.
xx	30.	lurida, Lowe	
		var. α	P.º S.º
		var. β	P.º S.º
	31.	laciniosa, Lowe	Des. Bor., D. Gr.
xx	32.	depauperata, Lowe	P.º S.º
x	33.	Squalida, Lowe	Mad.

[p. 51]

	34.	latens, Lowe	Mad.
xx	35.	oblecta, Lowe	P.º S.º
xxx(x)	36.	paupercula, Lowe	Mad., P.º S.º, D. Bor., D. Gr., D. Aust.
	37.	pulchella, Müll.	Mad.
	38.	pusilla, Lowe	
		var. α (annulata)	Mad., Des. Gr.
		var. β (scricina)	Mad.
	39.	Gueriniana, Lowe	Mad.
	40.	cellaria, Müll.	Mad.
	41.	Scintilla, Lowe	Mad.
	42.	crystallina, Müll.	Mad.
x	43.	bifrons, Lowe	Mad., Des. Gr.
x	44.	calathus, Lowe	Mad.
	45.	deflorata, Lowe	Mad.
	46.	Armitageana, Lowe	Mad.
(x)x	47.	actinophora, Lowe	Mad., D. Gr., D. Aust.
	48.	arridens, Lowe	Mad.
	49.	lentiginosa, Lowe	Mad.
	50.	stellaris, Lowe	Mad.
x	51.	arcta, Lowe	Mad.
		var. α	Mad.
		var. β	Mad.

xxxx	52. arcinella, Lowe	
	var. α	Mad.
	var. β	Mad.

[p. 52]

x	53. capsella, Lowe	Mad.
x	54. *fausta,* Lowe	Mad.
x	55. obserata, Lowe	
	var. α	Mad.
	var. β	Mad.
x	56. calva, Lowe	Mad.
xx	57. rotula, Lowe	P.º S.º
xx	58. consors, Lowe	
	var. α	P.º S.º
	var. β	P.º S.º
	var. γ	P.º S.º
	var. δ	P.º S.º
xx	59. calculus, Lowe	P.º S.º
xx	60. compacta, Lowe	
	var. α (vulgaris)	Mad.
	var. β (major)	Mad.
	var. γ (Portosanctana)	P.º S.º
	var. δ (pusilla)	P.º S.º
xx	61. commixta, Lowe	
	var. α (major)	P.º S.º
	var. β (pusilla)	P.º S.º
xx	62. abjecta, Lowe	
	var. α (conulus)	P.º S.º
	var. β (conuloidea)	P.º S.º
	var. γ (subdepressa)	P.º S.º
	var. δ (bicingulata)	P.º S.º

[p. 53]

xxx	63. sphaerula, Lowe	
	var. α	Mad.
	var. β	P.º S.º
	var. γ	P.º S.º

	xx	64. bisarinata, Sow.	P.º S.º
		65. echinulata, Lowe	P.º S.º
	xx	66. oxytropis, Lowe	P.º S.º
		67. turricula, Lowe	Cima
	xxxx	68. vermetiformis, Lowe	P.º S.º
	xx	69. mustelina, Lowe (var. of 70)	P.º S.º
	xx	70. cheiranthicola, Lowe	
		var. α (spilospira)	P.º S.º
		var. β (taeniospira)	P.º S.º
		71. polymorpha, Lowe	
		var. α (pallescens)	Mad.
72		var. β (nigricans)	Fora
	xx	72. pulvinata, Lowe	P.º S.º
	xx	73. attrita, Lowe	P.º S.º
		74. tabellata, Lowe	Mad.
	(x)x	75. syenilis, Lowe	
		var. α (nebulata)	Mad., Des. Gr.
		var. β (vittata)	Des. Gr.
		var. γ (pusilla)	Des. Bor., Des. Aust.

[p. 54]

72		76. lincta, Lowe	
		var. α (rosea)	Mad.
		var. β (cinerea)	Mad.
	xx	77. papilio, Lowe	Baxo
	xx	78. discina, Lowe	P.º S.º
	xx	79. testudinalis, Lowe	P.º S.º
		80. Lyelliana, Lowe	Des. Gr.
	xx	81. Albersii, Lowe	P.º S.º
		82. *Bulweriana,* Lowe	
		var. α	P.º S.º
		var. β	P.º S.º
	xx	83. tectiformis, Sow.	P.º S.º
	xxxx	84. Delphinula, Lowe	Mad.
	x	85. tiarella, Webb	Mad.
	xxxx	86. coronula, Lowe	Des. Aust.

xx	87. coronata, Desh.	P.° S.°
	88. *compar,* Lowe	Mad.
	89. Maderensis, Wood	Mad.
	90. spirorbis, Lowe	Mad.
	91. leptosticta, Lowe	Mad.
(x)	92. micromphala, Lowe	Des. Gr., Des. Aust.

[p. 55]

	93. dealbata, Lowe	
	var. α	P.° S.°
	var. β	P.° S.°
	var. γ	P.° S.°
	94. fictilis, Lowe	P.° S.°
	95. lenticala, Fer.	Mad. P.° S.°
xxxx	96. lapicida, L.	P.° S.°
xx	97. Webbiana, Lowe	P.° S.°
xx	98. Wollastoni, Lowe	P.° S.°
Bulimus		
	99. ventrosus, Fer.	
	var. α	Mad. P.° S.°
	var. β	Mad. P.° S.°
	100. decollatus, L.	Mad.
Achatina		
	101. Maderensis, Lowe	Mad.
	102. folliculus, Gron.	
xx	103. gracilis, Lowe	
	var. α (terebella)	P.° S.°
	var. β (subula)	Cima
	var. γ (vitrea)	P.° S.°, Des. Gr.
	104. acicula, Müll	Mad., Des. Gr.

[p. 56]

xxxx	105. eulima, Lowe	P.° S.°
[x]	106. producta, Lowe	Des. Aust.
	107. mitriformis, Lowe	
	var. α	Des. Gr., Mad.

		var. β	Des. Bor., P.° S.°
x	108.	tornatellina, Lowe	
		var. α	Mad.
		var. β	Mad., P.° S.°
xx	109.	melampoides, Lowe	
		var. α	P.° S.°, Cima
		var. β	P.° S.°, Cima
	110.	tuberculata, Lowe	
		var. α	P.° S.°
		var. β	P.° S.°
xx	111.	oryza, Lowe	P.° S.°
xx	112.	triticea, Lowe	P.° S.°
	113.	Leacociana, Lowe	Mad.
xx	114.	ovuliformis, Lowe	P.° S.°
xxx	115.	cylichna, Lowe	Mad.
Pupa.			
	116.	limnaeana, Lowe	Mad.
	117.	microspora, Lowe	Mad.

[p. 57]

xxxx	118.	linearis, Lowe	Mad.
	119.	Fanalensis	Mad.
	120.	anconostoma, Lowe	Mad.
	121.	cheilogona, Lowe	Mad.
	122.	vincta, Lowe	Mad.
	123.	*irrigua,* Lowe	Mad.
	124.	laurinea, Lowe	
		var. α	Mad.
		var. β	Mad.
x	125.	*sphinctostoma,* Lowe	
		var. α (rupestris)	Mad.
		var. β (arborea)	Mad.
	126.	laevigata, Lowe	Mad.
x	127.	*recta,* Lowe	Mad.
	128.	macilenta, Lowe (var. of 127)	Baxo, Mad., Des. Gr.
	129.	fusca, Lowe	Mad.

(x)x 130.	millegrana, Lowe	Mad., Des. Gr., Des. Aust.
131.	Ferraria, Lowe	P.º S.º
xx 132.	monticola, Lowe	P.º S.º
xx 133.	calathiscus, Lowe	P.º S.º

[p. 58]

x 134.	cassida, Lowe	Mad.
x 135.	cassidula, Lowe	Mad.
136.	concinna, Lowe	Mad.
xxxx 137.	abbrevista, Lowe	Mad.
x 138.	gibba, Lowe	Mad.
x 139.	lamellosa, Lowe	Mad.
x 140.	saxicola, Lowe	Mad.
141.	seminulum, Lowe	Mad.
Balea		
142.	perversa, L.	P.º S.º
Clausilia		
x 143.	crispa, Lowe	Mad.
xx 144.	deltostoma, Lowe	
	var. α (raricosta)	Mad., P.º S.º
	var. β (crebristriata).	Mad., D. Bor. (D. Gr., Des. Aust.)
145.	exigua, Lowe	Mad.
Craspedopoma		
xxx 146.	lucidum, Lowe	Mad., P.º S.º
x 147.	Lyonnetianum, Lowe	Mad.

[p. 59]

Limnaeus		
148.	truncatulus, Müll	Mad.
Ancylus		
149.	fluviatilis, Müll	
	var. α	Mad.
	var. β	Mad.

Fifteen Living Species Common to the Madeiras and to the South of Europe

None of them fossil

Helix cellaria	Mad.
" crystallina	Mad.
" pisana	Mad., P.° S.°
" pulchella	Mad.
" lenticula	Mad., P.° S.°
Achatina acicula	Mad., Des. Gr.
Zua Maderensis (lubrisa)	Mad.
" folliculus	Mad.
Bulimus decollatus	Mad.
" ventrosus	Mad., P.° S.°
Balea perversa	P.° S.°
Lemnula	Mad.
Truncatula fluviatilis	Mad.
Testacella N.° 6 & 7. p. 48	Mad.
Living spec. formed fossil	
Helix lapicida	P.° S.°

[February 8, 1856]

[C. Lyell to T. V. Wollaston[45]]

My dear Sir, [p. 60]

I have just traced outlines of the different sizes of a well known Mediterranean species in my collection as found in Italy and Africa and on the identity of which E. Forbes who gave me the African individual never doubted.

Still it may fairly be said that as we cannot without a violent hypothesis imagine the climatal & geographical condition in P.°S.° to have changed suddenly, we have a right to ask where are the intermediate & transition states between the extinct & living varieties (if such they were) of H. Lowei, & its smaller counterpart and of H. Bowditchiana. It is a very singular question of evidences. Perhaps no island has altered so much in size as P.°S.° judging by the extent of the 100. f[atho]m line area. The Caniçal district also is a fragment, not only the corresponding slope *a*, but the axis *B* having been carried away by the Atlantic during, I presume, several oscillations of level & a vast number of centuries. These changes may have made some species more local & have exterminated others, & have starved some and dwarfed them, and they may also explain why a small [p. 61] narrow ridge like the Desertas, may have far more than its fair share of peculiar species as some of them may have belonged to the wider area of which the Desertas is certainly the last remnant, for the exertion of the creative power may (& I imagine must have) a relation to extent of area. For the continuance or endurance of a species must depend, *inter alia,* on extent of range & if one of the laws of species be that they are to last for many generations, they will not be formed except where there is space enough, actual or about to be.

Your criticisms on the 14 species were very necessary, H. lapicida being a living species elsewhere, the number of positive cases is reduced to about 6, if the giants were eliminated, & your discovery of H. tiarella is indeed a warning. Yet I am inclined still to believe that according to the doctrine of chances where some, once prevalent & rich in individuals, are now more confined in range & even in several cases all but extirpated, there must be a few quite lost. There has been a vast revolution in the small world inhabited by them. The Desertas have suffered [p. 62] as if Ireland were cut down to one county. But the glacial period proves how small a variation in the percentage of species, whether of marine or terrestrial shells, answers to a vast amount of change in Phys. Geogr.^{y} or of alteration in the mammalia.

C. Lyell

[February 11, 1856]

T. V. Wollaston, Esq.[46]

To the same Feb. 10.

If the smaller representation of H. Lowei and H. Bowditchiana exist fossil in the shapes of H. Portosantana & H.

punctulata in Caniçal etc., then in the absence of intermediate vars. we must conclude that they are true species? But grant them to be only permanent varieties, these vars. have become extinct, after having persisted to exist for ages, fixed by certain long-prevailing conditions which then ceased.

The disappearance of these large shells seems certain. [p. 63] The Natchez loess land shells[47] come nearer to those of the plains of the Mississippi than do the Caniçal & P.°S.° shells to the mollusca now inhabiting the Madeiras. Yet the Natchez shells are above 100,000 years old & were contemporary with the Megatherium & Megalonyx. So the Grays Thurrock shells of the valley of the Thames do not depart farther from the living British fauna (the land shells perhaps less far) than do the Caniçal from the existing land shells of Madeira. But the Grays' beds contain Rhinoceros, Hippopot.ˢ, Eleph.ᵗ to say nothing of Macacus pliocenus.

If we accept Heer's[48] estimate, 1/4 of the plants of the fossil leaf-bed of S. Jorge are not now living in Madeira. In order to reduce the island to the size which it had when those leaves were green, we [p. 64] should have to take away 1100 or 1200 ft. of lava from S. Jorge & 2500 perhaps from some parts of the central region. Still the island may have been as large as ever for to say nothing of local movements the level of the floor of the whole Atlantic, relatively to the ocean, may then have been very different.

But the Flora of S. Jorge seems clearly to prove that Madeira was then as now covered with laurel & myrtle & heath & ferns & other plants belonging to an island of the Atlantic & not to a continent.

There have been no land shells found in the old tuffs of S. Jorge. [p. 65] The list of 13 S. European species sent (by Wollaston), all of them not as yet met with fossil at Caniçal & P.°S.°, seems to imply that they have all been introduced in modern times, unless perhaps H. pisana be an exception & Balec perversa f.ᵈ on the top of P. de Facho & [?] ft. high in P.°S.°

It would follow from this, that in tracing back the history of the fauna of Madeira, the signs of its connexion with a continent become less strong, (altho' Helix lapicida not now living occurs fossil, at ——). Its separation from P.°S.° is also implied by the fact that of the fossils 42 are [p. 66] peculiar to P.°S.° & 32 to Madeira & only 4 common to both. The first step therefore in our retrospect leads us to infer complete insularity. The next step to the S. Jorge beds probably embracing a period many times as great as the long interval which separates the Caniçal beds from our time still implies an island.

Again the S. Vicente & Baxio marine limestone with their well-rounded becah-pebbles imply the vicinity of land consisting of volcanic rocks & the Las Palmas beds in the Grand Canary prove a littoral deposit formed [p. 67] on the shores of a pre-existing land formed of volcanic rocks.

The G.^{d} Canary therefore & Madeira were apparently islands in the Atlantic in the Miocene period as well as in the S. Jorge Pliocene era.

When, therefore, was there any land communication between the islands or between them and any continent? Is it indispensable in order to explain the immigration of insects, plants & shells common to the different islands & to the continents? Would it suffice if instead of a continuous land communication we merely imagined such movements in the floor of the Atlantic ocean as w.^{d} cause at certain periods of the past a great projection of promontories from S. Europe to the W. and S. W.—& the laying [p. 68] dry of numerous shoals between the present archipelagos of the Azores, the Canaries, Salvages & Madeira.

The insular character of the fauna & flora of each island in several past periods, the non-fusion of the species peculiar to each, would all be explained by such an hypothesis.

Certain winged species might in the course of ages, Sphinx stellatarum etc., be carried by the winds.

Dr. Hooker observes that as the destruction of the forests in St Helena may have rapidly exterminated some land shells, so many may have perished from the same cause in Madeira. I may add that still more the immigration with man of 13 species of land [p. 69] shells & of innumerable cultivated plants, some of them not fitted, or so fitted, as food for the indigenous shells, but suited to the new comers may have caused great havock among the aboriginal inhabitants.

Locally also the climate has been altered & is less damp. The rivers have diminished in dimensions, as the Soccoridos.[49]

Then we have the garden flowers brought in for ornament & the numerous weeds of wh. species are supposed to have come in around Funchal alone.

In proportion as the range of a species is limited, as in small islands like Madeira & P.°S.° & the Desertas, the influence of new plants & animals as exterminating causes is great & rapid. Therefore [p. 70] the combined effects of climate & of the changes in the Flora & Fauna & of Man himself & the cattle is so enormous that a few centuries w.d produce in the Madeiras as great alterations as the same number of 1000.s of years in Europe. There is 1.st the loss of native plants fitted for certain helices as food or shade, 2.dly the inroad of foreign species suitable as food to newly imported slugs & snails, 3.dly the loss of dampness & change of climate.

For these & other reasons we must not apply the ratio deduced from the pleistocene beds of the Thames & Mississippi to the rate of change which may have occurred in P.°S.° & Madeira, altho' the loss of land seems to have been so great, since the sand-dunes of both [p. 71] islands was formed, that I doubt not that the lapse of time has really been very vast.

The fossel lichens of P.°S.° ? were shown to Dr. Hooker who thinks that they may be specimens of Cenomyce.

Feb. 11. C.L.

February 12, 1856

Palaeoteuthis of Bronn[50], of Devonian (Paleontogr. V. Meyer)[51] Salter[52] says a fish bone—letter Woodward

February 17, 1856

Geographical Distribution

Woodward *Manual of Conchology*, p. 390.[53] E. Forbes seems to have assumed that S.t Helena's land shells imply the former proximity of Africa & America because certain *forms* not species [p. 72] are common in the Bulimus family to Brazil or Tropical America & St Helena. This view may be consistent with E. Forbes' view of specific centres, for a greater proximity of land w.^{d} give rise to a greater analogy of conditions when the spec.^{s} now extinct were born.

Immigration of Land Shells

Woodward^{pt} vol. 3. p. 391. (top)[54] Helix cellaria & H. Pulchella being found at the *Cape* of G. Hope, where all the other spec. (60 or more) are peculiar, is in favour of these same shells being also imported into Madeira.

Copy [of Letter] To Mr. Wollaston[55] [p. 73]

Feb. 12. /56

I am much obliged to you for the list of fossil shells, and more especially as you have reduced some of the species, so that I have it in an amended form. In this list H. coronata (which was given as from *Bugio* in the small box among the shells you lent me) is given as from P.^{o}S.^{o} and H. coronula is from Bugio. I presume that the list is correct. (yes—Feby. 20.th).

Unless I hear from you to the contrary I shall follow the list & not the naming of the specimens, but I [p. 74] mention it as I commented on the species in reference to your discoveries in Bugio.

I cast many a wistful glance at the Desertas, and feel what it is to have mislaid or lost your Bugio fossils, for I was told it was madness to attempt to land. Even Vidal lost all his nautical or surveying instruments, being upset in a boat there when attempting to land—no small pecuniary loss, as well as hindrance in other ways.

I have been trying to get at the proportional number of extinct and recent land-shells, deducible from your lists. According to the last corrections I find in Madeira Proper six extinct species of shells (the slugs must be excluded in such calculations): [p. 75]

1. H. Bowditchiana	2. H. arcinella,
3. H. delphinula	4. H. cylichna
5. Pupa linearis	6. P. abbreviata.

In your first list I find 85, including the above fossils, as Madeira shells (Madeira proper) which would give between 7 and 8 per cent of extinct, a large number, and of vast importance in geological chronology, unless we deduct largely from its value from considerations gone into my last letter, and derived from the probability of recent extermination by man & by the plants & animals accompanying him, or following in his track.

But perhaps a fairer estimate of the change of the mollusca would be arrived at if we take the 35 fossil shells [p. 76] of Madeira proper, as representing the fragment of a fauna of which 6 are lost. If we could, by dint of searching find 70 sp. of fossils, we might expect to have 12 extinct, and so on, in the ratio of about 17 per cent. Would not this be a fair view of the divergence of the recent and fossil fauna?

But in P.°S.° there are 42 fossils of which 4 only are extinct (excluding the lapicida) and this gives only a proportion of about 9 per cent, as if the change had been greater in Madeira, or as if the extirpation of living species had been twice as great in the more cultivated island? and the island into which 3 times as many exotic species have been introduced.

[p. 77] Does not the comparison of Madeira and P.°S.° rather favour your theory of the recent extirpation of the lost fossil species—even though there may be no reasonable hope of ever detecting all of them in a living state.

A careful examination of the Desertas would be valuable to ascertain in an island, with which man has least interfered, the proportional number of individuals of each species fossil and recent. I cannot but imagine that when the first colonists landed in Madeira they would have found H. tiarella, very rare in a living state, tho' it once abounded as the fossils prove.

Do you happen to remember anything about the circumstances as to depth [p. 78] below the surface, or locality, or companionship or other conditions in which H. lapicida was found in P.°S.° When Humboldt talks of tropical plants from

W. Indies, or even canoes drifting to the Canaries, one hardly knows what may not have happened in the course of ages, to carry a S. European shell to P.°S.° [56]

As H. Pulchella and H. cellaria occur at the Cape of G. Hope, you may fairly assume that they were imported—and if so, *a fortiori,* into Madeira. Suppose Man, modern as he is geologically, to have existed at least 20,000 years, as some philologists suspect, he may have aided in importing shells involuntarily at remote periods into Atlantic islands. [p. 79] The wreck of a canoe may have reached an island long before the rude people by whom they were constructed got there. Balea perversa is a remarkable case, certainly, and a grand discovery.

If a large proportion of the 13 South European shells, now living in Madeira, have been introduced by Man, is it not remarkable that so few of the P.°S.° species have been naturalized in Madeira and vice versa. The Baxio barges laden with limestone for the Funchal kilns, might have been expected to carry Helices, to say nothing of smaller boats. But altho' the fact of the small amount of fusion is at first sight very striking, if not perplexing, you may perhaps draw a conclusion from it, favourable to the [p. 80] activity of human intervention, as the chief importing cause. For I suppose there have been 10 or 20 vessels arriving from Europe at Madeira, for one P.° Santan boat or barge. If I read your first table correctly, there are 136 species of living shells in Madeira, & P.°S.° taken together, including the Desertas, & not counting the slugs except Testacella, which has a shell. Deduct 13 S. European species most or all of which might have been imported, and we have 123 species.

Now I find 4 of these No. 36—63—115—& 146, common to Madeira & P.°S.°. If I read aright we have thus only 4 in 123[57] as [p. 81] representing the amount of amalgamation whereas in the fossil fauna, we have 5 in 72 common to Mad. & P.°S.°, the 5 being H. Bowditchiana, H. paupercula, H. spherula, H. compacta, & Cyclostoma lucidum.

The different & greater degree of approximation thus indicated in the latter, unless you can explain it (in some way which I cannot as yet) is of very considerable significance, for in the case of the 2 fossil faunas, we have 7 per cent in common, & in the recent, scarcely have the 3 per cent. I should infer from these facts that there was a land of passage between P.^{o}S.^{o} & Madeira at a time when the two islands were not united with S. Europe, [p. 82] then a separation, and a certain coming in & going out of species, after the disunion.

Suppose the facts had looked in an opposite direction & the 2 recent faunas had had 7 per cent in common, I should then have suspected that the early human settlers had imported shells from one island to another for a long period antecedently to the springing up of a more active intercourse between Europe and Madeira. I had anticipated some such a result and therefore attach more meaning to the facts, 1.^{stly} the absence among the fossils of [p. 83] living European species, except the H. lapicida in one island, and 2.dly the greater affinity of the ancient faunas to one another than the recent ones of the two islands.

In this way, if I mistake not, you could reconcile the migrations of insects at which you hinted. Of course the mountain tops must have lowered if there has been an extensive subsidence within the period of existing species. Geology establishes great general upward & downward movements, extending over wide areas, accompanied by subordinate oscillations. Within the last two years, one island in New Zealand [p. 84] has gone up, and the southern island gone down 12 feet.

Dr. Hooker tells me that almost every plant of many thousands, 800 leguminosae alone of S. West Australia differs specifically from the plants of S. E. Australia. The intervening tracts are low, & as I know by fossil shells, sent to me of recent marine species, have emerged in the modern tertiary or post-tertiary period. We cannot compare this phenomenon with the former union of Mad. & P.^{o}S.^{o} but we may compare it with a supposed union by a low intervening land, of one of the distinct [p. 85] Atlantic groups of islands, the Azores & Madeira, or Madeira & the Canaries, & it proves that a long continued land communication may exist consistently with a continued separation of the floras, even where there is no difference of latitude. This consideration may perhaps aid you in regard to the migration of Coleoptera. Preoccupancy is evidently a powerful barrier against foreign invasion, and such cases as the recent progress of Anaeris alcynostrum in this island are exceptional. Have you attended to this prodigy? Hooker suggests a strange remedy to stop its alarming rapidity of propagation, [p. 86] namely, the introduction of the otter or male sea[l]. I have re-examined the lizard's bones, may I show them to Prof. Owen.[58]

On looking over your first great list again I fear I have omitted H. compacta from the species now common to Madeira & P.°S.° This would make the proportional number of recent shells common to Mad. & P.°S.° nearly 5 per cent, the fossil being 7 per cent. The only shell once common to both islands & now lost in one of them is H. Bowditchiana which together with H. Punctulata [p. 87] has disappeared from Madeira. All the others which can be shown by your fossil list to have ever got a footing on both sides of the channel have stood their ground. H. cylichna is the only species now common to both islands & only found fossil in one, on Mad., but this is a negative fact. It may turn up fossil in P.°S.°

C. Lyell

February 19, 1856

Mr. Wollaston mentioned a friend in Funchal having received in a flower pot, in which a plant was sent, (8?) species of helices alive—from Europe?

February 20, 1856 [p. 88]

Cretaceous

Mr. Barrett (L)[59] says that in the Cambridge Museum they have Teleosaurus from the chalk. The more the large reptiles enter into it, the more are the Stonesfield and Purbeck mammals included in the Age of Reptiles. According says Mr. B. to the progression system, Cetacea ought to have appeared before terrestrial mammals.

Mr. B. with Mr. MacAndrew[60] has lately dredged up tessellated star-fish of ½ inch diameter from seas of Norway, far north. These were suppos.^{d} to imply warm climates of Chalk. Also Schizaster (fragilis?) allied to Micraster in 150 f.^{ms} also beds of terebratula with a long loop.

Coralline Crag[61]

Mr. L. Barrett also says that in the region of Limopsis pygmida coralline-crag-like zoophytes were [p. 89] common. They also found Buccinum Dalei of the crag common in some of the northern regions.

They found the ear-bones of fish but no bones of same.

Progressive Creation

Dana's[62] Article, (see Index Book, 1.[63] p. 45) is another powerful adhesion to Agassiz,[64] Hugh Miller,[65] Owen,[66] Salter,[67] Sedgwick,[68] Dawson[69] & others—Bronn[70] and Heer[71] also. The "*6* days" of Dana & Dawson & H. Miller lessen their weight somewhat as judging on purely independent evidence—& the Stonesfield mammals are said by Dana to come in as a prelude to the Tertiary ones as tho' they immediately preceded in [p. 90] time.[72] Mr. Barrett says that he has a Green Sand vertebra that may be cetaceous, but that assuming Cetacea to be absent, it is a fact ag.^{t} the develop.^{t} that they do not appear *before* the Oolitic mammals.

Every year that the Lower Silurian, tho' searched in the U.S. & in Europe, refuses to yield fish, adds to the weight of negative evidence, but fish, says Barrett, except in genera affording ear-bones, are not brought up by the dredge.[73]

Every year that adds to the Oolitic mammalia, without any discovery of Birds or Mammalia in the Cretaceous & [p. 91] Wealden, adds to the argument in favour of the possibility of arriving at lengthened periods in which certain classes are absent or starved altho' in anterior times they flourished.

The recent discovery of reptiles in the Connecticut Red Sandst., if identifiable with the supposed footprints of birds, is important as showing—like the negation of perpetual motion—that the negation of the ancient existence of Birds leads to truth, or guards against hasty generalizations.[74] Leidy[75] & Agassiz suspected reptiles & Owen hesitated because of the great antiquity of the rock—so did H. v. Meyer—(see Paleontogr.[y] [p. 92] Oeninghen bird fig.[s]).[76] All the footprints, according to the number of toes, may be reptilian, says Owen, & cannot be mammalian—for then no toe c.[d] have more than 3 joints. If mammalia existed, when so many creatures walked over the Connecticut & Corncockle Muir sandstone, why do we find none of their tracks?

On the other hand it sh.[d] not be forgotten that, as Prof. Hitchcock[77] always knew there were reptiles (quadrupeds) as well as supposed bipeds (ornithic[nites] ?) in the Conn.[t] [p. 93] rock; we might always have anticipated the finding of reptiles first; as the bones of birds in the Tertiary strata, when we know that birds did exist, can so seldom be detected in comparison of the contemporary reptiles.

It is the pterodactyl affinities of the Conn.[t] bones, first met with, which favours the progression system rather than their reptilian character.

Dana speaks now of the carbonif.[s] era as the Age of Amphibians only 12? or 13? years after the first indicat.[n] of a batrachian in the Coal.[78]

Connecticut Sandstone

H. Rogers[79] & his brother[80] incline to place it between Trias and Lias, but the [p. 94] proofs from (Etherea?) or Possidnomys do not seem conclusive of the identity of the Richmond, V.[a], beds & the Connect.[t] New Red—nor are the Durham fishes & the Richmond ones of same genera, tho' H. D. Rogers hints at some recent addit.[n], of Redfield [81] of Pompton fish like the Richmond V.[a] fish.

February 20 [1856]

Red Crag

Owen finds chiefly Miocene quadrupedal teeth in it, Rhinoceros Schaueneaucken & another not R. tichorinus & leptorhinus, Mastodon angustibus, Tapires, etc. none of them Pleistocene species—Xiphius like the Antwerp? many like Eppelsheim.[82] Now as the marine shells are more & more pliocene (Limopsis pygmida) then mammalia may have been washed into the sea from old cliffs? Morris[83] remarks that there is no Miocene now in Holland & asks where [p. 95] the bones came from. But when the levels were different & when no loess covered up the Hesbayan plains, there may have been some continuation of the Rhenish Miocene farther north, some perhaps utterly destroyed by denudat.[n].

Prestwich says that his investigations from Cromer to Ipswich [along the eastern coast of England] have led him more & more to identify the Red & Mammaliferous crags, but qy, [?] are the mammalia the same? The more modern he makes the Red Crag by assimilating it to the Norwich, the more impossible is it that the mammalian remains described by Owen can belong to the period of the Red Crag Shells to which that of Antwerp belong.

February 21, 1856 [p. 96]

Gastornise, Sables de Rilly

The position of this bird was ascertained when the femur was found after the tibia. It was as placed in C. d'Orbigny's[84] list of Paris form.[s], at bottom of plastic clay & mottled clay at Meudon. Over the conglom. with Coryphodon,[85] A. [See Fig. 3.]

Prestwich seemed almost to incline to think that this great wader inhabited the land at the period of the Thanet Sands. But not sure that the bird is really older than the bottom of Plastic Clay, in which case it might be contemporary with [p. 97] the Kyson & Woolwich mammalia. The Rilly beds, Prestwich thinks, [are] not where Ch. d'Orbigny has placed them, but higher up, [that is they were formed in] a pond on the surface of the chalk, filled with land shells washed into it & all distinct & peculiar species.

Bowerbank[86] said that the oldest known bird in England was the Wealden tibia? of Mantell's[87] where is it? Owen says he never saw it.

Fig. 3

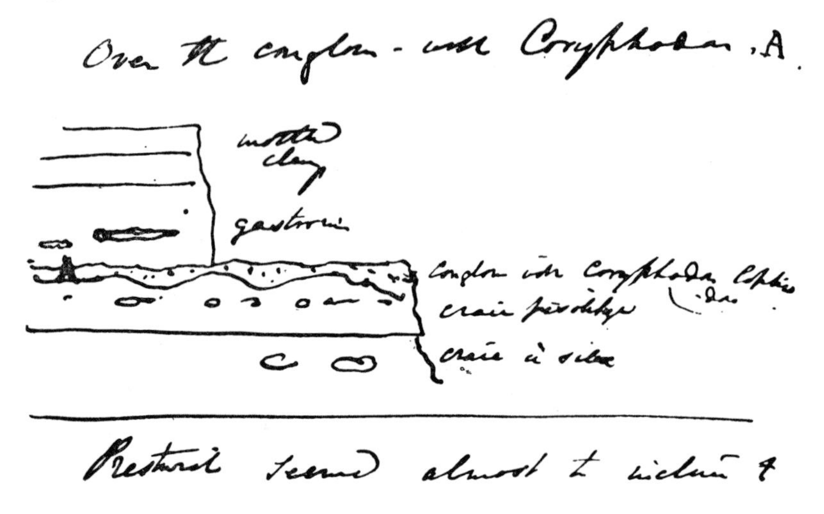

February 25, 1856 [88] [p. 98]

To Mr. Wollaston

I shall have to prepare illustrations & set artists to work for a lecture which I have promised on March 7.th at R. Instit.n on Temple of Serapis. I must, therefore, most unwillingly suspend my examination into the case of the Madeira land shells, fossil & recent. But before I stop, I will send you some of the numbers as I believe they will stand after the late revision. I hope you will some day show me the vars. of H. polymorpha or several of them. I will hunt up my own Madeira shells, which I have mislaid.

Madeiran shells of your original list, omitting slugs except Testacella	144
[p. 99]	
Deduct 23 species, as have varieties including H. Loweii & H. Bowdi[t]chiana	121
Deduct 15 living S. European species	106
Species extinct in the Madeira exclusive of H. lapicida	9

Living in Madeira Proper	77
Fossil in " "	35
Extinct in Madeira Proper	6
Living in P.º S.º	55
Fossil in P.º S.º	37
Extinct in P.º S.º excluding H. Loweii & including H. lapicida	3
Common to Madeira & P.º S.º of living spec.	9
" " " " " " of the fossils	8
" " " " " " of extinct	1
Living in Madeira excluding S. Europ.[n] species	63
Living in P.º S.º " " " "	51
Common to Mad. & P.º S.º excluding living S. European	6

[p. 100] I have not time to finish the list and include the Desertas & some other calculations. The most important result at present, unless I have made some mistake, is the greater connection which seems to have existed at the period when the fossils were accumulated, between the fauna of the two great islands & that which now exists, if we throw out of consideration the S. European invaders. For I make out 108 as the number of living shells in Mad. & P.º S.º, only 6 being common to both or 5½ per cent, whereas the proportion of fossil species is just double or in the ratio of 11 per cent! For [p. 101] there are 8 in 72 common to Mad. & P.º S.º among the fossils.

35 Mad. Proper }
37 P.º S.º } 8 common

Now unless we explain this by a former remote period of connection, or geographical union by land, of Mad. & P.º S.º, we might have looked for a result exactly in the opposite direction, inasmuch as the interference of Man in the last 3 centuries w.[d] tend so far as it acted at all to fuse the indigenous faunas of the two provinces. Can you not detect in the geog.[l] distrib.[n] of any species imported from P.º S.º into Mad. or from Mad. into P.º S.º. [p. 102] such a radiation from Funchal of the Porto Santan species, or from Baxio of the Madeira [species], as might agree with the idea of their having been carried by the Limestone barges?

Such discoveries, however, are not required; the enigma being to account for the marked dissimilarity & not the agreement of the living faunas.

I throw out these speculations for criticism. The greater extinction of Caniçal fossils, as compared to the extinction of P.º S.º fossil shells, is doubtless referable to the greater interference of Man in Madeira. Geologists overlooking this distinction might easily overstate the relative antiquity of the Caniçal beds. [p. 103] You will observe that we are brought, by very independent evidence, to the conclusion arrived at by MacAndrew from the littoral marine shells, namely, that the islands were once united, but at so remote a period that the creation and extinction of species has gone on very extensively since the date of that union.[89]

Copy of letter from Lord De Mauley[90] to Mr. Babbage—1851[91]

1841–2

The substructions of the Temple of Serapis at Puzzuoli, lately discovered, are formed of vaults or cloacae, [p. 104] upon which rests a pavement of coarse mosaic, which from the state of work & decoration, etc., the date of this part of the building is considered to be about 100 years before the time of Augustus. The height of this pavement is about 12 feet above the foundations & about 6 feet below the marble pavement, now visible, which consequently is about 18 feet above the foundations.

The marble pavement is evidently of the same date, as the three columns now standing, and connected with the Temple, of which they form a part. The friezes of the temple [p. 105] and work upon the columns, are judged to be of the age of Augustus both from the style of the work, & the quality of the marble, both of which are easily recognized by the Antiquaries. The cellar or vaults are of the Christian period. It may fairly be presumed, that for whatever purpose they were constructed, these arches were above water when built, as were likewise the two pavements above them.

There are then three Epochs at which the level of the water may be surmised, with tolerable precision, from the style of the building. The arched vaults or cloacae of [p. 106] the Christian age—the lowest mosaic pavement, above 12 feet above the ground level of the foundations of the time of the Republic—the marble pavement 6 feet above the other pavement of the time of Augustus—the standing columns, the perforation of which by the shell fish is at about 27 feet height, making with the 18 feet of the pavement, a difference of 45 feet between the ground level, and the height which it is supposed the water must have attained, so as to admit of the perforation of the columns. The water now covers the marble [p. 107] pavement, & Professor Nicolini[92] is of opinion that there has been a steady increase to the height of 4 or 5 feet during the last 4 years. But it is now (Feb. 23.[d]) flowing out constantly. The Guide however asserts, that that is always the case during the present month, from the melting of the snow on the mountains. There appears to be a slight difference in the level of the side of the Temple where the cellar & columns are, and that where the water escapes. The stones on the pavement are mostly cracked. Does this arise from the fall of the masses of marble or from the rising of the land? The boring of marine [p. 108] insects is observed at about the same level as Isola di Gaeta Terracina, etc., the formation there is limestone, here volcanic.

March 12 [1856]

With Wollaston[93]

1. Helix erubescens—large variety (ψ $\frac{1}{3}$[d] larger than the type of Mad. species) in the Bugio.
   H. phlebophora—
   var. from Ferro (P.° S.°)

Helix without umbilicus { undate—common in Mad.
vulcania—common to the 3 Desertas

Portosanctana
(H. labter like Loweii!) whether permanent vars. or species.

" armillate (qy striate Drap.?)
" (allied pisana)
" Pisana

[p. 109] ψ The large var. of H. erubescens in Bugio or S. Deserta (double the size? of Mad. type) is so far in favour of H. Loweii being the same as H. Portosanctana tho' this 4 times the volume?

Might not a question be raised whether even H. undata & H. Loweii be not the same?

[p. 110] Helix supplicate
" vulgate—(qy. H. Lurida same?) mouth different in P.° S.°—
" laciniosa (ragged) small or flat Desertas
" depauperate P.° S.° & H. squalida Mad. qy. if same, probably so—play the same part but diff.t permanent vars. from isolation

" latens
" obtinte } These are two of Lowe's species but one may be [of the] P.° S.° cole.t, whether [a] var. & seem to me more identifiable even than H. depauperata & squalida.
" pusilla
" gueriniana, from Lamarck, woolly like [the] Brit. H. radiate & tho' made genera in Ret. frio, may be like perversa introduced at a remote (continental?) period.

[p. 111] H. lurida has a constant difference in the shape of the mouth—in other respects the same, perhaps more different than any other of those united by T. V. Wollaston.

Such permanent vars. as H. vulgata, Mad.—& [H] lurida, P.º S.º,—or as H. depauperata, P.º S.º, & H. squalida, Mad., more distinct than same shell in two English islands.

March 12, 1856 [p. 112]

H. bifrons }
H. Calathrus } These 2 more like than any previously consid.[a] as distinct.

H. Armitageana (like H. hirsute English)

H. arridens & H. fausta a great deal alike, colouring in some like—umbilicus half closed.

H. arcta—many points of resemblance with H. lentiginosa, but the umbilicus more open in H. lentiginosa

Some might question whether H. stellares was not same as H. lentiginosa

1 arcte
2 lentiginosa
3 stellares
4 { arridens, fausta }
} or all one? preferable at present to make 4

are same { H. compacta, " calculus } living together & yet varying, difference in size chiefly.

March 12 [1856] [p. 113]

On going over half Mr. Wollaston's helices, [I] found several which after the reduction of 27 spec. might be sunk, rather than inclined to dispute those already merged. This tends to bring the fauna of the different islands nearer each other as far as species go, but not varieties which have become fixed by time, & breeding in & in [i.e., successive inbreeding]. May there not be now & then a natural hybrid causing difficulties & leading to merge 2 distinct species?

Coprophagous Beetles

There are some native over in Madeira which eat decayed vegetable matter before man & his domestic animals entered.

Therefore Prof. Heer[94] was mistaken in his argument even if 3 centuries were suffic.[t] to lead us to expect new species to be created.

[p. 114] If the two last were identified, it w.[d] be another case of fusion (qy see list)

H. bicarinata } identif.[d] in Catalogue double & single
" echinulata } kind
" turricula very distinct.

H. Polymorpha—if all which are sunk be identical, then comes the question whether H. cheiranthicola be not the same as some very conical vars. of polymorpha.

If H. tabellata were united, the polymorpha with H. cheiranthicola, there w.[d] be an end of species.

[p. 115] H. cheiranthicola, Wollaston thinks should be kept distinct, & perhaps so, but it proves the difficulties of species—the extremes are certainly very wide apart. H. cheiranthicola—& tabellata might almost be united if no check given to merging allied species.

March 14, 1856 [p. 116]

With Mr. Walter Mantell [95]

At Wellington, New Zealand, an old beach containing common mussel, (the pepe with the pecten), same as in Madeira sea-beach 12½ ft. high. Between W.[n] [Wellington] & on N. West shore of W.[n] harbour—N.E. of Wellington the old beach destroyed now—rolled pebbles & sand. C.[d] have got there in no other way—overgrown with small shrubs—above the road.[96]

No contemporan.[s] eruption in Tongariro as Col. Delamain & Mr. Babturbea of the Indian army testified—& no convulsion or extra action there always steam issuing.

Fissures in Wairarapa plain & in valley of the Hutt & threw up mud in Hutt valley & gravel several (4 ? ft.) high & 10 broad.

[p. 117] About 100 m. N.W. of Wellington is Whanganui [Wanganui] R. An earthq. (in 1845? or earlier?) after which, from houses back from the river, the river Whanganui was visible, from which previously it was not to be seen.

The 9 f.[t] perpendicular precipice was not at the foot of the hills but in the hills?

It may have been 90 miles. Old river courses deserted n.[r] Port Nicholson because of new cleft (qy. 100 yrs. old?) in clay slate, not fissile.

The feeling was as if you were standing on a table under which some man [was] staggering under the weight, during the upheaval.

Raised beaches everywhere if you look for them.

[p. 118] New Zeal.[d] country—sea covered with dead fish in Cooks Straits, vessels sailed thro!

March 28, 1856

C. Lyell to Mr. Wollaston

I have bought Albers.[97] I see that in his Preface he says, that he considers Bul. decollates as indigenous or not recently introduced. His first argument, that it is found in the Azores and Canaries is not worth much. His second, that it occurs fossil 500 ft. above the sea in Fuerta Ventura, is worth more, so far at least as that habitat is concerned. Is it always near Funchal in Madeira?

Albers also maintains, that H. lenticula is fossil in Madeira, and therefore not introduced, and H. pisana fossil in Porto Santo.

As to the latter case, one would like to know if it was superficial, or not.

He says, of 22 Madeira Pupas one only is common to the Canaries. If the latter had been well searched, it would be very interesting.

He says, that H. erubescens pauperenta and membranacea are common to Madeira and the Azores. This is in favour of a [p. 119] former union, if correct, although there is much, or some, commercial intercourse between Madeira and the Azores —both Portuguese?

Albers' H. Ludovici not enumerated by you, seems a variety of tectiformis. I have a suite of the latter, and could make other species in the same way.

Albers, I see, deals with H. polymorpha, as we have done and says, that Pfeiffer[98] goes always with him in all but H. attrita.

In regard to H. compar and Maderensis, he has, I should think, fixed an extreme variety for the latter.

He separates H. micronphale from H. leptosticte—fictilis from dealbata.

His own figures of H. depauperata and H. squalida are almost enough to satisfy one [that] it is a diff. of size chiefly.

H. obtecta and H. latens, which he separates, will, I suppose, be one of the most disputed of our assimilation. I have not yet gone into it again, but hope to find specimens; if not, I will look at yours again, though I remember them pretty well.

He admits, that H. echinulata is *"valde affinis"* to H. bicarinata, and that the former is even sometimes bicarinated. Even Lowe says in his last list *"nimis affinis"* etc.[99]

Even H. oxytropis seems to me to run the above pretty hard, [p. 120] if my specimens so named are the true H. oxytropis.

Albers places H. cheiranthicola very far off from H. polymorpha. I cannot see why, and he does not allude any where to H. musteliva and H. vermetiformis.

Under H. lentiginosa he sinks H. stellares.

Helix fausta and arridens he says might easily be confounded. His own figures show this fact very well.

H. capsilla not alluded to in Albers. H. calculus also not mentioned even as a synonym of H. compacta.

Next to H. consor, Albers inserts H. Hartungi, of which you and Lowe take no notice. Albers agrees, in sinking H. leonina Lowe into H. vulcania, but Lowe himself in his last list says that time may show them to be one & the same.

Helix advena is given by Albers as found by Hartung[100] in Porto Santo, a Canary species.

I see, that Lowe in his new list says that H. Bowditchiana may be a variety of H. punctulata.

Albers admits to the affinity of H. psammophora and H. phlebophora, but thinks them different. I should like to look again at your series which will come before us, when your fossils are inspected.

Albers distinguishes H. lurida (under the name of H. nitidiuscula) for H. vulgata. I will look again at this. (See Lowe p. 174 of List of 1854.)[101]

Upon the whole, the way in which Albers has dealt with [p. 121] the Helices, coupled with Lowe's admission in 1854, renders the new list less offensive to the latter.

He did not see any of your alterations, and when I pointed out a name, which he could not remember, it led him to remark, that 2 specimens might be united which are separated in your and his lists—but this I mentioned to you.

Achatina mitroformis is kept separate by Albers, but I see that Lowe says it almost passes into A. tornatillina.

Pupa irrigua and P. vincta. Neither Lowe nor Albers alludes to the possibility of their being the same.

Pupa recta and P. macilenta are united by Albers, as *olim* by Lowe.

The corrected statistics are now so much to be depended upon, that I do not fear reasoning and speculating upon them, but I must not enlarge on the subject in my volcanic paper on Madeira,[102] in which so many subjects of discussion occur, that I am considering what I can leave out, in order not to offer too long a memoir to the Royal Society. I shall merely allude briefly to the fossils, and in another paper treat fully of them—and the distribution of the recent, as ascertained by yourself and Lowe.

I consider the time indicated by the extinction of H. Loweii and H. Bowditchiana, and some other more undoubted specimen as vast, considered in reference to our historical notion of time, but as belonging to a period which if expressed in figures [p. 122] might be imagined. Yet vast as it has been, and including great changes in Physical Geography, it belongs to the times when the volcanic eruptions had ceased, unless possibly a few of the last, such as the lava which flowed into the S. Vicente Valley and down the bottom of it.

As the extinction of species, since the fossils flourished is very small, and the coming in of new species, doubtless equally small, the period alluded to comprehends a mere fraction of one geological epoch. Suppose 3 or 4 species now inhabiting the Madeiras to be lost, and as many new ones to come in, this would not alter materially the relation of the [p. 123] faunas of Madeira and P.º S.º, but (if Man's interference be assumed as nil) the introduction of new species would *pro tanto* cause the two faunas to differ more than before.

I rather think, from the numbers as they now stand, that if 8 species became extinct in the 2 large islands, & 8 more quite new ones came into being in their place, say 5 in Mad. & 3 in P.º S.º, this could only make the difference which we find between the state of the shells as they formerly existed and as they are now.

If species have only one birth place the divergence would go on increasing providing we can assume that no interchange by migration can take place [p. 124] between two such islands as Madeira and P.º S.º—& I believe you do not suppose it possible in the case of land shells. The fossil fauna have about 12 per cent & the recent about 11 per cent in common. The difference is as it should be & about as much as it should be.[103]

If we had a mass of helices in the S. Jorge leaf-bed we should no doubt find a much larger proportion of extinct species, and I sh.[d] expect a formation of corresponding age in P.º S.º to yield helices agreeing more with the S. Jorge, than do the fossils of Caniçal with those of La Bella in P.º S.º

[p. 125] Had the agreement of the latter groups of fossil shells exceeded the amount of agreement observable, between the living shells of Mad. & P.º S.º, as much as it seemed to do by our first calculations, it would have been perplexing.

I found helices under lava in Teneriffe. If some of the old soils or tuffs of Madeira could furnish them and also some old calc.[s] [calcareous] deposits in P.º S.º it would pour a flood of light on this subject.

C.L.

Queries

H. lentiginosa & H. arcta—differences of?

April 3, 1856 [p. 126]

Mr. de la Harpe[104] searched [the] Flysch, 1000 f.[t] thick above the nummulitic tertiaries, & in the lower part, where it became a thin—fine slate, as well as in the upper sandy portions, could discover no org.[c] remains except a few fucoids. This formation w.[d] constitute a good azoic slate if all the older rocks became metamorphic.

April 3, 1856

With Prof. Ramsay

The New Zealand Earthquake of Jan.[y] 23, 1856 affords an example of an *upthrow*—Ramsay has been in the habit of speaking of *downthrows* by preference on the opposite sides of faults.[105]

April 11, 1856[106] [p. 127]

C. Lyell to T. V. Wollaston

How can there be only 11 or 12 per cent of Pulmonif.[s] mollusks common to Mad.[a] & P.[o] S.[o] if, as Dr. Hooker says, nearly all the plants are common to the two islands?

Are the species of shells as true as those of plants & as well defined?

Or, is the longevity of species in plants greater than in land shells?

Or are the migratory powers of plants across marine channels from isl. to island greater?

Or are the land shells so much more limited in their range, even in a continuous tract of land, & is not the hypothesis in opposition to the known range of European helices [p. 128] & other land shells, as compared to plants?

Are there as many insects as there are plants (proportionately) common to Mad.[a] & P.[o] S.[o]?

April 11, 1856

If Helix Wollastoni [be] confined to one mountain in P.[o] S.[o], altho' fossil in the same mountain, its powers of migration [must be] so small that it cannot be expected to migrate from P.º S.º to Madeira.

Woodward remarks that he knows of no group of islands in which the species of plants are peculiar, if so are not the shells more limited because they cannot get from island to island [p. 129] as plants can, in consequence of seeds not being injured by salt water & being floated by currents.

But the land shells of all the 100 British Isles are the same & agree with Germany & France except a few Portuguese, as Lymnea involuta [of] Ireland. It is only very ancient islands which have their land shells peculiar—Miocene islands as E. Forbes said. Hence a geologist may infer the antiquity of islands by the isolation of their shells.

St. Helena had its own plants & the Galapagos [their own] but each isle of the latter only to a limited extent? Hooker, Linn. Trans.[107]

April 13, 1856 [p. 130]

With Darwin: Migration of Plants & Shells

Altho' seeds vegetate after immersion yet they sink in salt water & most of the plants sink soon. Their want of floating power is therefore the chief difficulty in the hypothesis of their transport by oceanic currents.[109]

The greater distinctness, of shells of the land & fresh wat.[r] in the islands of the same group, than of the plants, seems a general law all the world over.

No sufficient experiments seem as yet to have been made in regard to Helices & Pulmonif.[s] mollusks. Can they or their eggs resist salt water—are [p. 131] they readily drowned & how soon? The few experiments made by Darwin imply that they are soon drowned.

In the Galapagos isles the land shells collected were not numerous, but so far as the evidence went, it corroborates the Mad.[a] & P.º S.º case in regard to the distinctness of the fauna in each island.

St. Helena has been much changed since the extinct land shells accumulated there.

Frogs are not found in volcanic islands. Even New Zealand only provides one species lately discovered in island[110] & Darwin finds frogs' spawn to be very easily killed by salt water.

Letter from Mr. Lowe to Mr. Darwin [p. 132]

12 April '56

The Flora of Porto S.^{to} may be stated at from 270 to 280 species (indig.^{s} and perfectly naturalized).

1. Of these, 7 or 8 are certainly, and one or 2 others doubtfully, endemic—the doubt arising from difficulty of ascertaining identity of species—3 or 4 of the 7 or 8 are sufficiently striking and abundantly growing plants. 2 or 3 represent Mad.^{a} endemic sp.; 2 or 3 are more of the nature of 'weeds'.

2. Of plants common to P.^{o} S.^{o} & Mad.^{a}, but not hitherto found elsewhere, [p. 133] there are 20 or 21, perfectly certain, and 3 others doubtful.

3. Again, of plants common to P.^{o} S.^{o} and other countries (mainland) of Europe, but not *found at all* in Mad.^{a} there are 25 certain & 1 doubtful.

4. Lastly, of plants common to P.^{o} S.^{o} and other countries (mainland) of Europe, but *very rare* in Mad.^{a} there are 6, to which may perhaps be added 2, which are indeed only at this day occasional garden plants in Mad.^{a}, whence they were introduced in 1834, into P.^{o} S.^{o} which they now completely overspread! One of these is a [p. 134] Tamaria, (T. *orientalis* L. ?); the other is the Hottentot Fig. (*Mesembryanthemum edule* L.).

Of classes 3 & 4, almost all are common European 'weeds', proving of course little anyway. And let me add, that the very peculiar nature of the soil & climate of P.^{o} S.^{o}, as compared with Mad.^{a}, accounts of itself in great measure for the very different character & aspect generally of the vegetation in the 2 Islands. Many plants common in one Island, *can not* be made by any efforts to grow in the other.

[p. 135] This may be useful as a caution against attributing too much in this particular case to other possible modifying general causes or influences.

Hooker has I think considerably underrated the number of *good* endemic species in Madeira, which exclusive entirely P.[o] S.[tan] & Des.[tan] plants will be found I think rather to exceed than fall short of 100. But it would take a good deal more time & research than I can just now afford to speak positively & with accuracy on this head.

[p. 136] There are *very many common* endemic species in Madeira not occurring in P.[o] S.[o]

In the Desertas there are 3 very remarkable *endemic common* plants, one forming a new *genus* of Umbelliferae; another a new genus of Gramineae; the 3.[d] a new shrubby chrysanthemum, representing *C. Pinnatifidum* L. fil. of Mad.[a].

There is no freshwater fish in P.[o] S.[o] (as in Mad.[a]) but the Eel (*Ang. latirostris* Yarr.) I believe. Wollaston & I caught an arcelos[111] in a puddle last spring.

[Wednesday] April 16, 1856 [p. 137]

With Darwin: On the Formation of Species
by Natural Selection—(Origin Query?)

Genera differ in the variability of the species, but all extensive genera have species in them which have a tendency to vary. When the condit.[s] alter, those individuals, which vary so as to adapt them to the new circums.[s], flourish & survive while the others are cut off.

The varieties extirpated are even more persecuted & annihilated by organic than inorganic causes. The struggle for existence ag.[t] other species is more serious than ag.[t] changes of climate & physical geography. The extinction of species has been always going on. [p. 138] The number of species which migrated to the Madeiras was not great in proport. to those now there, for a few types may have been the origination of many allied species.

The young pigeons are more of the normal type than the old of each variety. Embryology, therefore, leads to the opinion that you get nearer the type in going nearer to the foetal archetype & in like manner in Time we may get back nearer to the archetype of each genus & family & class.

The reason why Mr. Wallace ['s] introduction of species, most allied to those immediately preceding in Time, or that new species was in each geol.[1] period [p. 139] akin to species of the period immediately antecedent, seems explained by the Natural Selection Theory.[112]

April 22, 1856

[Harley Street, London]

Why have we witnessed no new creations since Man existed? Perhaps because the time is too short for what ought to be considered a species to be introduced. If most of our species are varieties then genera or sub-genera at any rate are all that originate directly from the hands of the Creator. According to that view it would rarely happen that a new genus came into being. Man & the races of man which may be equivalent to one ordinary species may suffice [p. 140] to fill the chasm which the extirpation of mammalia by Man has caused. In this case we should have recourse less frequently to the acts of creation which appear to some minds inconsistent with the government of the organic world by general laws. The objection w.[d] be that if we concede the species of a genus so must we the genera of a family & the families of an Order.

The theory which requires acts of Creation or *saltus* in Nature or the springing up of new creatures out of nothing is said to be as hypothetical as the catastrophic in the inanimate world.

[p. 141] But how did Man begin? He was improved out of some anthropomorphus genus & is an *old world* form of that family. To this I object that his progressive power causes him to differ wholly from all the Brutes in kind rather than in degree. It is answered that a dog is progressive in a domesticated state, increasing in intelligence. The reply is that he has lost many of his instincts & attributes as a wild dog, possibly as much as he has gained.

You require, say my opponents, a perpetual series of miraculous interventions of the supreme or some unknown power to sustain the system of living creatures which Time [p. 142] & the changes of the animate & inanimate world have been always & always now are working.

You lay yourself open to Whewell ['s] objection that creation has ceased, but the Lamarckian machinery supplies a perpetual creation-power of new permanent varieties or species adapted to the new circumstances of climate, & geogr.[l] & contempor.[y] botanical & zoological conditions.

In the Ocean Man cannot have arrested a system which has been in force for 100 millions of years. Because *we* cannot explain its present working, shall we assume that the great world-clock [p. 143] has stopped? Was not this the first crude idea of those who advocated the doctrine of catastrophes?—in place of gradual & daily or annual changes. What revolution can marine life have undergone, by the presence of man, more important than that brought about by a new mammalian fauna replacing on the land some other, which may have died out?

On the other hand if we revert to the origin of things, how many original generic types do we require?

Must we not embrace the progression system as well as the permanent variety-making machinery? If Man is an indication of progress, there may have been antecedent [p. 144] links in the same chain. The chasm in the series of geological archives may be similar to those by which breaks in genera are produced.

Abortive wings, & organs called *rudimentary,* may be indications of a transition from one form or state to another.

The immensity of Time even in Pliocene & post-Pliocene periods may greatly facilitate the Lamarckian hypothesis.

The numerous types of gasteropods found in Bohemia by Barrande show how we may suddenly discover overlooked protoplasts of living & tertiary genera.[113]

April 29, 1856 [p. 145]

Origin & Reality of Species

After conversation with Mill,[114] Huxley,[115] Hooker,[116] Carpenter[117] & Busk[118] at Philos. Club,[119] conclude that the belief in species as permanent, fixed & invariable, & as comprehending individuals descending from single pairs or protoplasts is growing fainter—no very clear creed to substitute. Indefinite time & change may, according to Lamarckian views, work such alterations as will end in races, which are as fixed [as] the negro for example & unalterable for the period of human observat.[n], as are any known species such as the Ibis cited by Cuvier.[120]

The Oxslip, says Huxley, is not known on the continent.[121] It is a British permanent variety or race of the [p. 146] Primrose. Would it not be preferable to believe in the somewhat sudden development of new organs, say the sceptics, than the creation of a new species out of inorganic matter or out of nothing?

The successive creation of species is a perpetual series of miraculous interferences instead of the government of the organic creation by general laws.

Huxley shows that types intermediate between Mammalia & Birds & other great divis.[s] are not met with in Geol.[y] as they w.[d] be if there was a continual developm.[t] from one original type. Were there many first germs?

The best argument perhaps, in favour of the multiplication of new species on the permanent variety theory [p. 147] is this. Extinction has certainly been going on in the last 3000 or 6000 years. Therefore, as we have no reason to doubt that the living creation is as rich as any antecedent (some think it richer, which they cannot prove) then the lost species have probably been replaced. But in what manner?—We know of no machinery, except that of the gradual adaptation of species to new conditions, a change in their forms & organization which may be in progress & seems to be so in some.

If Man be modern, & if the negro & white man have come from one stock, & if such distinct races, if discovered in quadrumana, would have been pronounced species, then new species have been formed since the human pair originated.

[p. 148] Vast numbers of living species may be altered forms of extinct fossil forms, & the new climates & geographical position & contemporary fauna & flora may have made the present states of these animals, or plants the only ones which can survive in the great struggle for existence. Just as the black variety of man can alone flourish at Sierra Leone.

We have a right to question with Hooker one half of the recognized species of plants & if so, if naturalists are so little agreed as to the definit.[s] of what constitutes a species in the existing creation, they must be equally so in regard to fossil species.

Natura et fece et poi ruppa la stampa[122]

This might be equally true if the combined condit.[s] org.[c] and inorganic on wh. a permanent variety depending, past and present, can never in the doctrine of chances [p. 149] return again & this they can never do.—That it must be impossible is easily to be seen. Such impossibility however w.[d] equally explain why no species could be created now with all the same attributes & qualities as at some former period.

If in the course of the next 10,000 or 20,000 years any race of man should come to differ as much from the Caucasian as does this race from the Negro, the difference being in an opposite direction; as to form and mental power, there will remain scarce any doubt, that deviations equivalent to those which we usually estimate as sufficient to constitute species, will have taken place in the genus Homo. But the Negro and White Man or the Hottentot may represent the extreme possible divergence from a normal type? [p. 150] This is scarcely probable. The extremes of climate & of other condit.[s] now experienced are as nothing to those which may arise in an indefinite course of ages. It may require ½ a million of years or more to produce changes equal to those which caused the Glacial Period to be intercalated in the Modern Pleiocene. But if we could observe for ½ a million of years, & if as seems not improbable Man w.[d] not be exterminated by such an amount of change, then might we anticipate some greater departure from the common type than that now witnessed on the globe?

The idea that species after having been formed, in the place of others which are lost from time to time, [p. 151] after having for many millions of years or centuries succeeded each other, the idea that a total suspension of the mutability of the organic world sh.[d] have occurred in consequence of the presence of Man, & that too in the ocean as well as on the land, in the microscopic world as well as in the visible, this doctrine may seem plausible to those who view the planet & all other worlds as subordinate to Man & created in sole relation to him.

The lessons of Geology seem rather designed amongst other final causes to rebuke this vainglorious hypothesis. When we consider how nearly the races of Man approach to what are considered as equivalent to specific diversities in the inferior animals, & how perplexed the naturalist is to distinguish some [p. 152] varieties of dogs from true species, it does not seem reasonable to assign the character of fixity to the human period specially.

There seems no reason to affirm that animals & plants have not been varying in proportion to the amount of variation in climate & in the migration of animals & plants. The Physical Geography of the globe is altering & its inhabitants—species are becoming extinct. It is fair to presume that there is no diminution of organic forces of vital energy, no tendency to the depeopling of the land or of the sea, of the river or of the lakes—& if so, there must be some law still in [p. 153] force by which a succession of living creatures will be ensured, whether by the occasional introduct.[n] of new generic or specific types or by the modification of existing ones. If new genera spring up or the protoplasts of families, & these produce species by divergence, then these may be very rare occurrences. It may be argued that it is as likely that some of the extinct types may have deviated into new species, as that new ones sh.[d] arise out of nothing.

Hence the progression theory, which accounts for man being improved out of an anthropomorphous species, is natural the moment we embrace the Lamarckian view. It is not fair to avail ourselves of the theological support derivable from this objection. Yet strange to [p. 154] say there seems rather a disposition on the part of the orthodox to regard the progressive developm.[t] theory as more consonant with orthodoxy than what they term the uniformitarian doctrine.[123]

Yet the progressive theory must end in deriving man from the brutes & in teaching that the whole geological history of the globe is the history of Man.

Original Types

If there were protoplasts of plants & animals distinct from the first, why should we not grant original types of the leading division of Vertebrates & Invertebrates which we find in the lowest Cambrian or Bohemian primordials & ante-primordials.

The dignity of Geology w.[d] no doubt gain could we once conclude [p. 155] with any feeling of certainty, that the progressive development theory, the origination of species from modifications resulting from geological changes & a tendency to improvement, accompanying those modifications, were true—for then geology, ethnology and history would blend into one & the books of God's works w.[d] unfold to us by degrees the physical history of our species as well as of the beings which are now our contemporaries. But at present the mystery involving the subject does not appear to diminish. Our time of observation and of comparison of living & extinct nature is too short.

The origin of the organic world, like that of the planet, is beyond the reach of human ken. The changes of both may perhaps eventually be within our ken but ages of accumulation of facts & of speculation may be required.

INDEX [p. 156]

1. Creation of a "species to endure long requires many adaptat.[s] to org.[c] & inorg.[c] conditions (15)
4. Colour a protection—(be thereby modified.)
5. If glacial epoch has preceded, species will be thereby modified.
" Species introd.[d] by Man w.[d] often die out if unaided by him.
8. Rudimentary organs a mystery.
9. Self-imposed restrictions of the Creator & fixity or unity of plan render it intelligible to Man.

10. New species like new islands may die out soon.
11. New species without an antetype rare exceptions.
" Man greatest deviation.
" New species why allied to preexisting ones.
12. Relation of Ruminants to Man,—doctrine of Necessity.
13. Not necessary to suppose many originals of a species.
14. The theory invented from ignorance of migrations.
16. Insular species,—Salvages.
17. Shells common to Azores and Europe.
" Submergence of continent w.[d] leave peculiar species in islands.
18. Ireland neither was long united to England, nor was it the remains of an Atlantis, apart from England.
19. Absence of Eleph.[t] & Rhin.[s] in Irish & Scotch peat.
20. Enumeration of 12 works on distribution of species.
22. Sangatti & Brighton beds in favour of recent land-union of England & France.
23. N. America united with Europe in Newer Pliocene period —Sir. J. Richardson.

[p. 157]

25. If Alps have risen many thousand feet since Falunian period, so may an Atlantis have sunk.
" Species common to Atlantic isle & a continent thus explicable.
26. Neither ascidian nor Cephalopod naked in Silurian.
27. R. Brown, on no species common to S.[t] Helena & Congo.
29. Atoll theory need not imply building of corals for thickness of 10,000 f.[t]—volcanos may have raised mountains.
32. Volcanos in islands & their accompanying local upheaval may be consistent with general tendency to sink—volcanos, because of local rising & outpouring, being alone preserved.
35. Canaries & Madeira were islands in Miocene period, wh. is against a Pleistocene Atlantis.
36. Speculation on former union of the above islands, & long succession of changes uniting & isolating them.
41. Calder in 1794 speculating on recession of Niagara.

45. Alps high before Rigi—lost 7000 f.[t] then of height.
47. Successive upheavals (post-Miocene) & sinkings of.
" Stereognathus ooliticus not like Migale. Waterhouse.
48.–98. Mollusca of Madeira & P.[o] S.[o] List of Wollaston to show number of fossil & of species common to both & to Europe
60.–62. Letter of C. L. to Wollaston on giant varieties such as Helix Loweii & Bowditchiana not having intermediates.
" Changes in phys. geogr.[y] of Madeira & P.[o] S.[o] & Desertas.
63. Heer says ¼[th] of S. Jorge plants now extirpated.
64. Mad.[a] & S. Jorge flora insular & like the present.
66. Mad.[a] & Canaries islands in Pleo & Miocene times.
" Man & cultiv.[n] & destruction of forests extermin.[d] M.[a] species.
71. Paleoteuthis of V. Meyer, a fish-bone.

[p. 158]

" Forms not species in common led E. Forbes to assume former proximity of Africa, S.[t] Helena & America.
72. Helix cellarea & H. Pulchella in Madeira & C. of G. Hope imported.
74. Six extinct land shells in Madeira proper or 7 or 8 per cent—(see p. 86 correction)[124]
76. P.[o] S.[o] compared for proport.[n] of extermination.
78.–79. Canoes drifting may carry species—(age of Man)
78. If man 20,000 yrs. old he may have done much in modifying geograph.[l] distrib.[n].
79. Madeira barges landing in P.[o] S.[o] with shells.
81. Fossil faunas of M.[a] & P.[o] S.[o] have 7 per cent in common, the recent only 3 p.[r] c.[t].
84. Hooker's S. E. & S. W. Australian floras show the power of preoccupancy to resist foreign invaders.
87. Land shells migrating in a flower pot.
88. Teleosaurus in chalk. Barrett.
" Starfish recent in Norway—theory of climate.

89. Buccinum Dalei of crag found, far North.
" Progressive developm.[t] authorities in favour of—facts in favour of, & against.
92. Connecticut footprints may be reptilian, but not mammals for there [are] only 3 joints.

[p. 159]

93. Coal era called by Dana, Age of Amphibians.
94. Connecticut beds, why not between Trias & Lias.
" Mammalia of Red Crag, Miocene says Owen.
96. Gastornis of Plastic clay, age.
98. Madeira & P.[o] S.[o] shells in common & extinct.
100. Fossil fauna of the two isles was more like than the recent. Why: geograph.[l] connex.[n]?
103. Islands once united M.[a] & P.[o] S.[o] but long ago.
" Temple of Serapis old Mosaic pavement.
108. Size of Mad.[a] land shells with Wollaston.
110. Madeira [land shells] & change of species permanent vars.
113. " land shells, rather too many species still, & therefore islands more like, except in regard to Vars.
" Coprophagous beetles not wanting in M.[a] as Heer thought.
115. Madeira land shells illustrate the difficulty of deciding what species really are.
116. New Zealand Earthq. with Walter Mantell.
118. Albers' work on M.[a] shells & comments on—bearing of his identif.[s] on the difficulty of deciding on species.
122. Extinction of species in Madeira & Porto S.[o]
126. Flysch thickness & absence of fossils in Azoic.
" New Zealand earthq.—example of upthrow.
127. { Madeira & P.[o] S.[o] plants same, land shells different.
128. { Ireland on Lymnea peculiar (129)

[p. 1][125]

129. Miocene or ancient islands have helices peculiar.
130. Relative transportability of shells & plants.
131. Galapagos land shells distinct in each island.

132. Porto Santo only about 7 in 170 plants endemic.
133. Twenty three peculiar plants common to Mad.[a] & P.[o] S.[o]
134. Twenty-five common to P.[o] S.[o] & elsewhere not being of Madeira.
135. Madeira endemic plants 100—Lowe.
137. Darwin on extinction variat.[n] & Natural Select.[n] [1856]
138. " embryological argum.[t] for transmut.[n] & progression.
139. If genera alone created, we ought not often to witness birth.
140. Argument against 'selection' & transmut.[n], Man's origin.
142. Marine zoology must go on as before Man's creation.
143. Variety-making power draws as progression advances.
144. Rudimentary organs.
" Barrande's primordial trilobites supply antetypes.
145. Species, reality of, less believed, 1856.
" Oxslip, a variety not known on continent.
146. Creation theory implies miraculous intervention.
" No intermediates between Reptile, mammal, bird. Huxley.
147. Extinction goes on, no acts of creation; ergo Variation equival.[t]
" Negro & white man = species.

[p. 21]

148. Struggle for existence.
" *"E poi ruppa la stampa,"* true even with Darwin's theory.
149. Why sh.[d] Negro & European represent the extreme possible divergence in genus homo?
150. Geological condition varying, the divergence w.[d] also.
151. Man cannot cause suspension of Creation.
152. Races of the dog & Man against fixity of species in human period.
153. The orthodox incline to progressive development while hostile to transmutation.
155. Progressive developm.[t], if true, w.[d] raise geology.

Notes

1. Beginning in 1855, Lyell kept a separate series of notebooks in which he indexed books and articles he had read. He referred to these as his "Index Books." The entry beginning on vol. 1, p. 31, is as follows:

Nov. 26, 1855

Wallace, Alfred R. *Annals & Mag. of Nat. Hist.* p. 188—1855 (2.[d] ser.) vol. 16. Sept.[r] No. 93—

p. 188 No island of modern tertiary age possesses peculiar groups or species.

[p.] 189—Java, Sumatra & Borneo probably once united for the organic beings in great part are common to all—shallow sea between.

[p. 32] Wallace contin.[d]

p. 189 England separated from the continent in recent times has no species peculiar or scarce one—

" The Alps so modern that the species N. & S. of them the same as nearly as latitude & climate allow.

[p.] 190 In Geology closely allied species are found associated in the same beds.

" Progression from a lower to a higher organization in geology undoubted—

[p.] 192 Argument (at top of p.) proceeds far too much on the assumption that we know enough of the whole series at any former epoch to allow of assuming retrograde movements.

[p.] 193 Convulsions of Coal strata show the cause of great extinct.[n] of spec.[s] in the interval between the Paleo- & Neo-zoic periods—Qy as to supposed scarcity or poverty "divine idea" in that interval—

[p. 33] *Wallace* contin'd Nov. 27, 1855

[p.] 193 The greatest difference between the Paleo- & Neo-zoic faunas may possibly be due to a corresponding lapse of time or the vast duration of the interval, & if so the effect seems explained—Was the Permian Flora & Fauna poverty-stricken?

[p.] 194 Wallace asks, is the visible Silurian one ten 1000.th part of earth's surface.

[p.] 195 Rudimentary organs without any function—limbs hidden beneath the skin in snake-like lizards.

[p.] 196 The law "that every species has come into existence coincident both in time & space with a pre-existing closely allied species" goes far towards Lamarck's doctrine.

It may be that the old combination of Geological condit.^{s} still continues so nearly that as it never did before or since, yet some new condit.^{s} required a modific.^{n} beyond the variability of the old species so a new allied one is required.

These references are all to one article by Alfred Russel Wallace, "On the Law which has regulated the Introduction of new species," *Ann. & Mag. Nat. Hist.,* 1855, ser. 2, *16,* 184–96.

2. Lyell had set forth his concept of the nature of species in terms of geological succession, geographical distribution and place in the economy of Nature. He summarized his view in the statement that

> —Each species may have had its origin in a single pair, or individual, where an individual was sufficient, and species may have been created in succession at such times and in such places as to enable them to multiply and endure for an appointed period, and occupy an appointed space on the globe.

Charles Lyell, *Principles of Geology,* 3 vols., London, Murray, 1830–33, *2,* 124.

3. Charles Darwin had pointed out the South American character of the Galapagos flora and fauna:

> Most of the organic productions are aboriginal creations, found nowhere else; there is even a difference between the inhabitants of the different islands; yet all show a marked relationship with those of America, though separated from that continent by an open space of ocean between 500 and 600 miles in width. The archipelago is a little world in itself, or rather a satellite attached to America, whence it has derived a few stray colonists, and has received the general character of its indigenous productions.

Charles Darwin, *Journal of Researches into the Natural History and Geology of the Countries visited during the voyage of H.M.S. Beagle round the World,* 2d ed., London, Murray, 1845, p. 377.

Lyell is arguing that the similarity of the Galapagos forms to those of South America may result from a correspondence in climate and conditions rather than from colonization.

4. In the second volume of his *Principles* (1832), Lyell had quoted Buffon on the successive extinction of species. "They must die out because Time fights against them" (p. 176). He had also shown how the inexorable course of geological change gradually destroyed the particular set of geographical, climatic, and biological conditions on which each species depended for its existence.

5. Lyell had first read Lamarck's *Philosophie Zoologique* in 1827 and in the second volume of the *Principles* had given a thorough examination and critique of Lamarck's theory of the transmutation of species. Jean Baptiste de Monet de Lamarck, *Philosophie Zoologique ou Exposition des Considerations relatives à l'Histoire naturelle des Animaux,* 2 vols., Paris, 1809.

6. In 1839, Lyell had found the fossil remains of several species of mammals in beds of the London Clay, a formation of Eocene age, at Woodbridge in Suffolk. Richard Owen later identified one of the fossils as that of a Macacus monkey, a discovery that showed that the primates had appeared as far back as the Eocene. Charles Lyell, "On the occurrence of Fossil Quadrumanous, Marsupial and other Mammalia in the London Clay near Woodbridge," *Ann. Nat. Hist.,* 1840, *4,* 189–90. Richard Owen, "Description of the jaw of the fossil Macacus from Woodbridge," *Mag. Nat. Hist.,* 1839, *3,* 446–48. In 1862, Owen decided, on the basis of fresh fossil evidence, that the supposed Macacus was really a pachyderm. See Lyell, *Principles,* 11th ed., 1872 *1,* 162.

7. Samuel Peckworth Woodward (1821–65) was born at Norwich, England. In 1838 he became a subcurator at the Geological Society of London where he worked under William Lonsdale and Edward Forbes. From 1845 he was professor of geology and natural history at the Royal Agricultural College, Cirencester, and in 1848 was appointed an assistant in the Department of Geology and Mineralogy at the British Museum. Woodward identified the collections of fossil and living shells that Lyell brought back from Madeira and the Canaries in 1854 and sent Lyell corrections for the 1855 (fifth) edition of his *Manual.* The information given here appears to have been a verbal communication. The Salvages are a small group of uninhabited islands in the Madeiras.

8. Matthew Fontaine Maury (1806–73) was appointed in 1842 Superintendent of the Depot of Charts and Instruments of the U.S. Navy Department and Superintendent of the U.S. Naval Observatory. Woodward referred to Maury's *Physical Geography of the Sea,* New York, Harper, 1855, 274 pp.

9. Charles Darwin, *Journal of Researches,* 2d ed., 1845, pp. 486–91.

10. Edward Forbes, "On the connexion between the distribution of the existing fauna and flora of the British Isles and the geological changes which have affected their area, especially during the epoch of the Northern Drift," *Mem. Geol. Surv. Great Britain,* 1846, *1,* 336–432, p. 341. Edward Forbes (1815–54), naturalist and palaeontologist, had worked with the Geological Survey of Great Britain and had pioneered in the correlation of the geographical distribution of plants and animals with past periods of glaciation.

11. Ibid.

12. Joseph Dalton Hooker, *The Botany of the Antarctic Voyage of the H. M. Discovery Ships Erebus and Terror in the years 1839–1843,* 3 vols., London, Reeve, 1847–60, vol. 2. *Flora Novae Zealandicae,* 1855, "Introductory Essay," pp. 1–39. Joseph D. Hooker, M.D. (1817–1911) was the son of Sir William Jackson Hooker, director of the Royal Botanic Garden at Kew. His family had been long friendly with Lyell's, and after his return to England in 1843, he became a close friend of Charles Darwin.

13. Joshua Trimmer (1795–1857) was born in Kent, but at nineteen went to north Wales to manage a copper mine belonging to his father. In 1832, he became a fellow of the Geological Society of London and in 1841 published his *Practical Geology and Mineralogy* (London, Parker, 1841). After 1840 he lived as a farmer in Kent. "On the southern termination of the erratic tertiaries and on the remains of a bed of gravel on the summit of Clevedon Down, Somersetshire," *Quart. J. Geol. Soc.,* 1853, *9,* 282–96.

14. Joseph D. Hooker and Thomas Thomson, *Flora Indica: being a systematic account of the plants of British India, together with observations on the structure and affinities of their natural orders and genera,* 2 vols., London, Pamplin, 1855.

15. Alphonse de Candolle, *Introduction a l'étude de la botanique, ou traité élémentaire de cette science,* 2 vols., Paris, Roret, 1835. Vol. 2, book 4, pp. 250–318, deals with "Géographic botanique."

16. Sir Roderick I. Murchison and James Nicol, "Geological Map

of Europe," in Alexander Keith Johnston, *The Physical Atlas of Natural Phenomena,* 2d ed., Edinburgh and London, Blackwood, 1856, pp. 13–16, pl. 4.

17. This paper appears to have been Joseph Prestwich's "Theoretical considerations on the conditions under which the (drift) deposits containing the remains of the extinct mammalia and flint implements were accumulated and on their geological age. On the Loess of the valleys of the south of England, and the Somme and the Seine (1862–63)," *Phil. Trans. Roy. Soc. London,* 1864, 147–310. Sir Joseph Prestwich (1812–96), geologist, was an authority on the Tertiary formations of Great Britain, and in the 1850s became interested in early human remains.

18. Robert Godwin-Austen, "On the valley of the English Channel," *Quart. J. Geol. Soc.,* 1850, *6,* 69–97. Robert Alfred Cloyne Godwin-Austen (1808–84) geologist with the Geological Survey of Great Britain, was a friend of Edward Forbes and completed several of Forbes's works after the latter's death in 1854.

19. Edward Forbes, "Diagram of the distribution of British Phanerogamous Plants & Marine Mollusca," *Mem. Geol. Surv. Great Britain,* 1846, vol. I, pl. 6. Cf. Edward Forbes, "Map of the Distribution of Marine Life, illustrated chiefly by Fishes, Molluscs and Radiata: Showing also the extent and limits of the Homoiozoic Belts," in Johnston, *Physical Atlas,* 2d ed., 1856, pp. 99–102, pl. 31.

20. Edward Forbes, "Diagram illustrating the relations of the British Fauna & Flora before, during and after the formation of the Northern Drift," *Mem. Geol. Surv. Great Britain,* 1846, *1,* pl. 7.

21. Sir Henry Thomas de la Beche, *The Geological Observer,* 2d ed., London, Longman, Brown, Green and Longman, 1853, 740 pp., pp. 90–91. Sir Henry Thomas de la Beche (1796–1855), geologist, was appointed the first director of the Geological Survey of Great Britain in 1835.

22. William Hopkins (1793–1866), "Observations on Drift; on the causes of change in the Earth's superficial temperature; the doctrine of progression with respect to inanimate matter," *Edinburgh New Phil. J.,* 1852, *53,* 1–31.

23. Edward Forbes, "Diagram of the distribution . . . ," 1846.

24. From a list of shells found on the coast of Massachusetts, prepared by Augustus Gould, Woodward decided that more than half the species were common to Europe. Because these were littoral species that could not cross areas of deep water, he thought that

there must formerly have been a continuous line of coast between Europe and North America. Woodward also referred to Edward Forbes's discussion of the similarity between the shell species of European and North American seas in the northerly regions (Edward Forbes, "On the connexion between . . . ," 1846, p. 379; see also note 19, above) and to Sir John Richardson ("Report on North American Zoology," *Brit. Assoc. Repts.,* 1836, pp. 121–224). Richardson had argued from the similarity in species of bottom-feeding fish on both sides of the Atlantic and from the similarity in species of Owls in Europe and North America that the two continents had previously been connected. S. P. Woodward, *A Manual of the Mollusca; or, a rudimentary treatise of recent and fossil shells,* 3 vols., London, Weale, 1851–56, 486 pp., p. 358.

25. Sir John Richardson (1787–1865) in 1819 went as a surgeon and naturalist on Sir John Franklin's polar expedition.

26. Thomas Rupert Jones (1819–1911), palaeontologist, worked for the Geological Survey of Great Britain.

27. Daniel Sharpe (1806–56), geologist, visited the Alps in 1855 and later published two papers. Daniel Sharpe, "Structure of Mt. Blanc and environs," *Quart. J. Geol. Soc.,* 1855, *11,* 11–26; Daniel Sharpe, "On the last elevation of the Alps with notices of the heights at which the sea has left traces of its action on their sides," *Quart. J. Geol. Soc.,* 1856, *12,* 102–23.

28. Sir Andrew Crombie Ramsay (1814–91), geologist, was appointed an assistant on the Geological Survey of Great Britain in 1841 and, in 1847, also became professor of geology at University College, London. In 1871, he became director-general of the survey and, on his retirement, its director.

29. This estimate of the elevation of Scotland since the last glacial epoch was based, at least principally, on the mistaken interpretation of the parallel roads of Glen Roy given by Charles Darwin in 1838. Darwin took the parallel roads to be the lines of former sea beaches and therefore indicating the former sea level in relation to the land. See Charles Darwin, "Observations on the Parallel Roads of Glen Roy, and of other parts of Lochaber in Scotland with an attempt to prove that they are of marine origin," *Phil. Trans. Roy. Soc. London,* 1839, *129,* 39–81.

In 1862, Thomas F. Jamieson of Ellon showed that the parallel roads were the beaches of a former glacial lake that had filled Glen Roy. Thomas F. Jamieson, "On the parallel roads of Glen

Roy, and their place in the history of the Glacial period," *Quart. J. Geol. Soc.*, 1863, *19*, 235–59.

30. During the meeting of the British Association at Glasgow in September 1855, Lyell stayed with Mr. James Smith at his home Jordanhill outside Glasgow. Charles Lyell to Leonard Horner, September 16, 1855, Kinnordy mss.

31. Robert Brown (1773–1858), famous botanist, called by Alexander von Humboldt "facile princeps botanicorum," was curator of the herbarium at the British Museum.

32. James K. Tuckey, *Narrative of an expedition to explore the river Zaire, usually called the Congo, in South Africa, in 1816, under the command of Captain J. K. Tuckey R. N. To which is added the journal of Professor Smith; some general observations on the country and its inhabitants; and an appendix, containing the natural history of that part of the kingdom of Congo through which the Zaire flows (Observations on Professor C. Smith's collection of plants from the vicinity of the river Congo by R. Brown)*, J. Barrow, ed., London, Murray, 1818.

33. Lyell is basing these suppositions on Darwin's theory of the origin of coral reefs and atolls in the Pacific by a general subsidence of the sea bottom accompanied by a compensating growth of coral. Charles Darwin, *The Structure and Distribution of Coral Reefs*, London, Smith, Elder, 1842, 214 pp.

34. Lyell is here contradicting Darwin's theory for the origin of coral reefs and atolls, which assumed a large amount of gradual subsidence of the ocean floor and consequently a great depth of coral beneath an atoll.

35. Juan Fernandez—a group of islands off the coast of Chile, south of the Galapagos.

36. The Almendrado is a conglomerate formation about a mile northeast of Santa Cruz in Teneriffe.

37. That is, if the Canaries were islands in the sea during the Miocene period, it would be very improbable that during the succeeding Pliocene and Pleistocene periods they would have been surrounded by a continental land mass that had again sunk beneath the waves. Furthermore, during the hypothetical time when the Canaries were part of a continental land mass, continental plants would have spread out from Africa and Europe so as to destroy the unique character of their fauna and flora.

38. Lyell believed that valleys could be formed by the wearing of the sea against the land and the ebb and flow of the tide in estuaries.

39. Sir Joseph Banks (1744–1820), botanist, who accompanied Captain Cook on his first voyage, was president of the Royal Society from 1778 until his death.

40. Captain James Calder (1730–1808) was born in Scotland and about 1758 settled in Baltimore County, Maryland. On October 20, 1768, he was commissioned deputy surveyor of Frederick County, Maryland, and on October 16, 1771, deputy surveyor of Baltimore County.

41. Daniel Sharpe had believed that he had found a line of erosion caused by the action of the sea at an elevation of 9,000 feet in the Alps. On this basis, he argued that the Alps must have subsided to a maximum depth of 9,000 feet during the glacial epoch and then been reelevated. Daniel Sharpe, "On the last elevation of the Alps . . . ," 1856, p. 119.

42. *Clypeaster,* a genus of Coleoptera (beetles), renamed *Sacium* by LeConte. Clypeaster altus was a fossil representative.

43. Richard Owen, "On a fossil mammal (Stereognathus ooliticus) from the Stonesfield Slate," *Brit. Assoc. Repts.* 1856, *2,* 73. The significance of this fossil was that it was the first true mammal, as opposed to marsupial, to be discovered in beds older than the Chalk. It thus demonstrated that true mammals had been in existence much earlier than had been supposed.

44. George Robert Waterhouse (1810–88), naturalist, was appointed curator of the Zoological Society of London in 1836, and, in 1843, he became an assistant in the mineralogical branch of the department of natural history of the British Museum and, in 1851, keeper of mineralogy and geology at the British Museum.

45. Lyell first wrote to Thomas Vernon Wollaston (1822–78) on November 26, 1853 to ask for information on the natural history, particularly the Mollusca and Insects of Madeira. Wollaston, who was a nephew of William Hyde Wollaston and possessed independent means, had gone to Madeira in the autumn of 1847 for his health. During three long visits, he collected the insects of the islands and in 1854 published his *Insecta Maderensia.* Wollaston's replies to his letters from Lyell are preserved in the Lyell correspondence in the University of Edinburgh Library.

46. In reply to Wollaston's letter, February 4, 1856.

47. Lyell had visited Natchez, Mississippi, in March 1846 during his second visit to the United States and had studied the loess forming the bluffs of Natchez and Vicksburg. See Charles Lyell, *Second Visit to the United States,* 2 vols., London, Murray, 1849, *2,* 208.

48. Oswald Heer (1809–83) was a Swiss naturalist born at Niederutzwly, St. Gall, and educated at the University of Zurich where in 1834 he was appointed professor of botany and entomology. In 1850, Heer had gone to Madeira on account of his health and later published an account of his travels and a number of papers on the geology and natural history of the island.

49. Rio das Socorridos is one of the larger rivers draining the southern side of Madeira, entering the sea about two miles west of Funchal.

50. Heinrich George Bronn (1800–62) was a German palaeontologist of Heidelberg. The genus Palaeoteuthis had been established by Alcide c'Orbigny in 1849.

51. Christian Friedrich Hermann von Meyer (1801–69) was a palaeontologist of Frankfurt-am-Main. Lyell may have been thinking of his article, "Palaeomeryx eminens," *Palaeontographica,* 1852, *2,* 78–81.

52. John William Salter (1820–69), geologist, worked on the Geological Survey of Great Britain from 1846 until 1863. He wrote a paper "On the remains of Fish in the Silurian rocks of Great Britain," *Quart. J. Geol. Soc.,* 1851, *7,* 263–68, and carried on other studies of fossil fishes.

53. S. P. Woodward, *A Manual of the Mollusca,* 1851–56.

54. Ibid.

55. In answer to Wollaston's letters of February 4 and 7, 1856.

56. Friedrich Heinrich Alexander von Humboldt (1769–1859), German naturalist. See discussion of ocean currents in his *Cosmos: A Sketch of a physical description of the universe,* E. C. Otté, trans., 5 vols., New York, Harper, 1870, *1,* 306–09.

57. See letter from Wollaston to Lyell, February 10, 1856.

58. Richard Owen (1804–92), anatomist, was appointed in 1836 as the first Hunterian Professor at the Royal College of Surgeons, London, and, in 1856, became superintendent of the natural history departments of the British Museum. He was the leading authority on comparative anatomy in England.

59. Lucas Barrett (1837–62), geologist and naturalist, in 1855, was appointed curator of the Woodwardian Museum at Cambridge and, in 1859, became director of the Geological Survey of Jamaica, where he died while deep-sea diving in 1862.

60. Lucas Barrett and Robert MacAndrew, "List of the Mollusca observed between Drontheim and the North Cape," *Ann. & Mag. Nat. Hist.*, 1856, *17*, 378–86.

61. A Pliocene formation, of limited extent lying between the rivers Alde and Stour in the county of Suffolk in eastern England. It is a white calcareous marl or soft limestone very rich in fossil shells and bryozoa or coral-like mollusca that gave the formation its name. True corals, however, are very rare in it.

62. James Dwight Dana (1813–95), American geologist and naturalist, was born at Utica, New York, and educated at Yale College, New Haven, Connecticut. In 1837, he went as a geologist and mineralogist on the U.S. exploration expedition to the Pacific Ocean under the command of Captain Wilkes. After the return of the expedition in 1842, he spent the next years writing his reports. In 1849, he was appointed professor of natural history at Yale, where he spent the remainder of his career. The article referred to was James D. Dana, "Science and the Bible," Art. I, II *Bibliotheca Sacra,* 1856, *13,* 80–129, 631–56.

63. The entry in Lyell's Index Book, vol. 1, p. 45, is as follows:

Dana — James D. "Science & the Bible"
Bibliotheca Sacra Jany 1856 Art. III p. 80

p. 94 — Six Days of Creation.

[p.] 121 — Warm climate once general.

[p.] 122 — The absence of strata in the U.S. between the Coal & the trias or lias is called a suspension of life whereas a continental period w.[d] be to geologists a more natural hypothesis.

[p.] 123 — Successive creation of perfect species.

" — Law of progress from simple to complex as Agassiz has explained.

[p.] 124 — Insect life enlarged at successive period.

" — Creative force beyond Man's conception.

[p.] 125 — Stonesfield mammalia small & mere insect eating.

" — They are represented to have been just antecedent to the tertiary instead of having just preceded the Wealden & Cretaceous.

" — Close of Cretaceous vast chasm, 5th Day!

[p.] 126 The continental era i.e. the tertiary—But the carbonif.[s] must have also been continental.
" No proof of a species created since Man was on the earth. God rested.

64. Louis Agassiz (1807–73), Swiss naturalist, who came to the United States in 1846 to deliver the Lowell lectures at Boston Massachusetts, and, in 1848, was appointed professor of natural history in the Lawrence Scientific School of Harvard University. Lyell had read Agassiz's article, "Sketch of the Natural Provinces of the Animal World and their relation to the different types of Man," in Josiah Clark Nott, *Types of Mankind: or, Ethnological Researches, based upon the ancient monuments, paintings, sculptures, and crania of races, and upon their natural, geographical, philological and biblical history,* Philadelphia, Lippincott, Grambo, 1854, pp. 58–76.

65. Hugh Miller (1802–56), Scotch geologist and writer, began life as a stonemason. In 1822, he noted ripple marks in a quarry and became interested in geology. His collection of fossil fishes was described for him by Louis Agassiz, and Miller presented the results in 1840 in his book *The Old Red Sandstone.* In 1847, in another book, *The Foot-Prints of the Creator,* Miller argued against the theory of evolutionary development presented in the anonymous *Vestiges of Creation* (written by the Edinburgh encyclopedist Robert Chambers) in favor of a theory of successive divine creations.

66. Richard Owen, *On the Nature of Limbs,* London, Van Voorst, 1849, 119 pp., pp. 83–6.

67. John William Salter (1820–69), "On the remains of fish . . . ," 1851.

68. Adam Sedgwick (1785–1873), geologist, was appointed in 1818 to the Woodwardian professorship of geology at Cambridge University. In 1850, he published a fifth edition of his *Discourse on the Studies of the University of Cambridge,* London, Parker, 1850.

69. John William Dawson (1820–99), geologist, and a native of Pictou, Nova Scotia, studied at the University of Edinburgh in 1841–42 and again in 1845–46. In 1842, he accompanied Lyell on a visit to the fossil cliff at South Joggins, Nova Scotia. In 1855, he published his *Acadian Geology* and the same year was appointed professor of geology and principal of McGill University of Montreal.

70. Heinrich Georg Bronn (1800–62). See note 50, above.

71. Oswald Heer (1809–83), naturalist of Zurich, Switzerland. See note 48, above.

72. The Stonesfield mammals were discovered in 1814 when a stonemason from one of the quarries of Stonesfield, a village lying twenty-eight miles to the northwest of Oxford, England, brought a remarkable fossil jaw to William John Broderip (1789–1859), then at Oriel College, Oxford. There were two specimens of the jaw. Broderip sold one to the Rev. William Buckland (1784–1856), then reader in mineralogy at Oxford, and lent him the other more perfect one. When Buckland returned it, Broderip misplaced the specimen, and it did not come to light again until 1828 when Broderip published a description of it (William J. Broderip, "Observations on the jaw of a fossil mammiferous animal found in the Stonesfield slate," *Zool. J.*, 1821, *3*, 408–12). In the meantime, Buckland had shown the other specimen to Georges Cuvier, who had identified it as the jaw of a Didelphys, or opossum. Thus, it demonstrated the existence of mammals in a geological period when only reptiles were thought to exist.

73. Lucas Barrett (1837–62). See note 59, above.

74. Jeffries Wyman, "Notice of fossil bones from the red sandstone of the Connecticut River Valley," *Amer. J. Sci.*, 1855, *20*, 394–97. Jeffries Wyman (1814–74), American anatomist, was professor of anatomy at Harvard University from 1847 until his death.

75. Joseph Leidy (1823–91), a naturalist. A native of Philadelphia, Leidy studied medicine at the University of Pennsylvania (M.D. 1844) and after a period of teaching at the Franklin Medical College, in 1853, was appointed professor of anatomy at the University of Pennsylvania. Much of his work dealt with vertebrate fossils.

76. Oeninghen, a town in the Rhine Valley between Constance and Schaffhausen, where the Molasse, a Miocene formation of Switzerland, is best represented. Lyell is referring to Christian Friedrich Hermann von Meyer, "Palaeomeryx eminens," 1852.

77. Edward Hitchcock (1793–1864), geologist, a native of Deerfield, Massachusetts, carried out the first geological survey of Massachusetts from 1830 to 1833 and then continuing it in 1837 published his final report in 1841. In 1836, Hitchcock published the first description of large bird-like footprints (actually dinosaur tracks) discovered in the Connecticut River Valley. See Edward Hitchcock, "Ornithichnology—Description of the footmarks of

birds (Ornithichnites) on New Red Sandstone in Massachusetts," *Amer. J. Sci.,* 1836, *29,* 307–34.

78. James Dwight Dana (1813–95). See note 62, above.

79. Henry Darwin Rogers (1808–66), geologist, directed the geological surveys of New Jersey (1833–40) and of Pennsylvania (1836–42). In 1855, he was appointed Regius professor of natural history at the University of Glasgow.

80. William Barton Rogers (1804–82), geologist, was the elder brother of Henry Darwin Rogers. From 1828 to 1853, he was professor of natural philosophy and chemistry at the University of Virginia and, in 1862, was appointed president of the new Massachusetts Institute of Technology. For Lyell's reference, see William B. Rogers, "On the relations of the New Red Sandstone of the Connecticut Valley and the coal-bearing rocks of Eastern Virginia and North Carolina," *Amer. J. Sci.,* 1855, *19,* 123–25.

81. William C. Redfield (1789–1857), businessman and scientist, had collected fossils from the beds of New Jersey and Connecticut.

82. Richard Owen, "Description of some mammalian fossils from the red crag of Suffolk," *Quart. J. Geol. Soc.,* 1856, *12,* 217–30. Lyell seems to have read Owen's paper in advance of its publication.

83. John Morris (1810–86), professor of geology at University College London from 1854 to 1877.

84. Charles d'Orbigny (1806–76).

85. *Coryphodon,* a genus of extinct amblypod mammals from the lower Eocene of Europe and America.

86. James Scott Bowerbank (1797–1877) was an amateur geologist who in 1847 suggested and founded the Palaeontographical Society for the publication of descriptions and illustrations of British fossils.

87. Gideon Algernon Mantell (1790–1852), surgeon, first of Lewes, Sussex, and later, 1839, of Clapham, became famous for his discoveries of the fossils of the South Downs and Weald. Mantell was a personal friend of Lyell.

88. This letter seems to have been in reply to Wollaston's of February 19, 1856.

89. Robert MacAndrew, "On the geographical distribution of Testaceous Mollusca in the North-east Atlantic and neighbouring seas," *Liverpool Lit. Phil. Soc. Proc.,* 1853–54, *8,* 8–57.

90. The Rev. Maurice John George Ponsonby (1787–1855) was created fourth Baron de Mauley of Canford County, Dorsetshire, in 1838, by Queen Victoria.

The date 1851 here presumably refers to the letter, while the observations on which it was based were made in 1841–42.

91. Charles Babbage (1792–1871), mathematician and inventor of a calculating machine, was a personal friend of Lyell. In 1824 and again in 1838, he had written papers on the temple of Serapis near Naples and provided Lyell, who had used an engraving of this temple as the frontispiece for his *Principles,* with information concerning it. The geological significance of the temple of Serapis was that it demonstrated the occurrence of considerable and measurable changes in level of the land within historic time. It seemed, therefore, particularly relevant to the question of large-scale elevations and subsidences of land in the Tertiary period. Lyell gave a lecture on the temple at the Royal Institution. See Charles Lyell, "On the successive changes of the temple of Serapis," *Roy. Inst. Proc.,* 1854–58, *2,* 207–14.

92. This may have been Guiseppe Nicolini (1788–1855), Italian educator and writer and professor at Brescia and at Verona.

93. After their discussion on March 12, Wollaston wrote to Lyell on March 14, 1856.

94. Oswald Robert Heer (1809–83). See note 48, above.

95. Walter Mantell (1820–95) was the eldest son of Gideon Algernon Mantell (1790–1852). Walter Mantell had emigrated to Wellington, New Zealand and on a return visit to England was giving Lyell information concerning the effect of a recent earthquake in New Zealand in raising the level of the land.

96. In 1856, Lyell interviewed three men who had been eyewitnesses to the great New Zealand earthquake of 1855. They were Edward Roberts of the Royal Engineers, Frederick A. Weld, "a landed proprietor in the South Island," and Walter Mantell. Lyell was especially interested in this earthquake because it had been accompanied by many permanent elevations of the land. He later gave a full account of it in the *Principles,* 11th ed., 1872, *2,* 82–89.

97. Johann Christian Albers (1795–1857), *Malacographia Maderensis sive enumeratio molluscorum quae in insulis a Maderae et Portus Sancti aut viva extant aut fossila reperiuntur . . . ,* Berlin, 1854, 94 pp.

98. Ludwig Georg Pfeiffer (1805–77), German naturalist and physician, traveled in Cuba in 1838–39 and wrote on natural history.

99. Richard Thomas Lowe (1802–74), English naturalist, went to Madeira for his health in 1828. From 1832 until 1854, he served as the English chaplain there. Lowe published a number of works on the natural history of the island.

100. Georg Hartung was a young German geologist whom Lyell met during his visit to the Canary Islands in 1853–54.

101. R. T. Lowe, "Catalogus Molluscorum Pneumonatorum Insularum Maderensium: or a list of all the land and freshwater shells, recent and fossil, of the Madeiran islands: arranged in groups according to their natural affinities; with diagnoses of the groups, and of the new or hitherto imperfectly defined species," *Proc. Zool. Soc. London,* 1854, *22,* 161–218, p. 174.

102. Lyell never published his long paper on the volcanic geology of Madeira and the Canary Islands. Instead he discussed it in his next (sixth) edition of *Elements.* See Charles Lyell, *Elements of Geology,* 6th ed., London, Murray, 1865, 794 pp., pp. 621–47.

103. Lyell here seems to be stating a principle of species divergence through time.

104. Philippe de la Harpe, geologist of Lausanne, Switzerland, had written on Swiss tertiary formations.

105. Sir Andrew Crombie Ramsay (1814–91). See note 28, above.

106. This letter seems to have been in reply to Wollaston's of April 1, 1856. Wollaston replied on April 14, 1856.

107. Joseph D. Hooker, "An Enumeration of the Plants of the Galapagos Archipelago; with Descriptions of those which are new," *Trans. Linn. Soc. London,* 1841, *20,* 163–233. See note 12, above.

108. Sir Charles and Lady Lyell went to visit Charles Darwin at Down House in Kent April 13–16, 1856. They seem to have arrived at Down the morning of Sunday, April 13, and to have departed the afternoon of Wednesday, April 16. According to his calendar (Notebook 213, p. 1), Lyell had an appointment at Buckingham Palace, apparently in connection with the business of the Commission of the Exhibition of 1851, to which body he belonged, on Saturday, April 12, and he was to attend a wedding, which must have been in London, at 8:00 o'clock in the evening of Wednesday, April 16. This diary has against the entry April 13 the single word

"Down," and the entries in the Scientific Journal No. I (see pp. 52–53 and 54–55) record discussions with Darwin on April 13 and 16. Notebook 213, on the other hand, records a discussion with Darwin on Madeira on April 14, so that their scientific talks on three of the four days of the visit are indicated fairly clearly. On April 15, Lyell wrote a letter to Charles Bunbury from Down.

It was Lyell's custom to take his books and papers with him to Down and to spend part of each day working just as if he were at home. This habit made him a congenial guest for Darwin who also liked to spend at least part of every day at his work. It also suggests the easy and intimate atmosphere of their friendship.

In his Notebook 213, p. 29, Lyell recorded what he intended to take with him to Down.

Take to Down, April 13

1. Note B.[k] N.[o] 1 Scientific notes
2. Ordinary N.B. No. 213
3. Illus. Madeira paper—long parcel
4. Notes on Hartung & C. L. in envelopes
5. Hartung's letters
6. Manual, last ed. C. L.
7. Yellow notebook on Madeira notebooks.
8. Vidal's Map of Madeira

109. Charles Darwin, "On the action of Sea-water on the germination of Seeds, 1856," *Linn. Soc., J.*, 1857, *1*, 130–40, p. 134.

110. Darwin had carried out a series of experiments on the effects of salt water as well as on the eggs of snails and other land molluscs.

111. We have not been able to identify what Lyell and Wollaston found in the puddle.

112. Lyell is here again referring to Wallace's 1855 paper "On the Law which has regulated the introduction of New Species."

113. Joachim Barrande (1799–1833) was born in France and educated at Paris, but in 1820 went into exile with the French royal family who ultimately settled in Bohemia. Barrande, who had private means, spent the rest of his life in forming a great collection of the fossils from the palaeozoic rocks of Bohemia. He even opened quarries for the specific purpose of obtaining fossils from them. In 1852, Barrande had published the first volume of his great work on the Silurian system of Bohemia. Joachim Barrande, *Système Silurien du centre de la Bohème,* Prague, 1852.

114. John Stuart Mill (1806–73), philosopher of utilitarianism.

115. Thomas Henry Huxley (1825–95), zoologist, had, after completing his medical studies, gone as surgeon aboard *H.M.S. Rattlesnake* on a voyage to Australian waters. In 1854, Huxley had been appointed lecturer at the School of Mines, and in 1855 naturalist to the Geological Survey. His post at the School of Mines had allowed him to marry Miss Henrietta Heathorn in July 1855.

116. Joseph D. Hooker (1817–1911). See note 12, above.

117. William Benjamin Carpenter (1813–85), zoologist, had studied medicine at University College, London, and the University of Edinburgh Medical School (M.D. 1839). He became known for his work *The Principles of General and Comparative Physiology*, London, 1839. In 1844 he was appointed Fullerian professor of physiology at the Royal Institution, which, with other lectureships, he held until 1856 when he was appointed registrar of the University of London.

118. George Busk (1807–86), naturalist and surgeon, studied medicine at St. Thomas' and St. Bartholomew's hospitals and served as a surgeon on board the seamen's hospital at Greenwich until 1885 when he retired to devote himself to scientific research.

119. The Philosophical Club had been founded in 1847 by twenty-seven fellows of the Royal Society, of whom Lyell was one. The purpose of the club was to strengthen the scientific activities of the Royal Society and to provide for social interchange between the actively working scientists in the society. It met once a month on Thursdays except for the anniversary meeting, which was held on the last Monday in April. In 1856 this was held on Monday, April 28, so the meeting to which Lyell refers had occurred two days earlier.

120. Georges Cuvier, "Mémoire sur l'Ibis des anciens Egyptiens," *J. Phys.*, 1800, *51*, 184–92.

121. Huxley had delivered a public lecture on the oxslip.

122. Trans: "Nature made it and then broke the mold." Ludovico Ariosto (1474–1533), *Orlando furioso*, canto 10, stanza 84.

123. This paradox, which affects all discussion of ideas of progressive development in the living world from the 1820s until 1859, arose from the fact that the advocates of a theory of progressive development thought that the progress that they detected in the geological past arose from a gradual development in the mind of God of his concept of his creation. The successive races of

plants and animals in this view represented successive creations; they derived not one from another but each directly from the creator. The intoxicating feature of the theory of progressive development was that it seemed to allow scientists to look into the process of creation and to see visions of the developing mind of God.

124. On p. 86 Lyell noted that he had omitted Helix compacta from the living species of shells common to Madeira and Porto Santo.

125. Lyell ran out of space at the back of the notebook and continued the index on blank pages he had left at the beginning.

Scientific Journal
No. II

May 1, 1856

Dytiscus marginalis living, according to Mr. Charles Prentice of Cheltenham, a friend of Mr. Woodward's, was found with Ancylus Velletia lacustris, attached to him. The Ancylus had treated him as he would have done a stone. Suppose a gale to aid the Dytiscus when on the wing, might he not carry the Ancylus from one pond or river basin to another or from island to island. The seeds of water plants are made to bear long immersion in water & may be carried to the deltas of rivers easily & perhaps [p. 2] by marine currents to great distances. If aquatic plants can be transported by water more easily than the seeds of land plants, if they float better, this would make them more cosmopolite even with the advantage of the more equable temperature of lakes as compared to that of the air & the dry land. See N.B. 213 p. 56 with Woodward.[1]

May 5, 1856

Migration of Species

Colymbetes (coleopt.[a]) flew on board the Beagle, (Darwin Letter May 4.th)[2] when she was 45 m. from shore.

New Species [p. 3]

If conditions alter less in one region than in another the variation of species would be less. A moderate amount of fixed or permanent variation in 50 species or varieties of the Genus Squirrel, may amount to as much change in specific characters, in the way of gain, as the extinction of one species may amount to in the way of loss.

The change in the human period may be to the aggregate fluctuation, when two successive Geological periods are compared, as is the alteration of our continents. All may seem stationary. If the origin of the existing races of Man dates from 50,000 years ago, so the origin of a great many [p. 4] of the domestic animals may be as modern, if so the formation of species has gone on in human times or of many reputed species.

The number of generic types, from which reputed species have branched, is much reduced in number if we accept Carpenter's estimate in regard to Orbitolites as a test on Foraminifera & Hooker's (N. Zeal.[d] Flora)[3] in regard to plants. The more we diminish the number of original types, the rarer will be the birthdays of such stocks; according, therefore, to the doctrine of [p. 5] successive creation, the fewer will have been the opportunities for man to witness the first starting into being of protoplasts.

On the other hand the concession of a generic type prepares the way for admitting a prototype of an order & then all we require is the original creation of a certain limited number of vertebrate & invertebrate archetypes, whence the long series of geological assemblages of species were evolved by Lamarckian transmutation.

Wallace's hint as to the likeness [p. 6] of each succeeding geolog.[l] formation, in reference to its palaeontology, is the more deserving of attention when we allow for the multitude of missing documents.[4] We may reply to the question, where are the intermediate types. They are hidden from us, not yet found, many of them destroyed & irrecoverable. This is inevitable. Nature's system is not that of preserving a perfect history of her past creation, but only a broken & imperfect one.

The popularity of the *Vestiges* arises from any theory being preferred to what Grove[5] calls a [p. 7] series of miracles, a perpetual intervention of the First Cause as constant as is the dying out of species & as it has been from the beginning—so far as we can see back into the past.

But we must be prepared like Lamarck to include Man in the same system of change? If we exclude him the only sound argument for the popular theory of progress in the organic World is gone. If we include him, the great book which the

Geologist is trying to decypher becomes at once identified with Natural Theology as well as with Natural History. Mind & the Soul of [p. 8] Man will be found to be a development of the instinct of Animals? If so, why deferred so long after the Quadrumana & the Elephantine mammalia flourished. Have we gained one step towards reconciling us to the alternative in the last 50 years. Is not the dilemma really owing to the want of philosophy in species-making?

A species according to the Lamarckian hypothesis is a variety of a genus on which certain constant condit.[s] of the organic & inorganic world have been impressed for so very long a time that new condit.[s] cannot easily modify it, but may cause it to die [p. 9] out. Even the old conditions, if revived too suddenly, would produce extinction.

A race is an imperfect exemplification of the same result of conditions acting thro' a lapse of ages. The time, says Grove, to uncoil must be as great as to coil. To make a tropical race fit to inhabit arctic regions will take as long as to fit the arctic to live under the line.

When a species enters a new province by migration, it ought to spread at first rapidly because it finds fit soils, not cropped before; it may be years, even centuries, before there arrives such a combination of circumstances as is fatal to nearly all the individuals which have established [p. 10] themselves in the newly colonized region. At length the unusual multiplication of an inimical species, animal or plant, or a change of temperature, the prevalence of some wind or epidemic, annihilates all but one or two or a few individuals stronger than the rest, or provided with some exceptional attribute.

These individuals then constitute the stock from which the species in the new region will radiate & these will be better able to stand the recurrence of the adverse combination of causes.

If there be then an immigration of new species from the original country they will not spread because the fit stations are preoccupied by the [p. 11] stronger variety adapted to the new province.

But in a group of islands the new batch of settlers may find an isolated rock or isle, a new volcano perhaps, or a newly upheaved shoal unoccupied. Then we sh.[d] meet with the continental type unaltered as a representative form of the same genus. It may be centuries before the contingency recurs which was fatal in the island, first colonized, to the original settlers.

One island of a group may be twice or three times as ancient as another. In Volcanic islands especially, the intermittent nature of the igneous action favours successive production. So does upheaval of a submerged bed of the sea of originally unequal altitude. Volcanic eruptions also increase the chances of great fluctuations & irregularities, in [p. 12] the partial extinction of varieties & species.

The doctrine of the identity of the Human & the brute stock is only lowering to those who believe in the cessation of individuality after death, for the capability of progress compensates for past inferiority, but the Lamarckian doctrine would admit of deterioration of types as well as of amelioration. In short I see no parallax in the mystery of mysteries. It becomes every naturalist to grapple with the whole difficulty & to try & define *species,* if he cannot he must entertain doubts as to the signification of the words [p. 13] *race, variety, species* in Man & this implies a profound ignorance of the position of the Human Race in the system or its relation to the coexisting & past organic World, or beings revealed to us by Geology.

May 7, 1856

Dignity of Man

If there have been several (100?) millions of born & adult idiots & every gradation between them & men of every humble capacity & between the latter & men of genius, if there are every intermediate steps between the sensible or rational & the insane, why claim such dignity for Man as contrasted with the brutes. When does the sucking infant attain the rank of an intelligent [p. 14] dog? Has the child an hour before its birth a soul? or an hour after?

If a race of savages of the lowest capacity exist for 1000 years without progress, why not a race intermediate between them & the Chimpanzee?

May 7 [1856]

Progressive Theory

If the simplest forms succeeded each other in regular gradation at each successive Geological Period till Man came next to the anthropomorphous brutes, this theory most favored by Dana,[6] Owen?,[7] Hugh Miller,[8] Sedgwick?,[9] Agassiz[10] & the more orthodox, treads close upon Darwin's & Lamarck's. It brings Man into the same system of progressive evolution [p. 15] on which developed the Orang out of an oyster.

The Hindoo incarnation was at least a more spiritual theory. The idea, which I have advocated [in the *Principles*], of a uniform system of change to which Man forms an exception & which was uniform among other reasons because thereby alone could Man interpret it & which remained uniform with the sole exception of Man's reason & its effects, such a theory was most consistent with the popular notions of Man's distinctiveness and superiority. He may have arrogated too much in regard to his relative-rank & the force of facts must be presumed to be overpowering that c.ᵈ overcome the unwillingness of the many [p. 16] to infer that Man is a part of the one same & indivisible evolution of the complex out of the more simple, which evolved a mammal out of a reptile & a reptile out of a fish—which made H. Miller calculate the relative capacity of the brain to the skull in successive geolog.ˡ creations of Vertebreta.[11]

A law of progressive developm.ᵗ is the only way of reconciling the successive appearance of higher and higher beings on the stage, but can any such law evolve the rational out of the irrational. It is here that the analogy fails. The equality of many successive tertiary groups of mammalia, [p. 17] the Eocene Macacus (see L.ᵈ Brougham's argument of future archangel),[12] is a fact against Man being the latest step, unless in tropical conditions intermediate species between Man & the Orang sh.ᵈ be found fossil.

The intellect of Man 2000 years ago (Socrates, Aristotle, Plato) is quite equal to that of the 19.ᵗʰ century.

The relapse of the individual of the highest moral & intellectual power to 10 to 20 years of second childhood may be cited as a proof of individual annihilation for a term at least of suspension of existence, but it is yet an instant of infinitesimal [p. 18] duration in eternity.

The number of born idiots, of children born dead, of insane, of the lowest & most animal-like of savage races, of infants cut off before their capacity equalled that of the instinct of the Elephant or the Dog, has probably exceeded all the millions of the white races or Asiatics of the most civilized eras. The failures have been counted by Millions, so entirely does Nature subject Man to general laws—Epidemics, Earthquakes, Pestilences, Wars are allowed their full sway.

Why then should the law of Creative power which supplies the havoc [p. 19] & the chaos made by the destructive agencies cease to operate because of Man's appearance on the Earth? In poetry Milton might make all Nature shudder at the sin of man, the arresting of the creative or species-making power w.ᵈ be a greater shock.

But the soul's revelation (see W. Greg,[13] the grand Enigma) may be equally true in reference to a future state.

Species

If they are not real then Man can only be distinct as a genus. Linnaeus thought genera as real as species & Darwin would class Families & Orders in the same category, out of which permanent varieties by *selection* are formed.

Orthodox Theory [p. 20]

It might have been anticipated that the orthodox would most have shrunk from progressive develop.ᵗ—w.ᵈ most have wished to find a triassic mammifer. It has been rather the contrary—perhaps from a notion of adapting the 6 days of Creation to the crowning advent of Man. Care should be taken not to involve this question with preconceived notions about the Time or manner of the introduction of the original stock of the human race. But it is strange how much the 6-days theories have led towards C. Darwin. Hugh Miller is more of an advocate for the evolution of Man out of preexisting inferior grades than I have been. The absence of [p. 21] fish in the Lower Silurian, of reptiles in the Upper, of birds in Paleozoic, of Mammalia in the same, of quadrumana in the Secondary, of the Chimpanzee, & Orang even, in the Tertiary; all these negative facts, which according to my view, are of no real weight, yet they would all, as reasoned upon by the gradual development theorists & anti-uniformitarians (as by Whewell & others),[14] so identify man with the geological genealogy of the vertebrata as to prepare the way for the reception of Darwin's views, by which the perpetual-intervention hypothesis is superseded as much in the case of Man as in that of the inferior Creatures.

Idiots [p. 22]

They may be as numerous in the most advanced & full peopled parts of Europe as were the hunter-tribes of the same area. If these tribes were preceded by others less advanced it w.[d] not be more of a mystery, metaphysically consid.[d], than the existence now of so many failures or beings not coming up to the grade of Man nor of the Orang.

The insane form a 1/1000.[d]? part of the human race? or a million, hence we need not go back to primeval times nor to a Lamarckian hypothesis to establish the indifference of Nature (if some general law require it to be so) in regard to the forms, time or existence of a large number of beings wearing the lineaments of Man & yet not being gifted with reason—if such exceptions amount numerically to as [p. 23] many as could be supported by the globe when Man was in the earliest of hunter states & if this multiplicat.[n] of idiots, madmen, & human-beings in their first & second childhood be a constant condition of humanity.

May 12, 1856

Woodward, Part III, *Manual of Mollusca*[15]

p. 412—remarks that the genera of America suffice to identify the Coal, etc. tho' the species are different. Here we seem to have a community of condit.^{s} which according to the Lamarckian hypothesis must give rise to similar generic modificat.^{s} of the family type in the E. & W. hemispheres.

The migration of species (ib.) must take so much time that the same in very distant regions w.^{d} rarely imply synchronism of existence. [p. 24] The development of genera in time follows the same law as that of species. Woodw.^{d} ib. p. 411.

The creation in a single spot Woodward thinks intelligible, the dying out a mystery! to me it is just the reverse—p. 411. *ib.*

May 14, 1856

Species by Transmutation

If species were made by the effect of certain powerfully acting conditions of climate or coexisting species, it ought to follow that when several of the descendants of one stock reached several islands of an archipelago, they w.^{d} under the influence of these causes be converted into the same new species or into species closely allied. And this Darwin [p. 25] may affirm has occurred in regard to the vars. of Helix polymorpha in the Madeiras where doubts arise as to the distinctness of the species.

Species—Madeiran Shells

The resemblance of some individuals of Helix coronata of the Bugio to some of H. tiarella, the spira of the former being nearly of the same height as in H. tiarella, raises the question as to the identity of both & whether H. coronata of P.^{o} S.^{o} be really different also.

On the other hand H. tiarella recent & fossil & H. coronata recent & fossil show how those minute characters, by which each can be distinguished from the other species, have been constant, not only for a vast period (10,000 or 50,000? years) but through.^{t} vast changes of conditions [p. 26] for it is fair to presume that Caniçal & the north coast, where H. tiarella occur foss. & living, or P.^{o} S.^{o} in the two periods differed as much or more than Bugio & Madeira, or Bugio & P.^{o} Santo. All the species from the 3 isles might be mixed & again separated without fear of confusion.

The same it is true might be said of what Wollaston would admit as varieties of Helix polymorpha—so that this argument alone would by no means be conclusive.

H. Scabruscula & H. Wollastoni

The Sicilian species is the same in *size,* but the lip is not entire in the adult as in the adult P.° S.° species; the colour is not the same. Are colour & the integrity of the lip suffic.[t]?

May 24, 1856 [p. 27]

Transmutation of Species

Edw.[d] Forbes, according to G. Austen,[16] speculated on the probability of the Eocene mollusca pulmonifera of Central France having lived on in a region never invaded by the sea from the time of the old (Limagne) lakes to our days & having been gradually changed from the Eocene forms to ours.

If this could be admitted it is but a small step to go the length that Darwin wishes.

Paludina lenta may be P. unicolor—but in this & in the case of Helix labyrinthica, the species migrated from Europe where they became extinct & spread in the E. & in the West.

May 26, 1856 [p. 28]

Centres of Creation

O. Heer's letter receiv.[d] today alludes to certain areas from which the tertiary species radiated into the islands of the Atlantic & first from America into the Atlantic & then from Europe or Asia after the glacial period into W. Europe.

These ideas coincide with those of Wollaston that N. Africa & Sicily were the starting points of part of the land-shells of

P.º S.º & Madeira, not because such species as H. scabriuscula is really identical with H. Wollastoni, but because it & others, H. Muzzilii, are so allied as to imply a centre of similar generic & subgeneric types.

E. Forbes p. 27[17] points somewhat in the same direction—all tending towards C. Darwin's view of indefinite transmutation in time. The only half-way which I can imagine between such speculations & the theory of the reality of species is this. [p. 29] Certain types generic & sub-generic, more flexible than others, are created which give rise to the larger proport.ⁿ of what we term species. These types only are created directly, & in regions fitted for them at the time & for them only. Hooker's sub-species are derived from them—other less flexible types produce more feebly represented genera? Perhaps a type inflexible at our period may become more so under new geological conditions.

If this theory be adopted the creation of a true species will be a very rare event. The appearance of Man may have been a recent exemplification & the Negro a sub-species which in one of the Quadrumana might have been describ.ᵈ as a true species.

May 27, 1856 [p. 30]

Heer's Letter on Migration of Plants & Animals into Europe

At the time of the Atlantis, America (N.) was united by the Atlantic with Europe but there was sea to the East—the Mediterr.ⁿ joining with the Indian Ocean. This may have been in the Mollassic (falunian) epoch.[18] The climate was milder not only because the tropical Indian sea flowed freely to the shores of Europe, but because the Alps were low. Afterwd.ˢ the submergence of the Atlantis—all but a few volcanic isles & the icification (Uebergletscherung) of a large area separated America from Asia & after this the flora & fauna of Asia came back & peopled Europe when they became one continent, while America was disunited by the Atlantic intervening.

The Rise of the Alps [p. 31]

Heer imagines the Atlantis to have sunk in the Pleistocene time but the depth given in Maury's charts is immense.

May 27 [1856]

Insular Varieties

A Helix on one of the Desertas may be fed for 20,000 years on plants quite distinct generically from another of the same species (or stock) on the isl.[d] of Cima. Small isl.[ds] separated by the sea are peculiarly favourable for giving rise to such variations as can be found by restriction of diet & contact with different coexisting animals, & plants.

On a continuous tract of land the same experiment cannot be made as to the flexibility of a type, if limited to certain organic conditions, i.e. to certain limitat.[s] of intercourse with contemporary friendly or inimical creatures or nourishing or less fit food to say nothing of the inorganic & physical causes. A group of [p. 32] islands, therefore, is the fittest place for Nature's trial of such permanent-variety making & where the problem of species-making may best be solved.

Plants & animals introduced into Australia where the climate, & the organic inhabit.[s] all differed so much is an argument ag.[t] the variability of species, but only so far as a few years or generat.[s] can be compared to millions of years. If the species in the British Isles have remained the same as contrasted with the Atlantic Islands, because of the difference in the time, then tens of thousands of years are not enough for such experiments—because the lapse of centuries required for the conversion of the Glacial into the postpliocene era is so vast that the historical era is as nothing to it.

But undoubtedly the period which divides the Helix Loweii period from the actual may compare with that which intervened between the Glacial & our times. [p. 33] And if the duration of time causes the differences between the distinct provinces of species in certain Atlantic isles of our group, or of the distinct groups when contrasted (of Mad., Azores, Canaries or of Mad. & P.º S.º), then we may solve, in the same way, the distinctness of the fossil land-shells of Mad. & P.º S.º, & the uniformity of the species foss. & recent in each isle, or their agreement with the living & 3.dly the difference of s.d species from the rest of the world. For suppose the Madeira mollusca to be Meiocene altered shells, they may have had a million or 2 millions of years to change, & the interval between 1856 & the era of Helix Loweii or H. delphinula may be only (if we take the lowest Caniçal beds) 30,000 years.

Yet one would have expected a shade of difference between the fossil & the recent, more than we find in Mad.a & P.º S.º, some fraction of the transmutation. [p. 34] Whereas, if species are created by some process or power distinct from the transmutation or variety-making power, then the discordance of species in islands, in proportion to their inability to migrate, is beautifully explained.

May 30, 1856

In two distinct localities & in two distant geological periods, a perfect coincidence of contrasting causes cannot concur & therefore, species will never—scarce ever, if created at remote times or remote places—be very closely allied.

June 1 [1856]

Genealogy of Species

Two hundred generat.ns of Man w.d carry back the history of many a living pair 2500 yrs., 400 generat.s 5000, & we might perhaps find the ancestors of the negro or the pair representing the [p. 35] white man as distinct as we see they were 3000 yrs. ago on the painted walls of an Egyptian temple. But if we could double or quadruple the number of generat.s & go back ten or twenty thousand years should we not, say some naturalists & ethnologists, begin to find an approximation between the ancestral pairs which would stand before us as the progenitors of their respective stocks. According to one hypothesis the negro type would recede indefinitely into the past until we sh.[d] arrive at the eventful era when there would be no antecedent pair which could be evoked from the tomb. Such a vision alone, if [p. 36] the figures could merely pass before us, as in the last scene of Macbeth, would in one way or another destroy many a favourite conjecture or hypothesis.

If we could trace back any single species to the exact period when it had not begun to exist & [had] seen the first link of the chain, it would afford an important insight into a great mystery or what is now a great enigma.

We might select some well known fixed type, some species considered as inflexible, not prone to aberration or variability & see whether, in following back its history, it remained as unalterable as in our own time till the year came when [p. 37] nothing stood before us. In the present state of Geology we can recall certain portions of the past, but never any consecutive chain which has not wide intervals or blanks. It is as if we asked to see how a species lived, what form it wore 5000 years ago or 10,000 or ½ a million—& then the next record may be thousands or millions of years or ages, we know not which, so that the grand question as to the extent of possible passage from one form to another can never be determined. Yet we may conceive that in the course of time much accurate inform.[n] may be accumulated on this head & while [p. 38] investigations are made in referring to the past, the future concealed from us will be more & more known to our posterity, so that we have no reason to imagine that this mystery will not be in part cleared up.

Whatever be the truth the explanation, if revealed to us, would be startling. It might confirm a bold conjecture & if so we sh.[d] be surprised at its realization. It might confound our anticipations. It must coincide with all known facts & legitimate generalizations respecting the actual geographical distribution & the past geological history of species or what we call species in the organic world.

The existing forms we know would [p. 39] not recede

indefinitely into the past. They would blend into preceding groups insensibly or abruptly, by transmutation or by successive creation of nearly allied species. Their habitation, climate & coexisting animal-relatives & plant-relatives, w.d vary.

Abrupt reappearances of animals extirpated by Man or extinguished by other causes w.d take place. Would the first appearance of their progenitors be equally abrupt? Above all sh.d we trace back Man to one original or to several stocks or starting points? & so of other species—& where w.d Man commence & in the form of what race, Red, White or Black? savage or civilized, superior in stature & form or inferior or of [p. 40] average beauty.

The measure of time w.d become more & more perfect as geologists observed & chronicled the series of changes now in progress, & thus they w.d obtain & will acquire a gauge for estimating the antiquity of fossil remains, destined to be hereafter brought to light, in the Indian Ocean volcanic islands & in the interior of Asia.

(p. 154. Wollaston) Index B.k 1. p. 54[19]

Natural Barriers & Specific Centres

Darwin's theory w.d assume correctly that the precipices of Madeira are modern, & when they were prod.d, they separated species of insects, coleoptera, & by isolation led to their divergence from the original types, so that the specific centres may be dispensed with.

June 1 [1856]

Species

Species [p. 41] may be in a particular genus, if transmutation be true, representatives of the permanent varieties resulting from distinct measures of Time, or of fluctuations in the organic & inorganic world. If a certain variety takes 1000 generat.s to make & it comes into contact with another, which has required for its aberration 3000 generat.s, they may differ widely but not so far as two others which diverged 500 generat.[s] back under circumst.[s] more discordant, very distinct climate or species of coexistents or isolation & breeding in & in etc.

June 4 [1856]

Races & Species

If races are forming in Man & in Plants & Animals, as fast as races are dying out ,there will be a constant succession of different races & there may be always as many in spite of the [p. 42] perpetual extinction of them. No race would ever be created, except the first. Might this be the type of all the others? According to Lamarck it would be a preexisting species transmuted.

If the greater part of our existing species are of the same value, only as are the races of Man, & differ no more than does the Negro & White Man then as no new Human Stock or allied type has entered the earth in historical times, so very few other new types w.[d] be looked for in the same period. In other words the variety of Creation may be chiefly kept up by the formation of races miscalled Species & true Species may appear at rare intervals, not by batches. This may help to explain the fact [p. 43] of the appearance of a new generic type or the extinction of one being very rare. Yet one w.[d] suppose the exterminat.[n] to go on most rapidly.

June 13 [1856]

A theory consistent with all that is now known ethnologically & geologically, whether it embraced the doctrine of our stock at a very remote epoch or of many stocks of different antiquity perhaps of higher & higher grade as time passed on, w.[d] give a shock to the opinion of nearly all men. The most orthodox view that can be conceived might close the chair of nearly every University ag.[t] its promulgation, & other views w.[d] ensure the expulsion of a Prof. already installed. The most orthodox [p. 44] creed must assign an antiquity to the human race so great that 3000 or 4000 years tells for little in the formation of a distinct race & in ascending the stream of time, we must reach an era of tens of thousands of years to revert to the departure from a common stock. There comes the question whether less civilized races deteriorated from a more highly gifted or more advanced race, or whether the first stock was of low capacity & improved into higher.

If many stocks, did those of the humblest capacity precede the rest?

If any one c.[d] have the truth revealed to him. or c.[d] know what will probably be known 1000 years hence, he could not publish it without giving such a shock to [p. 45] received notions as would exclude him from a professorship in most universities.

Was the original stock of moderate or average ability & their species some higher some lower?

Or did they (as some have ventured to conjecture) slowly emerge from the inferior anthropomorphic creatures into Man?

If not, an act of creation is implied in (geologically speaking) modern times, & whether there were one stock or several created, whether simultaneously or in succession, this (or these) creation w.[d] involve the admission of that interference which many deprecate, *Nec Deus intersit* is their maxim, Man perhaps may be then "*dignus vindice nodus*".[20]

The progressive theory has been on the whole rather the [p. 46] favourite among the orthodox & those who have advocated a six day theory.

The dog and the wolf, having the same period of gestation, seems the best argument in favour of many races like Man being equal to species as usually admitted.

We do not advance but we accumulate facts which makes old notions inadmissible. We cannot prove that the races of man all came from one stock, or that they had each distinct starting points, nor decide how many aboriginal races there have been, but to explain these away by referring all to one stock, we require time so vast that the historical era carries back no very appreciable distance nearer to the [p. 47] point of departure from the supposed common stock.

Scarce any perceptible convergence of the two lines to any point in the past is indicated by Egyptian monuments, yet the period of predominance of existing species in Geol.[y] is so long (the Glacial Period being a mere intercalation) that the lines might converge in an era, which may be as of yesterday, in the latest geological epoch. The continuity may have been nearly such as they are in their principal features & the Mountain-Chains whereas 9/10ths of the species, now coexisting with Man, have survived great geographical revolut.[s] in Europe & N. America & probably in the rest of the globe, if we had studied them as accurately.

June 13 [1856]

A supplementary vol. to *Principles* (& *Manual*?) *referring* to each [p. 48] and pointing to corrigenda & to old views still held & novel ones proposed or suggested—*Origin of Species & of Man*—C. Darwin—Wollaston[21]—Hooker, N. Zealand[22]—De Candolle[23]—Agassiz[24]—Species in Islands—Progressive theory. Dana[25]—Hugh Miller[26]—Grove[27]—Ethnologists—transmutation—*Vestiges*[28]—Whewell's[29] denial of the creation of species in the Human Period—E. Forbes,[30] how far he favoured this view.

If true (as W. [Whewell] in *Plurality of Worlds* opines[31]) it affords the best argument in favour of transmutation; for the system is now what it was in so many essentials, as we know as Geologists, & especially in the gradual extinction of species so that if there be no concomitant creation we may conclude that the causes of variation in the organic [p. 49] world are sufficient to cause new species & to vary as much, as Extinction confirms. The loss of a species & substitution of more of another is an assimilating of the fauna. It becomes more uniform—so the introduct. of cultivated plants lessens the variety in Nature, tho' culture producing new races of animals & monstrosities in plants, compensates somewhat.

Madeira & S.[t] Helena are examples of the destruct.[n] of native fauna & flora. New species come in, it is true, but the Vine or Sugar Cane or Banana cover very large spaces, perhaps larger than gregarious or—native plants? Vaccinium Madeirense is an argument ag.[t] this & Erica arborea?

Man's interference is limited in the water especially.

The new races of Man are a source of variety.

In place of native stocks, reduced [p. 50] in Madeira or limited, we have domestic animals—races of man, cultivated plants strangely altered—new helices more than the percentage for extinct. If these races undergo modification, as great as the difference between Negro & White Man, they w.[d] be good species for future conchologists.

The tendency of the known causes of extirpation is not to lessen the amount of organic beings, but to reduce the number of types, unless the total amount of variation in Man & domestic animals & cultivated plants compensates for the loss. Here we come to the old ambiguity as to the nature of a Species.

If the Dodo be merely replaced by a new race of birds of an old species, & the Dodo be a true species [p. 51] & not a race of some living type, then such an extinction is a loss of a kind not to be set off ag.[t] the gain, or the gain of a marked & distinct race of dogs or pigeons, which is not an equivalent for the extirpation of the wolf.

Is there then a tendency to uniformity in the specific character of animals & plants? Are the types destined under Man's rule to diminish in variety & number?

As one dominant race may expel several races—

Dana says, Anniv.[y] Address, 1856,[32] that the creation was never so rich & varied as now. How can this be proved or supported? Probably he wd. appeal to negative geological evidence of no value.

June 13 [1856] [p. 52]

As the growth of 1000 trees in a forest may just equal in one day the growth of one tree 1000 years old which a new settler may cut down, so the minute, & imperceptible deviation of all the living species of animals & plants in a century, may equal the extinction of a species in a century. True of all ages or the acorn & seed to others 1000 yrs. old.

This loss of a large tree is a marked event, the growth of all the rest is in a day imperceptible or in an hour.

The coming in may be by the deviation from preexisting types so imperceptible, when spread thro' a million [p. 53] of what we term species, or half a million, that in a hundred years is inappreciable, whereas the extirpation of a variety or species well established, the growth of ages is a marked, unoverlookable, unconcealable event. According, therefore, to a Lamarckian view the formation of species is an act as invisible to Man as the growth of a continent or a mountain chain or much more so.

Was the origin of Man of this nature? It is unfair to bring down upon the advocates of such an hypothesis all the theological prejudice which such a question may raise. Answer the question how we may, the truth, could it be known, must when it embraces all established facts be such as can be reconciled to no popular creed.

June 14, 1856 [p. 54]

The genealogy of man would be as dignified & exalted if he were traced to a "clod" (Young's Night Thoughts)[33] as if he had been transmuted out of mammal highly organized & almost as wonderfully made as Man.

The vertebrate type, the anthropomorphous archetype preexisted for ages. Instinct, not unlike that of a child a year old, powers of memory & some forethought, parental affection, courage, anger, jealousy, industry, acquired knowledge, preexisted. Were the human faculties superadded to this, or some form of this already created & ancient machinery, or was it a new start? Was it an evolution out of [p. 55] what most nearly resembled it or out of inorganic, lifeless matter? Was it a modification of what was most akin to it, or a conversion of the material into the spiritual by direct fiat, without any intermediate gradations?

Can a rational sensible Man be born of two idiotic parents, if so is it a greater wonder than if a savage race came of the highest of the Brutes? If a poet or philosopher of the greatest genius can be born of his ordinary commonplace parents, it lessens the wonder of a brute race producing a rude human savage race.

The German Prof.[r] [34] [p. 56] regarded the inferior animals as he considered somnambulists who can speak & even write correct grammar, unlock doors, use their eyes & exert their memory of places & yet be unconscious actors, & so insects & elephants can display an acquaintance with mechanical & mathematical laws & yet be acting like automatons.

The record of Man's origin may be written in the stoney framework of mammoths of the globe—his antiquity, thro' what phases the human species has passed, how many races, if any preceded those known to us, which were the earliest, whether of higher [p. 57] or lower grade, where they first originated, whether their organization even varied much more from the present races than does the Negro from the Caucasian; whether there was a passage from some very dissimilar form or species into the immediate progenitor of Man or none whatever. It will require Time to investigate such a problem, but the discoveries of Geology w.[d] have seemed more marvelous if foretold 500 years ago than w.[d] all these results if now predicted.

If a retrospect of 200 generations or even 400 (5000 years) finds the White Man & the Negro as wide apart as now, it may merely mean that 2000 or 4000 generations w.[d] be required before [p. 58] we should ascend to a common stock. But after carrying back our investigation 400 generations & finding the White Man in cold countries & black man in the tropics not materially different, we might begin to see in a small number of generations a great change as we sh.[d] have reverted to a period of migration & first settling & change of conditions.

During all this time the geography of the Earth might be

unaltered & thousands or tens of thousands of generations may be required in order to get back to an era when the original stock, whether of a vertebrate or invertebrate creature, was subjected to geological conditions, organic & inorganic, very dissimilar from those now [p. 59] prevailing or when the climate was as distinct as that of the Glacial epoch.

The migration southwards of many crag species & their return to repeople the British Seas, after the cold of the Glacial Period had mitigated, is a strong exemplification of the resistance of species to changes & of the obstinate retention of specific characters.

Woodward says that as the horse died just when he had been taught to live without food (accord.[g] to the old Greek saying) so just as a species w.[d] be on the point of transmutat. into another it w.[d] perish.

The reply is that many do die out but others succeed & survive [p. 60] & the likeness of each geol.[l] fauna to that which had preceded it proves the passage.

Against this is the doctrine that species must be created like those immediately forerunning them because the zoological, climatal botanical, geographical & other coincident conditions never c.[d] have been so like at a more remote epoch as in the one just antecedent.

If the Deity can allow so large a portion of the entire human race 1/5th? to remain for several 1000 years in the state of the African negros & Hottentots, there may have been races even more rude than these or the Fuegian or [p. 61] Australian for a long period, so that we cannot speculate on the ground of probabilities.

If we ever came to the conclusion that each successive geolog.[l] group of animals & plants was evolved out of another by descent & without the intervention of the first cause or only by Laws appointed by Him like those governing the succession of individuals then we have to decide whether the creation of Man is to be instanced as exceptional.

The incarnation of Vishnu[35] embraced the form of a fish, a tortoise, a serpent, but never of an Ape, & the form most like the human w.[d] be most repugnant to the imagination. Yet [p.

62] the idea of the transmutation of a human or divine soul or spirit into some one of the inferior animals is not evidently repugnant to the mind.

It is by this intimate connexion of the reasoning faculty with the organic world that we acquire knowledge not so easily obtainable if our sensations & structure physiolog.l & anatomical were more foreign & distinct. We interpret nature as she exists & her past archives by this intimate identification.

The transmutation theory w.d be favoured by the progressive & w.d in turn favour it & we might look [p. 63] forward to future progress & to men who w.d regard us as we look upon the Fuegians. The absurd confidence of those who rely on negative facts in Geol.y throws discredit on the progressive system & gives to it in the hands of its principal abettors the appearance of a theory founded on a preconceived idea. We must be on our guard against being too much prejudiced against it because of the unaccountable manner in which Dana, Agassiz & others assume that each genus began where we happen to have first met with it or that each did not exist where not yet found fossil. Arguments drawn [p. 64] from finding no terrestrial remains in marine strata, or in shifting sands, must not conceal from us the positive evidence of abundance of reptiles, ferns & other forms now rare in the same latitudes.

[June 17, 1856]

Letter to C. Darwin,[36] June 17, 1856

Even if D. Sharpe's 9000 ft. of submergence of the Alps in the Glacial Period be rejected, the Jura blocks imply that the Jura & the Alps may have been islands within the Glacial epoch?

The glacial theory w.d perhaps throw as much of European land under water as was submerged in post-Eocene times in the Map of Europe in 1.st Ed.n of *P. of G.*

If the tropical conversion of sea into [p. 65] land & land into sea has been as great in the Glacial & post-glacial times as the Arctic & Antarctic revolut.s in Geography, then in a fraction of one Geolog.l Period w.d a great alteration of land & sea occur.

The Atoll theory causes continents to sink in the period of existing coral species.

The Central Alps [are] partly of Eocene strata, therefore, our continents [are] post-nummulitic.

Littoral species of MacAndrew[37] bear on the land shell argument in regard to P.° S.° & Madeira having been once united, but very long disjoined.

Identity of land shells in British Isles owing to [union] within Glacial Period? If so, it agrees with Madeira & Canaries having been disunited for a comparatively indefinite era.

New Zealand earthquakes give [p. 66] great mechanical movement in the life-time of Washington.

How came the Crag shells, when they returned to the British Seas after their long expulsion by Glacial cold, to be unchanged specifically during so immense an interval? & after such migration & changes of condition & Climate & coexisting species.

June 25 [1856]

[Darwin's Reply]

If the human intellect had been placed in the Elephant & the human animal had only been endowed with just so much instinct as the Chimpanzee their position w.[d] have been reversed in regard to power & M. S. as minute as any now written might have been planned by aid of the proboscis. & cannon fired & ships built & navigated. But if any one of the [p. 67] numerous forms of Proboscidians, which preceded those now living, had been endowed with human understanding they would have left monuments of their inventive & constructive power. If therefore Man be placed here in one of the forms of the Quadrumana in order that he may better interpret the system, it is a sufficient finite cause, without seeking any other, of his subjugation to all the laws of the terrestrial fauna.

Extinction

The failure of species may be like the failure of individ.[s] If certain species be the effect of Time the past continuance will generally secure the future endurance of certain conditions & then no species will be formed except to last very long. [p. 68] There will never be time to produce a marked deviation unless a volcano on our island or a desert last so long that the species shall be worth fitting for such new conditions.

If introduced species like Vanessa Atalanta be alike over the American continent it may arise from the recent emergence of that continent after the Glacial Epoch. Just as England has not had time to modify the Asiatic land & fresh water shells, having emerged only a few hundred thousand years.

The Negro may require absolute isolation & breeding in & in for 100,000 years & may thus people a large area with a fixed race which may not readily intermarry.

[p. 69] If there be a disinclination to sexual intercourse in proport.[n] to the remoteness of races, this may explain the coexistence of 2 permanent varieties for a few thousand years—especially if there be some local condit.[s] capable of reviving the exceptional race or refining it as a mountain, a desert, a morass—heat—the sea—air—etc.

It is remarkable how all the hypotheses for the separate & independent Creation can be turned in favour of the Transmutation theory also—& if the latter has the advantage of introducing a known general Law, instead of a perpetual intervention of the First Cause, it may assert a superiority unless we suppose a law as [p. 70] subtle as that of Life with its power over matter to modify our species into another & to fix it for a time, until the condition will no longer admit of the species continuing. But such a power or law, it will be said, w.[d] be exerted over preexisting organized matter rather than over inorganic.

A good illustration w.[d] be V. Atalanta as in Brit. & N. America. Then this seems as ancient in Madeira V. Callirhoe? See Wollaston, who considers this of the same species, & why not, if the American be the same.[38]

H. Tiarella

So marked is the divergence of each species of Helix in each separate Atlantic island that Wollaston feels confident that he shall know at once [p. 71] whether Berthelot's[39] specimen of the living H. tiarella came from Madeira or the Canaries. Whether it be correct or not his faith shows the singular extent to which permanent varieties have been formed in species long settled in such islands, and the varieties of H. polymorpha in different parts of the Madeiras have no counterpart in any British shell, perhaps in no European species?

Wheat & Aegilops

Aegilops caudata & Aeg. ovata have been said by ——[?] to have been converted into Wheat. They are annuals & by selecting the seed in each generation wh. comes up most like wheat in 4 or 5 generat.[s] the plants are hardly distinguishable. But these experiments [p. 72] made on Sicilian specimens of Aegilops in S. of Europe have not, it is said, answered in the N. Bunbury[40] has in 2 years got a luxuriant & altered plant, but not yet like wheat nor essentially altered. Henslow is experimenting.[41]

28 June [1856]

Philippine Islands

Mr. Cuming,[42] says Mr. Geo. Waterhouse, collected a species of Curculio from 4 islands & in each it differed, 2000 individ.[s] yet all distinguishable, metallic lustre in one, black in another & made by a French entomologist into separate species.

Waterhouse wrote a paper to show they were all one.[43]

Thibet Shells [p. 73]

[Are] nearly all British species, tho' made different by some because of geographical position. Himalayas & Thibet collected by Dr. Thomson.[44] The Cyrene consobrium or torquinta made into 4 or 5 species by authors but all are with the Grays Thurrock fossil.

Brain

Carnivora [are] above insectivora. These [are] above Rodentia & Marsupials, says Waterhouse, below all.[45] The corpus callosum evanescent in the marsupials, prominent in carnivora.

June 30, 1856 [p. 74]

Volcanic Areas & Upheaval

Volcanoes distend & uplift the earth's crust independently of their effect in building up by ejectamenta & outpouring of lava. Hence they check submergence within a limited area but probably have no wide effect in arresting the downward, or accelerating the upward, movement of large areas. The action of subterranean heat & volcanic action may be the cause on a great scale of Ocean Beds & of Continents, but the superficial volcanoes may be ineffective except as local agents. Hence if one area be rising [p. 75], as South America, it will rise faster & higher in the Andes if there are volcanoes there. If it be sinking in the Pacific & Indian Oceans it will not sink so deep, nor be so entirely submerged, if there be active volcanoes, but around such volcanic vents there will be deep sea as soon as we get beyond the local & limited area of the volcanic influence.

Sicily would have risen in the Newer Pliocene Period even without the aid of Etna. That portion, where the great volcano is, has been elevated higher than it would have been had there been no lava, but may not for all that be so high as Castrogiovanni.

[p. 76] There has always seemed to me a difficulty in reconciling two facts in Darwin's theory of volcanic & Coral areas—namely that Volcanoes are the upheaving power and yet, that nearly all the islands in the middle of great oceans are volcanic, whereas there are not many active, nor an extraordinary number of Tertiary volcanoes in continental areas.

If we suppose a subterranean power indifferent to the casual site of superficial outbreaks, to uplift certain extensive tracts & cause others to sink, the effect will be that we shall find some volcanoes in large islands as in Java & Sumatra, some in [p. 77] continents as in Mexico, but if in spite of the eruption & outpouring force & the expansive action of heat, we find no general coincidence between volcanoes & continents, but rather the contrary (or a geographical connection between oceans & volcanoes) then there can be no truth in the idea that volcanoes belong to areas of Elevation more than to areas of subsidence.

If there be a growth of coral there will be three reasons for volcanic islands, even in areas of subsidence—outpouring, local upheaval, accumulation of organic matter. To a certain extent shells supply this 3.^{d} cause even where slow-building corals are wanting.

[p. 78] In the Atlantic one fourth, as in Madeira, of the whole height, or less than that in the Grand Canary (1/8th or 1/6th?), may be due to local upheaval & hence we find submarine volcanic form.^{ts} upheaved 1000 or 1400 feet, but in the case of the G.^{d} Canary & in Madeira the rise is post-miocene, but the accumulation [of] Miocene [beds] probably [occurred] during subsidence?

Darwin's rule, p. 142,[46] that volcanoes are absent in areas subsiding, or which have recently subsided, w.^{d} be contradicted by the above.

Darwin's best argument is that there are no volcanoes in the areas of atolls, p. 141,[47] but may not this be explained by saying that volcanoes when extinct may sink & their site be only known by atolls [p. 79] or by *encircled* islands as w.^{d} be the case with a granite mountain or any other. But if a volcano break out in a subsiding area, then as in the Friendly archipelago, there will be upraised atolls near the volcanoes, & the volcano will turn the area into one of elevation. If, however, as D. says, there is no active volcano within several 100 miles of a *small* group, even of atolls, then the local upheaval must reach several 100 miles & it w.^{d} be strange that the ocean sh.^{d} be so deep between Madeira & Africa, for the volcanoes are in their origin Miocene volcanoes.

Let the areas of subsidence & elevation be equal & the volc.^{s} ought to be equally distributed, but with the difference that local upheaval w.^{d} give an excess in favour of elevat.^{n}, if in spite [p. 80] of the volc.^{s} be[ing] chiefly in oceanic tracts, it w.^{d} imply their connexion with subsidence—but this may prove that the areas of subsidence are in excess & therefore volc.[s] are chiefly insular or oceanic.

W.[d] not Java be a low island if all the volc.[c] matter were subtracted? Certainly Madeira & G.[d] Canary w.[d] cease to be islands. Etna w.[d] not be a lofty part of Sicily, nor Naples so high as the Apennine region—the upheaved volc.[c] recent masses are, as in Ischia, subaqueous volc.[c] tuff & lava & may only imply oscillation 2000 f.[t] of a shoal sinking & the re-elevation of the same shoal. If the upward & downward movements in the Canaries & Madeira have been balanced then [p. 81] we sh.[d] have the submarine portion accounted for by the stratif.[d] rock of the islands themselves, plus the organic remains. If upheaval was in excess, then some older tertiary or secondary peaks w.[d] peep up.

May not many of the Atolls be founded on Older Pleiocene & Miocene volcanic chain such as Mont Dor & Central France & Hungary.

Maury chart of 1855, Pl. xiv, makes a depth of 1875 fathoms between Cape Verde I.[s] & Africa more than 1000 f.[ms] 6000 f.[t] between Canaries & Africa & more than 1200 feet between Madeira & Africa! [48] Can it have been joined in Miocene period? The sea between M & Azores as deep—or above 2000 f.[ms].

June 1, [1856] [p. 82]
[See Fig. 4.]
[July 1, 1856]

"Whether Volcanoes Are in Areas of Elevation"— Extract of a Letter from C. Lyell to C. Darwin, July 1, 1856

[p. 83] Of course it is true, as you well show in your Coral volume, that the active volcanoes have recent deposits with marine shells uplifted in them. This is the case to some small extent in all the principal Atlantic islands except Palma, which I visited, & Palma has not been thoroughly examined & may somewhere exhibit signs of elevation comparatively therefore,

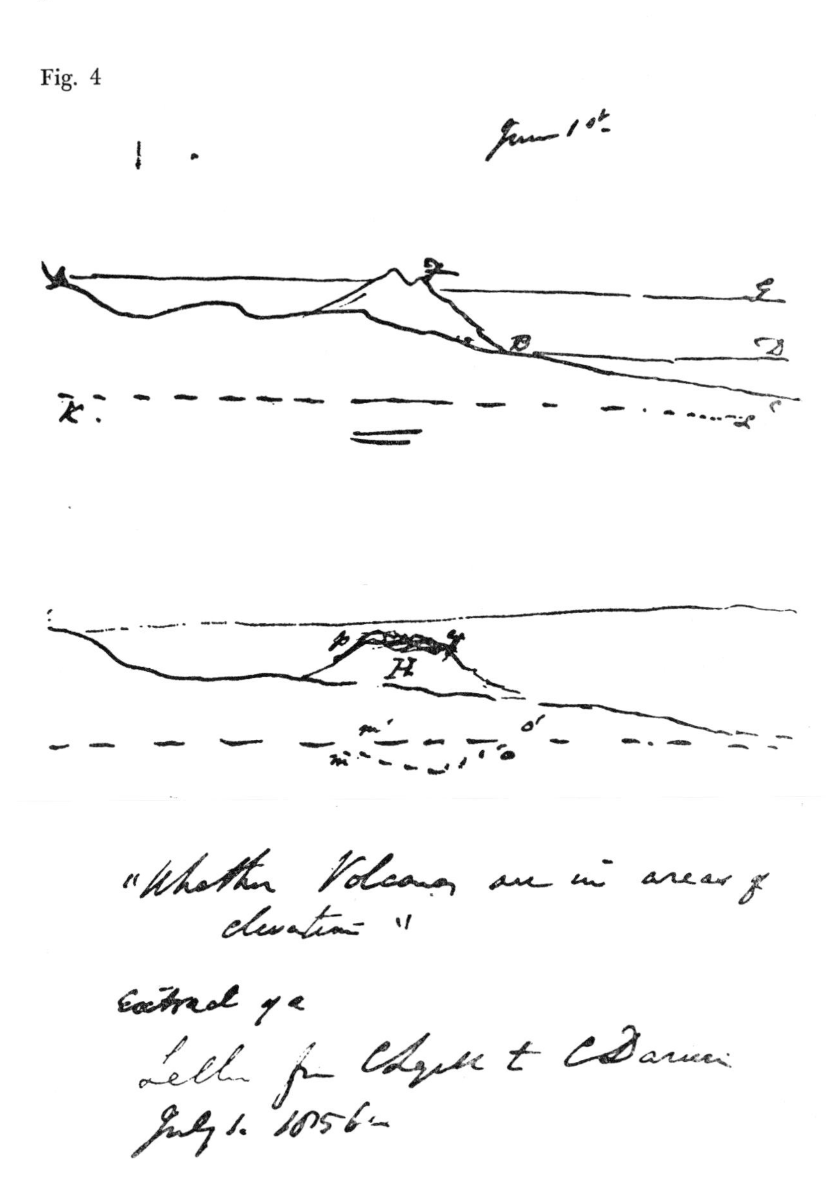

Fig. 4

and by contrast with the Atoll areas, you may represent the volcanic as rising.

Still I have always felt a little uncomfortable at being called upon to assume that in recent & Pliocene ages volcanic action has been and is connected with the growth of land. Were this the case should we not find that [p. 84] the continents would be the great areas of extinct Pliocene and of active volcanoes, and that the latter did not affect sea-side and insular and even mid-ocean sites.

If we find active volcanoes in Oceanic areas, & few or none of them in the middle of continental areas, it furnishes a *prima facie* case in favour of the doctrine that the grand uplifting power acts very independently of the accidental sites of existing superficial outbreaks. An argument might even be raised in support of the theory that active volcanoes are more connected with sinking on a great scale, however true it may be, that locally they tend to upheave as well as to form land by outpouring of lava & of ejectamenta.

[p. 85] Maury's last chart of the Atlantic makes the Atlantis hypothesis more bold than it appeared when E. Forbes proposed it, for the Canaries are separated from Africa & Europe by deep sea depressions of more than 6000 feet & Madeira by depths exceeding 12,000 feet! The data I fear are scanty however.

I find in Madeira & the Canaries upraised littoral deposits of the Miocene period, in my sense of Miocene, when there was a certain proportion of living species already in being. This, I think, rather increases the difficulty of the continental extension hypothesis.

But I want to ask you whether it may not be true that the bed of the Atlantic has been gradually sinking all the [p. 86] while the Canaries & Madeiras have been forming & that very slight local upheaval only has occurred even on the sites of these volcanic islands. I sometimes think that I can dispense with all excess even of local upheaval, over & above that of the adjoining deep sea spaces.

Thus, for example, suppose A. B. C. to represent the original Europeo-African continent and B. D. to be the level of the Atlantic. A gradual sinking down of 6000 feet takes place in a short part of the Miocene Period, (not occupying possibly above ½ a million of years).

The ocean has thus risen relatively to the land up to B. But in the meantime the volcano F. has been gradually built [p. 87] up 7500 ft. & is 1500 feet higher above the sea. A pause in the volcanic action takes place during which a subsidence of partial extent under I.[s] occurs, causing F. to lose 1500 feet of its height by slow depression during which every part of the subaerial mass of F. gets submerged & covered or faced with a marine littoral deposit, full of rolled boulders and pebbles with palettae & other littoral shells. [See Fig. 5.] The subjacent

Fig. 5

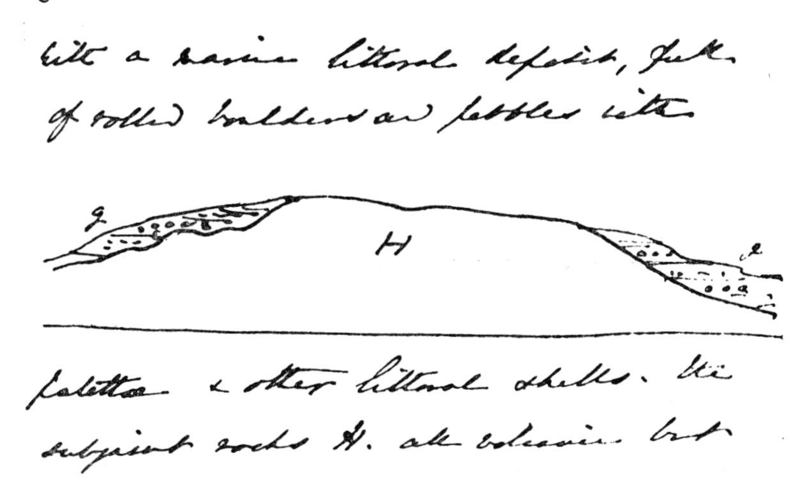

rocks H. all volcanic but as entirely free from marine remains as if exclusively subaerial.

We now have the original subterranean layer K. L. bending down at m, n, o. [p. 88] 1500 feet below the general depression of 6000 feet, & if it be then restored to the level m 'o' we have the volcano H. pushed up again 1500 feet with the marine beds p. q. abutting against the foundation of older subaerial rocks. This is what I observed in one part of the Grand Canary.

The 4000 or 5000 [ft.] of additional subaerial volcanic beds may be built up & you have Madeira. In the Grand Canary I suspect most of its height was attained before the submergence of 1500 or (1100) feet. [See Fig. 6.]

But my reasoning you see is the same as that which I adopted about the Atolls [p. 89] before you invented your theory, namely that oscillations occurring in a sea filling up with coral, or with volcanic matter, may cause uplifted marine formations provided subsidence & upheaval be just equal, the one to the other.

Take away all the volcanic matter from Etna, Ischia etc. &

Fig. 6

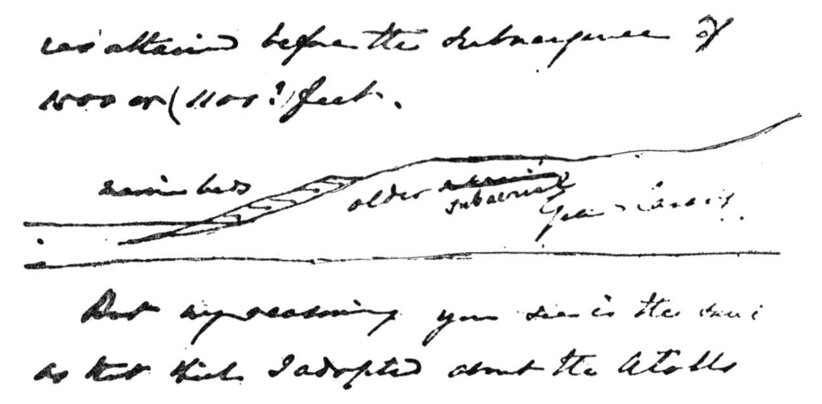

the marine shells could sink down below the sea level. All the marine beds in the Canaries & Madeiras are volcanic except the corals & shells themselves. If the active volcanoes were connected with a continent-making power, we should see secondary and non-volcanic rocks uplifted by them. I do [p. 90] not, however, want to contend that active & Pliocene volcanoes belong to subsiding areas, rather than to areas of elevation, altho' half inclined to that alternative in preference to the opposite theory. But surely they are so distributed as that they seem to belong quite as much to Pliocene & recent subsidence as to upheaval during the same period.

July 2, 1856

P. S. Not Sent to Darwin

The deepness of the sea round Madeira & P.º S.º & other Atlantic Islands is against Volcanoes being connected with upheaval for the upraising power w.ᵈ tend at least to render the sea shallow sh.ᵈ it fail to push up dry land in the neighbourhood of oceanic volcanoes.

Measure of Time [Darwin's Reply][49] [p. 91]

If the Elaphrus lapponicus migrated in Glacial period to the Grampians & is unaltered from the Lapland heath, it implies a vastly greater lapse of time since the British Isles were parts of the European Continent than the time since which Madeira & Porto Santo were separated.

Always assuming that the variations in the Madeira Islands afford a measure of time, floating ice, however, might have transported beetles as it has carried bears from Greenland to Iceland.—

By this means apparent universality of species may be explained & species which can obtain new [p. 92] conditions by enjoying greater powers of migration may be preserved unaltered & untransmuted & return when the conditions return. But ought there not to be all the intermediate varieties preserved, ought not the study of the fossils to be more puzzling?

Rosa Tomentosa

If it varies according to stations it may be adduced in favour of the flexibility of certain species. Lindley,[50] when asked how many British Roses there were, answered that there was at least one—(C. Bunbury); he also said there were 5 principal types perhaps. But the passages of one into the [p. 93] other are such as to make it no easy task to limit the species if more than one be admitted.

How like the question as to the Human Species & the principal types & the intermediate shades.

If on a sandy plain at Cape of Good Hope we find 10 distinct species of Heath, all under the same condit.[s] of climate & soil how, says C. Bunbury, can we suppose them to be offsprings or divergents from one stock?

One might imagine several islands in which in the course of ages they [the species] had become distinct or they had migrated in different directions & thro' distinct regions & climates & reached the Cape. It might require as much time to revert into one original typical form, if such there were, or to become [p. 94] thoroughly & indistinguishedly confounded as for the White, Black & Red Man, the European Negro & American Indian to get fused in Virginia. It has perhaps taken 20 thousand years or much longer for these races to form if ever they ramified from one parent stock & if so, a few centuries may do little towards their assmilation, unless race constitutes no disuniting cause, and it seems admitted that in proport.[n] as races are disunited the disinclination [*sic*] to sexual intercourse diminishes. Does it so in dogs?

We have two ways of testing the antiquity of races if the white man & the Negro be the same—3000 years is as nothing in the Egyptian monuments—& 300 as nothing in assimilating the two in Virginia. Subject them to the like conditions & several centuries do but little [p. 95] towards approximation. Ascend 3000 yrs. towards the supposed starting point & you seem to have made no step towards their junction.

July 5, 1856

Letter [to] Darwin

Icebergs & floating ice between latitudes 35 & 80 in each hemisphere may have been great agents of transporting species & have done much which is attributed to continental extension.

Take the glacial period as a unit & multiply this by all post-Miocene Time & the result may afford an amount of geograph.[l] change capable, with the additional aid of means of transport & migration of species, of carrying them everywhere even across the line by aid of cold periods & mountainous islands & [p. 96] floating ice & floating timber.

Species

If it be a mere metaphysical abstraction a certain amount of variation from a given type & that amount incapable of definition now & forever so that one naturalist may have as much right to adopt his own standard, provided he can point out certain differences, as another, the admitted want of precision & exactitude is such as to render it doubtful whether Zoology & Botany can deserve the name of science. But the naturalist may retort on us, can you define a Period in Geology. Is it not a mere abstraction? Are not format.[s] as unreal as species—one of them representing a longer, others a shorter lapse [p. 97] of years & breaks in the series, or in other words, our ignorance affords the only general.[y] received lines of demarcation.

If it be possible to call geology a science in spite of the incapability of the most skillful to define a period even zoologically, if some species are always dying out & others coming in, so if some variations are always in progress in the descendants of the same stocks in the organic world, it may be that we only can avail ourselves of accidental breaks in the chain to make good species & yet we may use them as realities.

If a species is a mere abstraction, is not a plant & an animal equally so? How many original starting points are to be assumed? These first stocks [p. 98] must have been the only realities—from these all others were evolved, even the rational out of the irrational.

No doubt many faculties—such as anger, memory, parental affection, courage, fear, hunger, thirst existed before Man.

July 8 [1856]

Zoological Society

J. E. Gray[51] says a great many of the species of Cyrena (20 Woodward thinks) will have to be united to Cyrena fluminalis, —— ——[?], C. trigonula, C. consobrina, etc., that such is the state of conchology that what with Dealer's species & those made to gratify the vanity of naturalists desirous of naming a new spec.[s] & the theoretical system of assuming all species from remote regions to be distinct, the multiplication of spurious species has [p. 99] been very great & the time has to come soon when, as the Botanist is already doing, numerous varieties will be united into one. But meanwhile the time has not been wholly lost & the distinctions of varieties belonging to localities & climates, stations & provinces, are of importance & may deserve names.

Mr. Stephens[52] observed that some butterflies of the Himalayas are identical with British & yet not found in the intermediate country.

[July 10, 1856]

Plato's Ideas

Our conception of the reality of specific types may border on the eternity of the 'divine ideas' which at certain periods, formed & predetermined, became united with matter or incarnate, according to Plato. If any resemblance exists between the modern [p. 100] doctrine of specific types & the metaphysical abstraction of Plato's genius it need not raise on that account any presumption ag.[t] them. For under any theoretical hypothesis or way of viewing genera & species, even if there be one perpetually changing system, progressive or non-progressive, we cannot, if we believe in the foreknowledge of the Deity or in such predestination as any pantheist will allow, doubt the preexistence of those immaterial conceptions, instinct, life, & finally the human reason & intellect—& then intimate union with certain material organizations. Is this one case, where we believe in nothing but the individual, we still [p. 101] have to conceive the changes from the embryo to the infant & so on to adolescence & age. All this we may call the embodiment of one idea, but if so, & if we thereby admit a succession of ever differing individualities, we arrive at something like Plato's [concept].

July 10 [1856]

Continental Extension

Whatever amount of sea has been in post-Miocene ages converted into land, we may & must presume that a like & equal space has been changed from land into sea.

More than this, if oscillation of level in post-Miocene ages have several times converted & reconverted the same areas into land & sea & land again, we must accept [p. 102] all this amount of fluctuations as having greatly affected the distribut.[n] of animals & plants. If we found no anomalies, no facts at variance with an assumed distribut.[n] of species in conformity with the present shape of land & sea, how great would be the contradiction, the discordancy of the results between geol.[y] on the one hand & natural history investigations on the other.

Linnaeus

When he invented generic & specific names he never suspected the real existence of the groups of beings which these names designated. Perhaps it may be said that in declaring that genera had as real an existence as species he unintentionally sanctioned the doctrine that both were unreal in nature. [p. 103] (See Wollaston on generic theory.)[53]

From the Lower Mammalia to Man

There are no proofs of superior sagacity of the anthropomorphous mammalia over the Pachydermata or Carnivora or Canine in instinct. Man, therefore, being an old-world form of anthropomorph, is no reason for presuming a transition, tho' if the race began in Asia, it may be imagined to have some connection. As to the metaphysical & theological difficulties & 'moral repugnance' felt ag.[t] the passage from brute to man, we must consider fairly & look in the face the corresponding ones in the existing state of the creation. The mystery here is as great. If it were proved that there had been every intermediate stage between irrational & rational in a [p. 104] long succession of ages, whether by successive races or species, & if it were asked when did a responsible soul begin, it is the repetition of a doubt, so often put as to the infant, newly born & never developed, so far as this earthly career is concerned. The number of infants who die before — months bears a large proport.[n] to the adults through.[t] the globe. It may be always supposed that their spiritual career is cut off to be continued hereafter, as may be that of idiots, but it is at least an enigma of the same kind as that of the imaginary case of a transition series of beings, from a mere animal existence like that of a new born infant, to a higher state.

If we do not encounter more perplexity by embracing this hypothesis than by rejecting it [p. 105] we must welcome it sh.[d]

the weight of evidence be on that side. The reason I presume of the continuance of the prevalence of the theory of progressive development, in spite of the frequent shifting of the dates of the first appearance of different classes of vertebrates & invertebrates, is the established fact that Man, the most highly developed, came last. It being taken for granted that Man as one of the animal Creations is the highest, an integral part of the system, not as Vishnu came down & was embodied & incarnate in the form whether of fish or reptile or ——, but as a real inseparable constituent part, intellectually & morally, as well as physically of the organic system of the globe & that this develop.[t] came last & was preceded by nothing comparable in dignity, we have this one sure step [p. 106] made good in a progressive series & may look for, first, the disappearance of the beings most nearly resembling Man, then of the next in dignity & so on till we arrive at the invertebrates & according to some, to plants. However, frequently the evidence gives way & the first appearances of birds, reptiles & fishes are made to recede into the past; still the fact of the crowning apparition of the most perfect family of the mammalia, the most highly advanced class of the vertebrates, being last in the order of time, is sufficient presumption in favour of an antecedent progression.

It may be said, why not, if it were simply like the visit of Vishnu; sh.[d] not some other order of animal have been chosen, some amphibious form [p. 107] but that may remain for the future. If this view be true, the progressive theory, coupled with that of the indefinite variation of species, will lead inevitably to the conclusion that the genus Man will in time give rise to many species, while the tendency to perpetual improvement will evolve something higher in power of mind & reason. The whole creation will thus come to be viewed in a new light & the relation of Man to the wonderful series of living beings & to the countless changes of the past will acquire an interest much higher than it is otherwise invested with.

It cannot lower Man more than the contemplation of the millions of infants cut off prematurely, the hundreds of millions of savage or semi-barbarous races, the long existence of negroes, [p. 108] the millions of idiots & of the insane.

Miller's gradual development of brain higher & higher is a

remarkable proof how far the orthodox are prepared to identify the human with the brute organization in the highest function.[54] In fact instinct & reason are undefinable as to their limits.

While the appearance of Man, last on the stage, is the real foundation on which the progressive theory rests after repeated defeats, so the fact of our witnessing no new creation, while extinction is going on as actively as ever, is the great fact in favour of the variation of species being the Creative Power.

Or rather, species are abstractions, [p. 109] not realities—are like genera. Individuals are the only realities. Nature neither makes nor breaks molds—all is plastic, unfixed, transitional, progressive, or retrograde.

There is only one great resource to fall back upon, a reliance that all is for the best, trust in God, a belief that truth is the highest aim, that if it destroys some idols it is better that they should disappear, that the intelligent ruler of the universe has given us this great volume as a privilege, that its interpretation is elevating, that if in deciphering records relating to many [p. 110] millions, perhaps millions of millions of past ages, we discover much that is irreconcilable with all the popular creeds which exist now, and all that have ever existed, it is no sign of our being false interpreters for it will not shake what has been common to the greater number of faiths in all ages & among all races, a belief in the Unity of the system, the intelligence, order & benevolence of the Deity. It will not alter our hopes of a future state—it cannot lessen our idea of the dignity of our race to gain such victories over Time. It renders the vastness of the system more apparent, it reconciles us to the much that is unintelligible & mysterious.

Rate of Change

[p. 111] If an ordinary earthquake is a sinking, it gives the rate at which a chain of mountains rise, if the deposits of the Ganges or Mississippi give the pace at which old stratif.[d] format.[s] accumulated, why sh.[d] not the rate of variation in each species be a measure of the rapidity of the passage from one geol.[l] assemblage of beings to another. If this be the real mystery of Creation it ought to be as non-apparent, as latent, as unsuspected, as incredible when first announced, as the theory of Gradual Causes or ordinary causation being sufficient for all geological purposes.

Nec Deus intersit[55]

The idea of Nature first trying 'her 'prentice hand' to borrow the playful expression of the Poet Burns[56] [p. 112] at making reptiles & then creating birds & more highly organized beings, may be less worthy than the notion of an original from which all the rest were to follow & to be evolved.

Each type or species, some have imagined, contained all the future embryos encased in it so we might suppose of the first series of types which were created when the planet was peopled the first time.

Creation

Rather than suppose that in the coming of Man creation ceased, the hypothesis of indefinite variability sh.[d] be embraced.

They who believe Man to be part of the system so integrally as not to be separable as rational, they who think [p. 113] his coming a link in the same chain as that of mammalia after reptiles may well grant that Man's advent did not disturb the creation & reproductive force which have subsisted unimpaired for so immense a period.

Least of all should the believers in the progressive theory incline to any momentous change in the order of Nature consequent on Man's appearance.

In a given time has any type varied more in physical character or intellectual power than Man, if the human race has started from one pair. If what we have hitherto called Creation is resolved into the power of organic beings to produce, by successive generations, other beings unlike themselves [p. 114] then the Human Race has already, if it started from a single pair, done as much in this direction, achieved as much as any other in an equal period unless the dog be thought an exception—the bushman, the negro—the Caucasian.

The power of plants & animals to give birth in a single

generation to plants & animals slightly differing in form, instinct & organization from themselves & in thousands of generations to a more marked deviation, the new forms, habits & instincts being hereditary, so that in millions of generations they may give rise to plants & animals exceedingly unlike, as unlike as what [p. 115] are usually termed distinct species by the best living naturalists & in several millions of years to beings generically distinct, in millions of ages, classically distinct & so on.

Besides this, according to some, there is a tendency to higher or more complex organization in the descendants of successive periods as well as deterioration.

Loss of Species

May be like the premature death of individuals, a sudden cessation of variation in a given direction, never to be fully recovered. It may be a setting of limits to certain developments in a given direction for the time being. The conditions may have become so unfavourable [p. 116] that in that line the flexibility or power of accommodation cannot keep pace with the alteration of climate, etc., but it may not prove that species are not indefinitely variable.

Thus a species may be cut off in Iceland & survive in Norway. The Musk ox may live in the Arctic regions & may come to vary slightly.

I have never doubted that whatever theory we may come to, the re-peopling of the globe in past times & the distribution of species *now* has been & is governed by *Laws* in the same sense as the Universe is governed by laws—that if New Species [p. 117] begin *de novo* by some cause quite distinct from the mere 'variety making' power. If the 'species-making' law be some other still it will be a law, a system, not a direct intervention of the First Cause, of a nature unlike that which allows of the extinction of what we call Species.

Progressive Theory

They who embrace this are bound in all other respects to level Man down to the mammalia in general. If the step from the higher mammalia to Man be part of the same chain of successive developments, which is believed to be traced in the [p. 118] gradual introduction of invertebrata, fish, reptiles, birds & mammalia, then we may as well cede all 'muir' questions, such as whether Man be a species like any other mammal, & the races of Man be equivalent to the varieties in the Dog. We have waived all objections when we have made the advance from instinct to reason, from the brute to the human intellect, from the brain of the Tiger or Elephant to that of Man, a step in the same progressive chain.

If the first progressive being indefinitely capable of progress [p. 119] as a species be no innovation, but part of one, & the same continuous chain of events, then may we safely reason as to minor points of affinity from brute to Man.

Creation

Whatever be the power which has for hundreds of times repeopled the Earth with tribes of plants & animals as fast as they became extinct, that power I have always held is still in full & unabated action as is its antagonist or destructive power. It has gone on thro' indefinite time all the causes for it remain the same, there is no reason to [p. 120] suspect diminished energy in this power, it may be occult, it may remain a mystery, but to believe it to be dormant, or in abeyance, or lost, because Man is in the World, or because Science has not yet established its working or the laws by which it is regulated, this w.^{d} be an unphilosophical hypothesis not borne out by anything we know of in the natural world.

Lamarck supposed that creative power was manifold in the capability of members of the animal & vegetable [p. 121] Kingdoms to generate offspring differing from themselves.

According to this view, or that of the transmutationist, no act of Creation in the ordinary sense, or the origination of a new and distinct type of plant or animal has taken place on this planet, in any other form than that of monads. But these Nature is always producing. This theory therefore, is not entirely the same as that which throws back the beginning of things in the organic world to the earliest period, when the earth was peopled by living beings. If any thing *de novo* is formed daily, the [p. 122] idea of an initiating power being always at work, though governed by general laws, is not abandoned. It is merely assumed that the implicit elements, as it were, of organic life are brought into being & out of these higher & higher forms are in the course of ages evolved. Such an hypothesis deviates somewhat less widely than the principle of hereditary variation from the old received ideas, by which each species was created as it is so long as it flourishes on the earth. It agrees however with it, in the idea that species [p. 123] have no reality and are not merely abstractions, but are abstract ideas not founded in truth for they presuppose some fixed model or standard admitting of deviation or being plastic within narrow limits, in some cases more & in others less narrow, but if the conditions are stretched beyond those limits, the die is broken & the vessel which will expand & contract when exposed to heat or cold, especially if reduced slowly, which will crack & fall to pieces if subjected to a sudden increase or diminution of temperature & which will not, however [p. 124] gradual the changes, endure beyond certain limits, but will be wholly annihilated.

Were we to ask the advocates of the indefinite variability of animals & plants when the last act of creation took place on the globe, they could place it beyond the period of the remotest geological formation hitherto discovered, containing any vestige of organic structure, unless they assumed, as is too often done, that the last detected of such fossils indicates the real beginning of life on this planet. By this they mean so vast a period that the array of figures expressing [p. 125] it could be inconceivably vast, if they enter into the views which are shared by a majority of living geologists.

Another hypothesis would make an exception in regard to the human race, admitting a special intervention in this one single instance. Others are content to disregard this problem which however is unphilosophical unless it be carried out consistently. If, for example, a progressive theory of development lean mainly for its support on the acknowledged fact that man came last into the globe or, if not last, long after a series of antecedent [p. 126] assemblages of animals & plants, then they who embrace this doctrine and who believe that the step from the irrational to the rational is an intimate part of the advancement from lower to higher grades in the scale, are bound to adopt for the creation of man and other mammalia some analogous natural plan or system. They may imagine the human spirit, soul, or intellect, to have come down at a certain period to visit the earth and to have become embodied in some vertebrate form enabling it more thoroughly to study and [p. 127] understand the surroundings and contemporary organisms and those of past ages, in their relation to the inorganic elements of the globe, being so assimilated as to cause the least possible disturbance to the uniform system. But if Man is comprehended in the same series of developments which according to many, explain the absence of vertebrata in the oldest fossiliferous rocks and the late appearance of mammalia in the stratified series, i.e. we then identify so completely with the whole vital system of this planet from the remotest past, that we cannot fairly admit creation, in its [p. 128] usual acceptation, of the first human pair and deny it for the first elephant or any other species.

July 16, 1856

The doctrine of progress has derived its main support from the comparatively recent advent of man. This great fact remaining, its supporters are ever ready to repair the fabric as often as one portion of the evidence after another, first relied upon, has given way and if a fossil mammifer were found tomorrow in the earliest rock in which we could reasonably hope to meet with one—the carboniferous or [p. 129] coal formation, as being the oldest in which terrestrial plants and insects are preserved in any number, it would not cause the abandonment of the progressive theory so long as man is avowedly or secretly regarded by geologists as one of the mammalia and considered as at the head of the mammalia.

July 18 [1856]

The infant, the boy of 9—youth of 18—25—50—70—middle-aged & old might be compared, defined & contrasted, & if a being be supposed to have seen no others, he would wonder, could he be told that they implied mere differences of age. All his previously acquired [p. 130] skill in discriminating w.^{d} appear to be at fault, he would distrust himself, & however real his previous knowledge, it w.^{d} no longer estimate as highly as before—every strong line of demarcation w.^{d} have vanished & his landmarks which he felt to be secure guides w.^{d} now seem to be the marks of ignorance or the evidence of lost links in the series, or links overlooked.

July 19 [1856]

With Hooker

See p. 4—N.B. 214 On Creation & progressive development.[57]

Hooker says that Ferns, Lycopodiums Coniferae & Cycads have not been developed or improved or advanced since the Carbonif.^{s} & Oolitic epochs wh. is ag.^{t} develop.^{t} theory or the transmutation line.

[p. 131] So Huxley says that Ganoid fish were as perfect as any other now living.

In regard to such cases as Helix Wollastoni & H. scabriuscula where the difference consists in the integrity of the lip, it becomes a question of the quantity of matter (calcareous) secreted by the full-grown individual, for many species which have, when young, incomplete lips, have entire ones, as H. Wollastoni, in the adult—See e.g. H. polymorpha.

Lysimachia Vulgaris

The common loosestrife lately found in the Australian Alps with a few other species of European plants—

How, if it migrated there, came it not to change its aspect & be a very marked variety?

Animal Origin of Human Race [p. 132]

If every individual is conscious of having in infancy played unconsciously the part of an animal with less intelligence than the Elephant & is aware, that if cut off in that early part of his career by an untimely death, he w.[d] have had, so far as this earthly scene is concerned, no higher a claim to superiority than that which a young mammifer of almost any species may challenge, why may not the whole race have had an early period of brute condition thro' which it passed leaving no record behind.

Berberis Vulgaris [p. 133]

Retains its character from G.[t] Britain to Thibet? where moist atmosphere, (also in New Engl.[d]?) but when it gets to dry climate around the Mediterranean, [it] acquires new characters wh. have been mistaken for specifics. If whatever happens in space may happen in Time a species in 2 successive eras may assume a sufficiently distinct form & habit to rank as a new species in the opinion of most botanists as does Berberis Cretica [?].

November 7, 1856 [pp. 134–135]

[Ms.] Notes on the Sewalik Freshwater Shells
by Edward Forbes

Recent	Fossil
Pulmonifera	
Planorbis 1. unnamed in Mr. L.'s collect.	Planorbis identical with No. 1
" 2. unnamed in Mr. Lyell's coll.[en]	
" 3.	
" 4.	
	(Pupa? not resembling any species I have seen in east)
Pectinibranchiata	Paludina unicolor
Paludina unicolor	" allied to Bengalensis but apparently distinct
" Bengalensis (in Mr. Lyell's col. unn.[d])	

Melania Thiarella	Melania quite distinct from any recent one I have seen
Ampullaria glauca?	Ampullaria glauca identical with specimens from the Jumna, differing somewhat from the recent Sewalik specimens
Lamellibranchiata	
Unio favidus (Mr. Lyell's spec. Sylbet)	Unio favidus—a var.
" marginalis Lam.	" new or Jumna sp.
" (undesc. ?)	" in Mr. Lyell's collect.[n]
" (undesc. ?)	"
	" Apparently undescribed fossil species
	"
	"
	"
	"
	"
Cyrena. small species	

Remarks on the List [p. 136]

The total number of fossil species belonging to freshwater genera is 15. The total number of recent species brought from the same country is 15. Four of the fossils are identical with recent species. The remainder are probably extinct forms.

They consist of:

1.st a Paludina which though extremely near P. Bengalensis must be regarded rather as its representative than its homologue.

2.d Of a *Melania* which is very distinct from any Indian form I can find a record of & which appears to have taken the place of the existing *Melania Chiarella.* (I have seen [p. 137] 10 Indian *Melania.* It is none of them.)

3.d Of a series of *Unios* which are different from any species of which I can find a record. In 1838 Mr. Lea enumerated 13 species as the total number of known Asiatic Uniones.[58] Of the 10 Sewalik fossil species one only is identical with any of those 13 species, so that 9 remain to be accounted for. The common

Unio of India appears to be U. marginalis of Lamarck which is not to be found among these fossils.

Of the species found fossil, which are identical with recent forms, the only Unio appears to be [p. 138] identical with one of the commonest existing species of India. The *Paludina* is the *unicolor* which is found among Paris basin shells, and also in the Mammaliferous Crag of England. At present this species ranges from Egypt to India. The *Ampullaria* is remarkable for bearing a more close resemblance to the individuals of the species from the Jumna than to those existing at present in the Sewalik.

Besides these fossil freshwater shells there is a species of Pupa or Bulimus with which I am [p. 139] unacquainted.

On the whole I am inclined to regard the evidence of the Molluscan fauna as indicative of the older Pliocene period *at latest.*

(Qy. 1846?)

Edward Forbes

Note on the Table by Dr. Falconer[59]—"N.B. the materials for this comparison in recent forms of India very limited & incomplete."

Species in Islands

If it be said that a submerged continent would produce in consequence of the partial distribution of species, a set of islands with peculiar species, it may on the other hand be equally argued that even that continent may have derived a part of its peculiar & [p. 140] partial range of certain species in preexisting islands. The Canaries & the Madeiras if united w.[d] have species which wd. be very long in fusing entirely, so long, that a geol.[l] epoch might expire without completely effacing the original specific centres. The Alps may still have some remaining signs of such an original set of insular specific centres—so if the creation of species goes on, the archipelago may become a continent & the continent a group of islands all within the life time of a majority of the same species continuing on the earth & such must sometimes have been the case & must not be overlooked in whatever manner we explain the origin of species.

March 29, 1857

Hugh Miller, Testimony of the Rocks, etc.[60] [p. 141]

p. 54 We ought not to contrast the existing Fauna & Flora of the globe with a Tertiary or Secondary or still older Fauna & Flora but compare the fossil remains of a peat-moss, a marl loch, the river banks of the Mississippi & its delta, with the fossilif.[s] rocks of antiquity. Even thus we have too great an advantage over antiquity; time would efface many a monument, some must perish daily—the cast, the faint impression are proofs of it. If the Middle Purbeck mammalif.[s] layer (93) is as productive of species as any known layer in a modern delta, it may be taken as a proof of at least [p. 142] as rich a fauna as that now inhabiting the globe.

How difficult is it to detect in peat, which is half turned to lignite, even a few insects—Why then infer that the Carbonif.[s] insects were few because we have only found 12 individ.[s]

April 1, 1857

Cartilaginous Fish

A dog fish, says Mr. Perley,[61] if seen on the coast [of] New Brunswick proves to him that abundance of such fish as he preys on, are there. Osseous fish may be presumed to exist because the species of cartilaginous fish of prey are found, instead of inferring that only cartilag.[s] existed. The one reason infers the non-existence from the very sign which to the other implies the abundance of osseous fishes.

April 6, 1857 [p. 143]

Species

A botanist, says J. Hooker, who knows only 1000 species will have a diff.[t] view as to their distinctness from the same man when he has studied 10,000. He will admit that some wh. were distinct till the intermediate spec.[s] were known are no longer so.

After 50 yrs. we are no farther in regard to the English Brambles whether there are many or only one species.

It is not very different with the roses.

The groups of species of a genus, as of Heath at the Cape, is very readily explained by the theory of their all coming from one typical genus.

Darwin does not develop from a monad, but has generic types created?

[p. 144] Almost all facts may be turned both ways, in favour of the Creationist or of those who develop by variation or selection.

May it not be with the great problem of Creation & of Species as with religion, the effort of the finite mind to comprehend the infinite must give rise to as many theories or diff.[t] opinions as diff.[t] religions. No two men, unless their faculties are equal, can entertain precisely the same estimate of the scheme of organic nature, actual & geological—or no two having equal powers, if they possess different amounts of knowledge of Nature.

The divine ideas in the genera [p. 145] & species of Animals & Plants are too numerous past & present for the human mind to fathom.

INDEX [p. 146]

5. Creation of genera only w.[d] explain why Man has not witnessed act of creation.
6. Absence of intermediate types because Nature's system is not to record.
7. Miracles as numerous as new species.
   Man must be included in progressive theory.
8. To uncoil takes as long as to coil, to fit a tropical for the pole as arctic for the line.
9. New plant immigrating finds fit soil.
10. Why marked insular varieties in an archipelago & on some isles the original type unchang.[d].
12. If individuality persist after death future progress guaranteed by past improvement.
13. 18. 22. } Millions of human intermediates between infant & adult, rational & irrational, sane & insane.
14. Man's evolution out of the Brute, H. Miller.
18. Man subject to same laws, Epidemics, etc.

[p. 147]

20. Orthodox most favour progressive developm.[t].
22. Idiots & exceptional cases as numerous as normal ones in rudest hunter state.
23. Genera of shells, Woodward, Coal period suffice to identify in 2 hemispheres tho' species differ.
24. Development of Genera in time following same law as that of species, Woodward—24.
" Allied species from one stock formed in different isles of an archipelago.
25. H. tiarella & coronula, Madeira & question of species & their constancy for ages.
26. Helix Wollastoni lip differing from nearest ally.
27. Edw.[d] Forbes s.[d] by Austen to have favoured transmut.[n] of Eocene land shells to living (?).
28. Heer in favour of radiation of *Generic* types.
29. Whether man one of the created genera.
30. Heer on sinking of Atlantis in Pleistocene period.

31. Insular varieties caused by diet in land-shells.
32. Species in the numerous British Isles remain the same—in Madeira are different.
33. 63. { But ante- & post-glacial as nothing to Miocene.
34. Endemic species in islands most in favour of special acts of Creation.
" Species of remote geolog.[l] periods can rarely be closely allied, never identical.

[p. 148]

37. In Geology such breaks that we cannot see a consecutive series of changes of species.
38. A true theory of the origin of species must reconcile all known facts of geology & actual geogr.[l] distrib.[n] & races.
39. Origin of Man, whether black or white.
41. Races, if forming as fast as others are extirpated, will keep up the variety of the whole.
42. No race ever suddenly created.
43. If *one* original stock of Man, its origin was very remote.
44. Was the first stock higher or ruder than the living?
44. 45. { If special act of creation be the *dignus vindice nodus* in Man, then other species also real.
46. Whether one or many starting points for man or animals an allied enigma.
47. 9/10.[ths] of living species have survived great geographical changes.
47. 48. { Plan of work on Origin of species & Man.
49. Variation of plants & animals caused by domestication balances Man's extirpating power.
" If cultivated plants in Madeira cover large areas so did gregarious wild plants.
50. Compensation of new varieties for lost species.
51. Dana (like Brown) says creation was never so rich as now.

[p. 149]

52. A tree cut down made up for by growth of 1000.[ds] compared to loss of species counterbalanced by new varieties.

53. Origin of a species by transmutation must be equally invisible as growth of tree.
54. Whether evolved out of a clod or mammal Man's genealogy equally dignified.
55. Man of genius born of commonplace parents.
54. Instinct, memory, industry, jealousy, before Man.
57. Negro & White Man do not converge in 4000 y.[rs].
59. Migration of Crag Mollusca, & their return northwds. unaltered, shows persistency of Species.
60. Whichever of 2 theories we adopt, species must be allied in two successive epochs.
" God who suffers negros to be stationary or other tribes to retrograde may have allowed rude savages.
61. No incarnation of Vishnu into an ape.
62. If we were not corporeally identified with the living type of animate creation, we could not interpret Nature past & present.
63. Abuse of progressionist doctrines must not blind us to their probability.
64. Geographical revolut.[n] in glacial & post-glacial times ag.[t] Darwin's permanence of land.
66. Crag species, why not altered by migrating south?
" Elephant's proboscis might equal human hand.
67. Failure of a species may be like infant cut off.

[p. 150]

69. Aversion to sexual intercourse of distinct races helps permanent varieties.
70. Vanessa Atalanta, is it a species?
" Helix tiarella c.[d] not come from the Canaries, s.[d] Wollaston.
71. Aegilops turned into wheat—ask Henslow.
72. Curculio, distinct vars. of one species in as many of Philippine islands, made species, but shown by Waterhouse to be insular vars.
73. Thibet land & freshw.[r] shells like British & Cyrena fluminalis example.

" Marsupial brain, says Waterhouse, wants the corpus callosum.

74. Volcanic action may not cause upheaval, but only locally retard the sinking of a subsiding, or exaggerate the rise of a rising area.

75. Etna not so high as Castrogiovanni minus the outpoured & outthrown matter.

76. Volcanoes, why often in oceanic islands?

" " & not numerous in Tertiary continents.

77. Volcanoes do not properly belong to areas of elevation; but rather the contrary.

Volcanic island in sinking area due to lava, coral—local upheaval.

78. Take away volcanic matter & volcanic regions w.[d] some be low land, others seas.

81. Deep sea near & around volcanoes.

[p. 151]

82.–90. { Volcanic areas not areas of Elevation.
letter to C. Darwin. 1856 July.
Madeira & Grand Canary }

91. Elaphrus lapponicus unchanged by migration.

92. Roses, if more than one (or 5?) British spec.?

93. Heath different species in one climate at Cape.

94. Negro & White not assimilated in 3 centuries tho' in same climate in Virginia.

95. Ice agency in transporting species.

96. Species only as difficult to define as are formations or chronological eras in Geology.

98. Time has not been lost says Gray & Cumming in making species or noting fine variations in shells.

99. As butterflies in Himalaya are identical some with British so Cyrena.

" Plato's ideas & modern theory of Species.

101. Oscillations of land in post-Miocene period proved by Geology require anomalies in geograph.[l] distribut.[n] of living species.

103. Man evolved out of inferior animals.

104. Number of infants cut off.
105. Man not an ideal of perfection, not amphibious.
106. { " crowning & last act of Creation, support of theory of Development, even with H. Miller.
108. }
111. Species if formed by transmutat.[n] ought to seem immutable like continents.
" Nature's "prentice hand" a playful expression.

[p. 152]

112. Development hypothesis as dignified.
113. Generation produces successive beings unlike the parent stock
114. Races of Man prove as much transmutation as any other thing.
" If in short period permanent vars., so in long, species, & in millions of years, orders—invariability also an assumption.
115. As extinction, so renovation [is] a law still in full activity.
117. The species-making law whether like Man's free will different from ordinary laws.
" If Man to prove progress we must level him down to a link in the chain.
119. The antagonist reproductive power quite as much required now as ever.
120. To suppose it in abeyance, unphilosophical.
121. Monads of Lamarck differing from Darwin's.
122. The progressionists imply that Man came in like other beings as a step in advance, not incarnation in preexisting form.
No finding of a mammal in Coal w.[d] alter their doctrine.
129. Ages of Man as distinct & definable in character as species if no intermediates
130. Hooker ag.[t] developm.[t] as exemplif.[d] by non advance of Carbonif.[s] ferns & Cycads.
131. Lysimachia vulgaris in Australian Alps.
132. As individuals so the human race may have had its irrational infancy.

133. Berberis vulgaris same in Thibet & G.[t] Britain but quite unlike in Mediterranean.
134. Sewalik freshwater shells, E. Forbes listing.
136. } These imply Older Pliocene (or older?)
138. }
139. Dr. Falconer says the materials of comparison incomplete so far as recent shells go.
" Species, centres of, in islands may be due to limited range on a continent now submerged or isolation on a continent to antecedent isles.
141. Hugh Miller sh.[d] have contrasted any old flora or fauna with peat moss fossils, not with living fauna & flora.
142. Carbonif.[s] insects not few because 12 only known.
" One species of a predacious fish, says Perley, implies abundance of their prey.
143. Plants, says Hooker, J., pass more into each other in proport.[n] as more species known.
143. English Brambles & Roses one spec. or many?
" Heath of Cape many derived from one?
144. Organic Nature so vast that no two men with different powers of mind, or no two with equal powers, but unequal knowledge, can think alike, no more than [they can think alike] of God.

Notes

1. The entry in Lyell's Notebook 213, p. 56, reads:

Velletri lacustris taken out of the water adhering to Dytiscus marginalis by Mr. Ch. Prentice of Cheltenham. It lays eggs.

Neritina lays small eggs with hard envelopes.

Gelatinous masses of eggs of Lycreris Narcellae & Neritinae are there in pools of freshwater—on the shore—w.[d] flock.

Woodw.[d] p. 384—land shells orderly distributed.

Nanina Macandrii Gray who supposed it to be from the Pacific, therefore a Nanina.

2. On May 3, 1856, Darwin wrote Lyell a letter (completed May 4) containing the following passage:

> Your cases of possible transportal beat all that I have ever heard of; & if anybody had put such cases hypothetically to me I sh.[d] have laughed at them. I have known Colymbetes fly on board Beagle 45 miles from land, which, by the way surprised Wollaston much. We had much, to me most interesting, conversation when he & the others were here. Wollaston strikes me as quite a first-rate man & very nice & pleasant into the bargain. It is really striking (but almost laughable to me) to notice the change in Hooker's & Huxley's opinions on species during the last few years (Darwin-Lyell mss.).

Darwin is referring to the discussion that took place the week before when Joseph D. Hooker, Thomas H. Huxley, and T. Vernon Wollaston visited Darwin at Down House, Kent.

3. Joseph D. Hooker, *The Botany of the Antarctic Voyage of H.M.S. Discovery Ships Erebus and Terror in the years 1839–1843*, 3 vols., London, Reeve, 1847–60; vol. 2, *Flora Novae Zealandicae*, 1855, passim.

4. This is a reference again to Alfred R. Wallace's article of 1855, "On the Law which has regulated the Introduction of New Species," *Ann. & Mag. Nat. Hist.*, 1855, ser. 2, *16*, 184–96.

5. Sir William Robert Grove (1811–96) was an English lawyer,

judge, and scientist, and was best known for his invention of the gas voltaic battery in 1839.

6. James Dwight Dana (1813–95), American naturalist and mineralogist, was professor of natural history at Yale University.

7. Richard Owen (1804–92), comparative anatomist, was Hunterian professor of anatomy at the Royal College of Surgeons.

8. Hugh Miller (1802–56), Scottish geologist and author.

9. Adam Sedgwick (1785–1873), geologist, was Woodwardian professor of geology at Cambridge University.

10. Louis Agassiz (1807–73), Swiss palaeontologist, had migrated to America where he was professor of zoology at Harvard University.

11. Hugh Miller, in *The Foot Prints of the Creator: or the Asterolepis of Stromness,* with a memoir of the author by Louis Agassiz, 3d ed., Boston, Gould and Lincoln, 1866, 337 pp., p. 308, wrote:

> The brain which bears an average proportion to the spinal cord of not more than two to one, came first,—it is the brain of the fish; that which bears to the spinal cord an average proportion of two and a half to one succeeded it,—it is the brain of the reptile; then came the brain averaging as three to one, it is that of the bird; next in succession came the brain that averages as four to one,—it is that of the mammal; and last of all there appeared a brain that averages as twenty-three to one,—reasoning, calculating man had come upon the scene.

12. Henry Peter Brougham, the first Baron Brougham (1779–1868). We have not been able to determine the source for Brougham's argument of a future archangel.

13. William Rathbone Greg (1809–81), essayist, wrote *The Creed of Christendom; its foundations and superstructure,* London, Chapman, 1851, 307 pp.

14. William Whewell (1794–1866), historian and philosopher of science, was master of Trinity College, Cambridge. He had been a rigorous critic of Lyell's views since his review of the first volume of the *Principles* in 1830 (*Brit. Crit.,* 1831, *9,* 180–206). See also William Whewell, *History of the Inductive Sciences,* 2d ed., 3 vols., London, Parker, 1847, *3,* 662–75.

15. S. P. Woodward, *Manual of the Mollusca; or Rudimentary Treatise of Recent and Fossil Shells,* London, Weale, 1856, vol. 3.

16. Robert Godwin-Austen, Preface to Edward Forbes, *On the Tertiary Fluvio-Marine Formation of the Isle of Wight,* London, Longman, Brown, Green, and Longmans, 1856, 162 pp., pp. xiii–xiv.

17. This is apparently a reference to a table in an 1852 paper of Forbes' in which he correlated various Mediterranean beds in Malta, Corsica, Greece, and southern Spain with those in the Azores. See Edward Forbes, "On the Fluvio-marine Tertiaries of the Isle of Wight," *Quart. J. Geol. Soc.,* 1853, *9,* 259–70, p. 270.

18. The Mollasse is a greenish sandstone, outcropping in the valley of Lake Geneva between the Alps and the Jura, of Miocene age and therefore contemporary with the formation known as the Faluns of the Loire Valley. See Charles Lyell, *A Manual of Elementary Geology,* 5th ed., London, Murray 1855, 655 pp., p. 180.

19. The entries in Lyell's Index Book *1,* 52–54, are notes on T. Vernon Wollaston, *On the Variation of Species, with especial reference to the Insecta; followed by an inquiry into the nature of genera.* (London, Van Voorst, 1856, 206 pp.) On p. 154, Wollaston concludes his discussion of the influence of natural barriers on the distribution of species and mentions that "the perpendicular edges of the ravines [in Madeira], which in many instances rise to an elevation of 2,000 feet, have acted (and ever *will* act) as impassable barriers to vast numbers of the insect tribes."

20. "Nec Deus intersit, nisi dignus vindice nodus, Inciderit; nec quarta loqui personna laboret." Horace, *Ars poetica,* lines 191–92. (Trans: "Nor let a God interpose unless the difficulty is worthy of his mettle, nor a fourth person strive to speak.") Horace is giving advice to dramatic writers not to use persons too exalted for the minor parts assigned to them.

21. Thomas Vernon Wollaston (1822–78). See note 19 above.

22. Joseph D. Hooker, *The Botany of the Antarctic Voyage,* vol. 2, 1856, *Flora of New Zealand.* See note 3 above.

23. Alphonse de Candolle (1806–93), *Géographie Botanique raisonée,* 2 vols. in one, Paris, 1855, 1365 pp.

24. Louis Agassiz (1807–73). See note 10 above.

25. James Dwight Dana (1813–95). See note 6 above.

26. Hugh Miller (1802–56). See note 8 above.

27. Sir William Robert Grove (1811–96). See note 5 above.

28. *Vestiges of the Natural History of Creation* [by Robert Chambers], London, Churchill, 1844, 390 pp.

29. William Whewell (1794–1866). See note 14 above.

30. Edward Forbes (1815–54), naturalist, was palaeontologist on the Geological Survey of Great Britain from 1842 to 1854.

31. [William Whewell] *Of the Plurality of Worlds, an essay,* London, Parker, 1853, 279 pp. (published anonymously). For Whewell, see note 14 above.

32. James D. Dana, *An address before the Alumni of Yale College, at the Commencement Anniversary, August 1856,* New Haven, Babcock, 1856, 28 pp.

33. Edward Young (1681–1765), *The complaint: or, Night Thoughts,* George Gilfillan, ed., Edinburgh, Nichol, 1853, pp. 282–83:

In nature's channel, thus the questions run:
  What am I? and from whence?—I nothing know,
But that I am; and, since I am, conclude
Something eternal: had there e'er been nought,
Nought still had been: eternal there must be.—
But what eternal?—Why not human race?
And Adam's ancestors without an end?—
That's hard to be conceived; since every link
Of that long-chain'd succession is so frail
Can every part depend, and not the whole?
Yet grant it true; new difficulties rise;
I'm still quite out at sea; nor see the shore.
Whence earth, and these bright orbs?—eternal too?
Grant matter was eternal; still these orbs
Would want some other father;—much design
Is seen in all their motions, all their makes;
Design implies intelligence, and art;
That can't be from themselves—or man; that art
Man scarce can comprehend, could man bestow?
And nothing greater yet allow's than man.—
Who, motion, foreign to the smallest grain,
Shot through vast masses of enormous weight?
Who did brute matter's restive lump assume
Such various forms, and gave it wings to fly?
Has matter innate motion? then each atom,
Asserting its indisputable right
To dance, would form an universe of dust:
Has matter none? Then whence these glorious forms
And boundless flights, from shapeless, and reposed?

Has matter more than motion? Has it thought,
Judgment, and genius? Is it deeply learn'd
In mathematics? Has it framed such laws,
Which but to guess, a Newton made immortal?—
If so, how each sage atom laughs at me,
Who think a clod inferior to a man!

34. The German professor to whom Lyell refers was apparently Johannes Müller (1801–58), professor of anatomy and physiology at the University of Berlin. In his discussion of sensory physiology, Müller compared sleepwalking to the instinctive behavior of animals. See Johannes Müller, *The Physiology of the Senses, Voice and Muscular Motion, with the Mental Faculties,* William Baly, trans., London, Taylor, Walton and Moberly, 1848 (pp. 849–1419 from Müller's *Elements of Physiology*) pp. 948, 1353, 1417–18. Cf. Johannes Müller, *Elements of Physiology . . . Translated from the German with notes by William Baly,* 2d. ed., 2 vols. (continuous pagination), London, Taylor and Walton, 1840–43, 1671 pp.; and William Baly and William S. Kirkes, *Recent Advances in the Physiology of Motion, the Senses, Generation and Development . . . Being a Supplement to . . . Müller's "Elements of Physiology,"* London, Taylor and Walton, 1848, 132 pp.

35. Vishnu is the second deity in the Hindu triad of Brahma, Vishnu, and Siva, the three principles of creation, preservation, and destruction. Vishnu has undergone a number of *avataras* or incarnations, the best known being those as Rama and as Krishna.

36. This is an abstract of Lyell's reply to a letter from Darwin written from Down, June 16, 1856. The original of Darwin's letter is among the Darwin-Lyell mss. and was printed, except for part of the postscript, in Charles Darwin, *Life and Letters,* Francis Darwin, ed., 3 vols., London, Murray, 1887, *2,* 72.

37. Robert MacAndrew, "On the geographical distribution of Testaceous Mollusca in the North-east Atlantic and neighbouring seas," *Liverpool Lit. Phil. Soc. Proc.,* 1853–54, *8,* 8–57.

38. T. Vernon Wollaston, *On the Variation of Species,* 1856, pp. 73–74.

39. Sabin Berthelot (1794–1880), French naturalist, published various papers on the natural history of the Canary Islands between 1826 and 1837. See Sabin Berthelot and Barker Webb, "Synopsis Molluscarum terrestrium et fluviatilium quae in itineribus per insulas Canarias observaverunt," *Ann Sci. Nat.,* 1833, *29,* 307–26.

40. Charles J. F. Bunbury (1809–86), botanist and country gentleman, was Lyell's brother-in-law.

41. John Stevens Henslow (1796–1861), rector of Hitcham, Suffolk, had been professor of botany at Cambridge from 1827 until 1839 and had exerted a strong influence on Charles Darwin while the latter was a student at Cambridge from 1828 to 1831. During his rectorship at Hitcham, Henslow remained an active botanist.

42. Hugh Cuming (1791–1865), naturalist, had traveled in South America, among the islands of the South Pacific and in the Philippines. In four years, he collected 130,000 specimens, which were placed in museums.

43. George Robert Waterhouse (1810–88), naturalist, was with Frederick William Hope in 1833 cofounder of the Entomological Society of London. The paper referred to seems to be his "Description of the species of the curculionideous genus Pachyrhynchus, Sch., collected by H. Cuming in the Philippine Islands," *Trans. Entomol. Soc. London,* 1841–43, *3,* 310–27.

44. Dr. Thomas Thomson (1817–78). Thomson was graduated M.D. at Glasgow in 1839 and in 1840 went to India as an assistant surgeon in the service of the East India Company. In 1847 he was appointed one of the commissioners to define the boundary between Kashmir and Tibet and traveled extensively in Kashmir and Tibet. See Thomas Thomson, *Western Himalaya and Tibet; a narrative of a journey through the mountains of Northern India during . . . 1847–8,* London, Reeve, 1852, 501 pp.

45. George R. Waterhouse, *A Natural History of the Mammalia,* 2 vols., London, Baillière, 1846–48.

46. Charles Darwin, *The Structure and Distribution of Coral Reefs,* London, Smith, Elder, 1842, 214 pp., p. 142.

47. Ibid., p. 141.

48. Matthew Fontayne Maury, *Physical Geography of the Sea,* New York, Harper 1855, 274 pp.

49. The original of this letter is in the Darwin-Lyell mss. It was published in part in *More Letters of Charles Darwin,* Francis Darwin, ed. 2 vols., New York, Appleton, 1903, *2,* 135–37.

50. John Lindley (1799–1865), botanist and horticulturist, was professor of botany at University College London.

51. John Edward Gray (1800–75), naturalist and physician, was an original member of the Zoological Society of London and had

been keeper of the zoological department at the British Museum since 1840. The statement here referred to appears to have been a verbal comment made at a meeting of the Zoological Society. Gray published nothing on *Cyrena*.

52. James Francis Stephens (1792–1852), naturalist, had held a position at the British Admiralty in 1818 but because of his skill as a naturalist he was given a leave of absence to assist Mr. William Leach to arrange the insect collections of the British Museum. After his retirement from the admiralty in 1845, Stephens began to catalogue the British Lepidoptera in the British Museum. At his death, he left a catalogue, first published in 1852 and again in a revised edition in 1856. James F. Stephens, *List of the Specimens of British Animals in the Collection of the British Museum, part 5. Lepidoptera,* 2d ed., London, The Trustees of the British Museum, 1856, 224 pp. Because this work contains no remarks on the geographical distribution of British insects, Lyell's statement is probably based on personal communication.

53. T. Vernon Wollaston, *On the Variation of Species,* 1856, pp. 160–64.

54. Hugh Miller, *The Foot-prints of the Creator,* 1866, p. 314:

> Nature in constructing this curious organ [the brain], first lays down a grooved cord, as the carpenter lays down the keel of his vessel; and on this narrow base the perfect brain, as month after month passes by, is gradually built up like the vessel from the keel. First it grows up into a brain closely resembling that of a fish; a few additions more convert it into a brain undistinguishable from that of a reptile; a few additions more impart to it the perfect appearance of the brain of a bird; it then develops into a brain exceedingly like that of a mammiferous quadruped; and, finally, expanding atop and spreading out its deeply corrugated lobes, till they project widely over the base, it assumes its unique character as a human brain.

55. See note 20, above.

56. From the last stanza of Robert Burns's (1759–96) poem, "Green Grow the Rashes," in *The Complete Poetical Works of Robert Burns,* James Currie, ed., New York, Appleton, 1855, p. 373:

> Auld Nature swears, the lovely dears
> Her noblest work she classes, O;
> Her 'prentice han' she tried on man,
> An' then she made the lasses, O.

57. Lyell's entry in Notebook 214, p. 4, is as follows:

No. 46 of Geol. Journ. lent to Hooker

With Hooker July 19, 1856

{Anagallis arvensis
Common Pimpernel

Great difficulty of producing hybridity between the blue & scarlet varieties of the plant whereas no difficulty of making one blue or one scarlet variety breed with another of same colour.

In plants even nearly allied races are unwilling to mix.

A good naturalist alone can feel the fixity of species.

Yet the geological argument of Time is logically powerful on the other side.

The number of species in Madeira is not as great as it ought to be according to the extent of ground in regard to plants.

Berberis communis

Lowe has made a new species of the P.^{o} S.^{o} Berberis saying it resembles the character of the Berberis cretica.

This last, says Hooker, is a mere diverg.^{t} var.^{y} of Berberis vulgaris of G.^{t} B.^{tn} & Europe. This plant when it reaches the Medit.^{n} region becomes dwarf & assumes forms wh. have acquired for it various specific names.

Hooker agrees as to Man as a progressive animal being distinct.

E. Forbes, says H.[ooker], had supposed certain waves of poverty & richness of Creative Power wh. his Polarity doctrine favoured, & tho' no creation has taken place since Man, yet he looked forward to periods when the Creative Power w.^{d} be active again as at certain past times.

Hooker agrees that the poverty of the fauna & flora at certain epochs sh.^{d} not be inferred from negative facts in the present imperfect state of our knowledge.

58. Isaac Lea (1792–1886), Philadelphia malacologist and publisher. Lea published a long series of papers on the genus *Unio.*

59. Hugh Falconer (1808–65), palaeontologist, who collected and studied fossils from the Siválik hills.

60. Hugh Miller, *The Testimony of the Rocks, or Geology in its bearings on the Two Theologies, Natural and Revealed,* Edinburgh, Constable, 1857, 500 pp.

61. Moses Henry Perley (1804–62), pioneer business man and man of science in the colony of New Brunswick. In 1849, he began a

series of reports on the fishes of the waters around New Brunswick and, in 1852, published *Reports on the Sea and River Fisheries of New Brunswick,* 2d ed., Fredericton, New Brunswick, J. Simpson, printer, 1852, 294 pp.

Scientific Journal No. III

July 25, 1856

Reality of Species: Letter to J. Hooker—Hamburg[1]

This kind of work will be very indispensable for some one of authority, seeing where we are drifting to, for whether Darwin persuades you & me to renounce our faith in species (where geological epochs are considered,) or not, I foresee that many will go over to [p. 2] the indefinite modifiability doctrine. If so it will not, or ought not, to make the slightest difference in regard to the rules you lay down, & in doing which you have with prophetic caution, anticipated the possibility of many of your readers embracing the transmutation theory. But the species-multipliers will be delighted with a theory, which sanctions [p. 3] to a great extent the conclusion that the boundaries of species are, in the nature of things, artificial, mere human inventions & therefore gives them a kind of right to affix their own arbitrary bounds. So long as they feared that a species might turn out to be a separate & independent creation, they might feel checked, but once abandon this [p. 4] article of faith & every man becomes his own infallible Pope. In Truth it is quite immaterial to you or me which creed proves true for it is like the astronomical question still controverted whether our Sun & our whole system is on its way towards the constellation Hercules. If so, the place of all the stars, & the form of many a constellation, will millions of [p. 5] ages hence, be altered. But it is certain that we may ignore the movement now & yet astronomy remains still a mathematically exact science for many a thousand years. You must go into the doctrine, laid down at p. xiii,[2] with fuller explanations & it must come to be understood that they who make species must be consistent with themselves & [p. 6] make them all of coordinate value.

If they will not admit your two cedars, p. xvii[3] to be our species they must then deny the authority of Linnaeus & other great naturalists who have made the primrose & cowslip one & the same. I cite this at random but some such inconsistency should be exposed & they should be set down as totally wanting in philosophical power & their synonyms [p. 7] never cited. I could furnish good examples in conchology of such other neglect of relative value in the species invented by the ignorant who at the same time adopt the names of their predecessors which comprehend a much wider range of varieties in one species of the same genus.

The speculation as to the quondam connection of the Antarctic lands now separated [p. 8] by oceans are very interesting. I think Darwin does not enough allow for a suggestion, which you advance, that the land from which species now common to A. & C. migrated may have been in the space B. now occupied by the ocean [see Fig. 7]. I always incline to the idea

Fig. 7

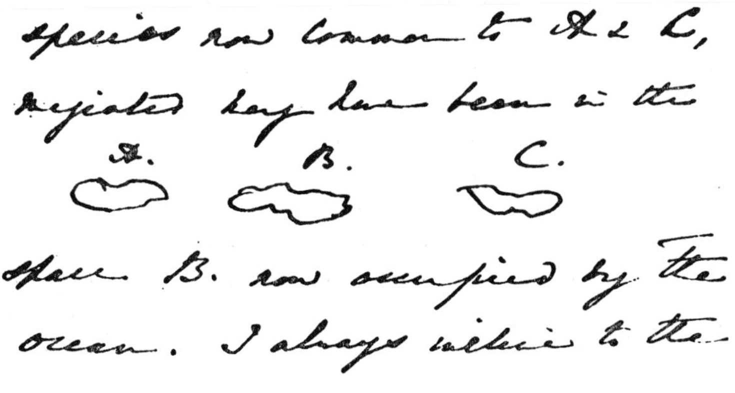

that existing continental areas are in great part of post-eocene date, as Europe, N. of Africa, & a large part of Asia best known, certainly are. For similar reasons spaces like B. may have [p. 9] been the compensating areas of subsidence.

Your idea of the Antarctic species going N. by the Andes,

& of the Panama Andes having once been loftier, is very grand & probable. The marine (living) shells, each side the narrow isthmus of Panama, are very distinct & it is most probable that in the older Pliocene period that the isthmus was higher & broader.

Your occasional criticisms of E. Forbes, to whom you have done ample justice, interested me much & one day I should like to talk over them, as well as a [p. 10] multitude of other topics. I expected you would have used ice-rafts more freely in the Antarctic, coupling them with oscillations of levels in the post-miocene ages, one might get a most complicated succession of states of the same area, geographical & glacial. I fear much that if Darwin argues that species are phantoms, he will also have to admit that single centres of dispersion are phantoms also, & that would deprive me of much of the value which I ascribe to the present provinces of animals & plants as illustrating [p. 11] modern & tertiary changes in physical geography.

I am also wholly at a loss to account for such facts as you told me of the Lysimachia vulgaris in the Australian Alps, if we embrace the indefinite variability speculation, doing away with all creation & substituting a power of individuals to produce offspring unlike themselves or a posterity referable to distinct species. For if such were the influence of external causes, one would think that the Lysimachia or Capsella bursa-pastoris could not retain [p. 12] its character after wandering over the globe & that it would be contrary to the doctrine of chances, that two individuals descending from common parents of very different species, should be found in the Southern & Northern hemispheres with identical characters, each having of necessity got into this new state or permanent variety (or species) by quite a different set of changes both as to climate & coexisting animals & plants.

If such results were possible I should expect the recurrence of the [p. 13] same species in distinct geological periods after they had been extinct or in abeyance for an intermediate period or two. But I must not run on. But please to remind me when we meet to talk with you on the question of the shells of Porto Santo & Madeira proper differing so much more than the plants, I mean the land shells, [p. 14] even after observing a large proportion of the Padre Lowe's species. I have seen the Zamia & Cycads of the Botanic Gardens here [Hamburg], but have not yet seen Lehmann,[4] but Otto, the inspector, showed Victoria Regina in flower & all sorts of Nymphaeas which are killed, root & all, every winter & yet are got up in 7 weeks in great [p. 15] beauty. I am just starting to see Wiebel, the geologist.[5] If ever you wish to write to me you can address me at Harley Street, but you will have more than enough to do before starting & perhaps we may meet in the Alps.[6]

C. L.

July 27 [1856]

Changed Constitution of Species

Hooker mentions in his N. Zeal.[d] Essay that some idivid.[s] obtained from mountain heights even of the same species are *hardier* tho' otherwise unaltered.[7] Now this corresponds to members of a family, being some of [p. 16] them consumptive & others not, the former may live to an ordinary term of life if they resort in time to a southern climate. They may be hardy there & some of the brothers or sisters may be weak & sickly if transferred to the same southern or more equable temperature. The negro in Canada is sickly, but not so in New Orleans, where he escapes the yellow fever & can live in the cotton marshes in the summer, wh. are deadly to the white man. The white man is not hardy in Sierra Leone. But if weak-lunged individuals transmit such physiological peculiarities to their descendants, so may species have offspring capable of surviving changes by gradual accommodation like the hardy mountain varieties of a particular plant. A race not materially different may be [p. 17] perpetuated.

July 30, 1856

Reality of Species

If the transmutation theory becomes more than a conjecture, there has been no creation on this planet for millions of millions of years—not even of Man himself unless the Human race be regarded as an exceptional case. If this latter alternative be embraced we must be carefully consistent & not reason from the superiority of Man, considered as an integral & inseparable part of the Brute or Vertebrate Creation, to a doctrine of progressive development. If the step from rational to irrational be such a saltus, as not to be a link in the same chain of events, then we have no right to presume that because Man succeeded [p. 18] the Chimpanzee or Elephant that therefore these last came later than Birds, & Birds than Reptiles, or the latter than fishes.

If on the other hand we unite the two doctrines & include Man in the same category, we arrive at consequences which throw a new light on his past history. The doctrine that different races of Man are not all derived from one stock, but have had distinct & independent starting points, suggests the question whether the inferior races doomed, as Cuvier said, to inferiority by the form of their skull, preceded the more highly favoured, as the progressive [p. 19] development theory would unavoidably lead us to expect. (See Lawrence, *Hist.y of Man,* for citation of Cuvier "Cette loi cruelle".)[8]

Dresden, August 14, 1856

When there are two plants in the same genus of allied species, the one very poisonous, the other innocuous or perhaps wholesome, & if these distinct attributes may last for 1/2 a million of years, it will not affect the practical necessity of distinguishing them with botanical precision, that a theorist may make out a probable case in favour of the one having descended from the same common stock as the other. The common mushroom, for instance, & its allied poisonous species.

Qy. the [p. 20] nightshade & the potato. It may be that these plants may outlast Man, himself, on this planet, as many species have already continued permanent in their characters throughout periods during which the geography of the globe has changed & still survive with Man a more modern creation.

Excelsior

A man of great genius born of parents of only ordinary, or even perhaps moderate & humble intellectual powers, is an example of an advance in intellectual progress, more vast than we have any examples of in respect to corporeal stature. No dwarf born of full-sized, or even gigantic parents, can compare, [p. 21] nor giant spring from a dwarf, if such a phenomenon ever appeared with the contrast in intellectual capacity sometimes exhibited between parent & child, or brothers & sisters. Let us only suppose several retrograde steps in the like direction & we are carried back to a being of as humble capacity as the lowest Australian savage & this without any miracle. If we could be assured that there were only 200 generations intervening between Shakespeare & two ancestors, or a pair as humble in intellect as the lowest of the human race ever known, being nevertheless of sound mind & deserving to [p. 22] rank as rational & responsible beings, we should not think the fact as so extraordinary as to require as much evidence to substantiate its truth as a new creation in the common sense of the term.

The appearance of two individuals, male & female, not produced in the ordinary way by generation but coming direct from the hand of their Creator, or by some unknown & hitherto undiscovered way ushered into the world, would, if asserted, demand proof of the same overwhelming & extraordinary kind which we have a right to ask when some supernatural event is related to have occurred. But we sh.[d] not insist [p. 23] on the same weight of evidence, not even a hundredth part of it, in favour of the derivation of men like Bacon, Newton or Shakespeare, from parents far inferior to an average Negro, or Fuegian or Australian or Bushman.

The question, therefore, which the opponents of the doctrine of New Creation have a right to put, is this, where will you stop—if the humble pair to which you have gone back in this hypothesis were themselves derived from ancestors as inferior in intelligence to them as they were to some of their posterity, would this be more of a supernatural event.

Creation

[p. 24] A German ethnologist told me at Berlin (1856) that he had satisfied himself that to speculate on a full grown Man having been created as such was contrary to probability. The germ, at the moment of conception, would be the moment at which the human rational, & intellectual spirit, or gheist w.[d] be given by the Deity. If in this way the spirit of Man was imparted to the embryo of a brute at the moment of conception, this w.[d] be a conceivable form in which consistently, with the general laws of the organic world as known to us, creation would take place. Still the progressive system might require that, as there are races of [p. 25] very different degrees of power, the least intellectual have preceded the more advanced.

Vienna, August 28, 1856

Letter to Barrande[9]

There are 65 species in the Colony of which 58 agree with the species in the base of E, while the remaining seven are peculiar, which gives a proportion of more than 10 per cent in the Colony E^1, which do not occur in the base of E. From this we may infer that if a larger number of immigrants, coming from the mother country [p. 26] of E^1, had established themselves in Bohemia, say 300, there would have been 30 species which would have become extinct before the fauna of the base of E came into existence, & an equal number of other species would be substituted for the lost ones in the mother country so that finally, when the fauna of E was at last thoroughly established in Bohemia, it differed from E^1 by one tenth of its species, a very considerable variance inasmuch as the same amount of variation, ten times repeated, represents the whole difference between the bottom of E & the top of H.

[p. 27] Seen from this point of view the colony E^1, when explained in your way by the temporary removal of a barrier soon afterwards renewed, ceases to be anomalous, especially if you find that the fauna of D, if taken from beds immediately above E, differs from the fauna of D near its point of contact with E. Have you data for determining whether there is not some perceptible difference in the fossils of the lowest & uppermost beds more than 3000 feet thick between E^1 & E. Some species of the region of the province D, probably disappeared & others came in during that [p. 28] long epoch.

A very important application, may I think, be made of your grand discovery of the colony E^1 to the question of the indefinite variability of species & their convertibility one into the other, a theory in defence of which some profound thinkers have recently been enlisted, & for the testing of which a Bohemian series of paleozoic fossils will be valuable in more ways than one. They who favour the doctrine of the instability of species admit that mere lapses of time, without concomitant changes of condition in the organic & inorganic [p. 29] worlds, would never give rise to material modifications in the specific character. But in this case we have first the revolution of the earth, which locally exterminated the colony E^1 so that they expelled the Fauna of D & established in its place the species of E on Bohemian ground. Besides other influences climatal, geographical, etc., the disappearance of a tenth of the Fauna of the old mother country & the [p. 30] appearance of 10 per cent of new species, must have placed a great number of individuals of the whole Fauna in new situations & relations. What was the effect of the struggle & with new circumstances? The answer is the total extermination of $1/10^{th}$ of the species between the era of the colony & the final establishment of the colony at the base of E^1. Those species which could not resist the pressures of the new circumstances gave way entirely & were exterminated, the surviving 90 per cent or 58 [p. 31] in 65 was not materially modified in their specific characters in spite of the revolution in the animate & inanimate Creation, of which Bohemia had been the theatre. If the indefinite modifiability were a true doctrine, might we not have reasonably expected that the 58 survivors, which passed from E^1 to E, would have been very marked varieties. Is this the case?

Vienna, August 29, 1856

The dogma of the author of the "Plurality of Worlds"[10] has said that all naturalists are convinced that no creations are now going on. On what ground can this be affirmed & by whom? [p. 32] If it were true that what had been going on for 1000.ds of years had been arrested, because of the appearance of Man upon the earth, his advent w.d indeed have been a revolution in the organic & physical worlds & the more inexplicable as the extinction of species is in full progress & must continue.

For those who desire to establish the creation of species by batches, this dogma is no doubt favourable.

If the changes now going on in the animate creation are the same as have ever gone on, the work of creation, or the coming in of new species, must be going on. Can it be so slow & imperceptible as to escape detection?

Millions die before birth in the womb of every generation of Man, millions soon after. To draw the line & to declare [p. 33] when the mere animal became a responsible & rational being with a human soul has been a question which theologians & metaphysicians have controverted keenly.

The ceremonies of the Roman Catholic church respecting the baptism of unborn infants, & the contempt of many protestant writers of such rites, sufficiently proves that no new enigma w.d arise if we c.d.

November 28 [1856]

Archiac,[11] Change of Species

It is going on now as much as ever tho' 6000 yrs. inappreciable—*Hist. des progrès*—Tome VI., p. 154.[12]

Lartet's Dryopithecus, 1856[13]

We have here a Miocene or falunian anthropomorph, so near the negro that an anatomist acquainted with its skeleton, tho' extinct perhaps for a million years & more, might pass a good osteological examination in the Coll. of Surg.[s]—as for human anatomy, & the physiology of such a being must have approached Man very nearly.

December 10, 1856 [p. 34]

All theories of the Chimpanzee having been develop.[d] into Man fall to the ground when we can prove that the Miocene epoch had an anthropomorphic species much nearer in form (& equal in size nearly) to Man. If Man was develop.[d] out of any Ape it was from the Dryopithecus & a million or more of years have been required for it.

December 15, 1856

Extract of Letter from Sir C. Lyell to Mr. Searles Wood[14]

I was in hopes that at the end of your treatise you would have given us for the whole of the Crag such a summary as you gave for the univalves, vol. [1] 1848, p. 199.[15] I am the more [p. 35] disappointed as I want to cite your latest results, immediately in a postscript to the French translation of my *Manual*[16] just coming out. You would oblige me if you would send me the total numbers of shells which you now make. I am afraid of falling into a mistake. On a hasty estimate I made,

Univalves	190
Bivalves	220
	410

but as I made the number about 425, or 15 more when I tried in my *Manual,* 5th Ed.[n], p. 170, to give your results. I am perhaps underrating the number now.[17]

If I understand your table p. 199,[18] [p. 36] the Univalves are thus:

Corall.[ine] Crag—127	of	which 58	recent
Red Crag 63	"	" 26	"

But is this your final result? It would, if correct, make the more modern or Red Crag diverge more widely (judging by the Univalves) from the existing fauna than the Cora.[lline] Crag.

This same inference you draw, or rather you point out that a similar conclusion would be derived from the application of the percentage system, if the bivalves described in your last Part are taken alone. But what I wish to know is whether you do not find, as I understand you [p. 37] to find in 1843, that on the whole the Coralline Crag contains a smaller percentage of living shells than the Red Crag.

I suppose we may safely conclude, however, from your work that the difference of age between the Red & Cor. Crags was not considerable, judging at least by the test of their relationship to the Fauna of the present seas. Also please give me the numbers of the Cor.[alline] & Red Crag univ. & bivalves & the propor.tn of recent in each, & tell me how many you make common to the Red & Cor. in univ. & also in bivalves.

Also, explain why in Table [p. 38] p. 306 Ap.x [19] the freshw.[ater] bivalves are omitted such as Cyrena, etc. Are there not 11 of these? You say at p. v.[20] univalves that the Grays' [21] Clacton[22] & Stutton beds[23] tho' having nearly all living species, are probably of age of Red Crag. You scarcely expect so many to occur recent, like Limapsis pygmaea, as to make the Red Crag mollusca approach the living Fauna as nearly as does the Grays' list. Do you not make 4 extinct species in Grays', etc.?

Deshayes[24] as to Unio, cf. Grays' distinct from U. pistorum. [p. 39] It must at least be a diff.t race if not species.

If a Helix like H. labyrinthica has survived several great geol.l periods, we have no right to expect it on that account, to have a wide range. See Wood p. [?] bivalves.[25] It may be dying out or past its maximum of extension, but we may expect to find it in one or more places very distant from each other or from its former fossil locality.

E. Forbes thought he had detected Pol. lenta among Falconer's Sewalik shells. If it has lived from Eocene times, it ought to occur in Miocene strata, & might have been transferred from there to existing Indian rivers. [p. 40] See p. 36— If we add to the Red Crag univalves 29 shells common to the Coralline Crag & the recent, we have then 92 Red Crag shells, of which 55 are recent, which gives a greater percentage of recent to the Red.

Cor. Crag	127	recent	58	or 45 per c.[t]
Red "	92	"	55	or

December 16, 1856

C. Lyell to S. Woodward

We cannot assume that all the shells common to the Coralline & recent periods, & missing as yet in the Red Crag Fauna, are to be taken as belonging to that fauna, or so counted as if they were [p. 41] present, when we strike a percentage unless we at the same time make allowance for other missing shells which must have lived on from the Coralline to the Red Crag periods.

It is a question of the relative value of lives & although these Cor. Crag species, which still exist in our seas, are now proved to have had the best lives yet we must admit, according to the doctrine of chances, that a certain number of the extinct species of the Coralline Crag, at present missing in the Red Crag, did nevertheless live on, & such [p. 42] a search as would be necessary to bring to light all the living species above alluded to, would also lead to the discovery in the Red Crag of some of the extinct species of the Coralline.

The fauna of each of the Crag formations is defective & when we allow for undiscovered species we may certainly allow most largely for those recent ones which we know to have been already established in the areas of Norfolk & Suffolk at an earlier period.

December 28 [1856] [p. 43]

With J. Hooker & Dr. Falconer[26]

The ovary in mammalia of the placental, surrounded with several envelopes in the marsupial, the foetus a small part of the time, 1 month in 9?, in the uterus, then extruded & placed in the outer pouch. In birds & reptiles the ovum goes thro' its metamorphoses externally by the heat of the body of parent or the sun's heat. In fishes the ova are impregnated externally by the male (not in sharks but in ordinary fishes).

So in plants, says Falconer, the gymnosperms approach more nearly to the case of fish where the ovary is naked. In the Coniferae the leaves are imperfectly developed, the flowers [p. 44] also. But, says Hooker, in Coniferae the pollen is more highly developed than in the Angiosperms, Monocotyl. or Dicotyled.[s]. It is composed of two sides—the embryo is developed in the seed?

In the Cycads, says Falconer, the leaves are crozier-like[27] but, says Hooker, they are only pinnules, not leaves, the axis is not as in ferns.

The wood in Coniferae is exogenous & has most of the complicat.[ns] of angiosperms.

On the whole, geologically considered, the occurrence of Coniferae in very ancient rocks, like that of Sharks & Ganoids among the vertebrata, or of Cephalopods among the Mollusca, is against degradation in older [p. 45] formations. We cannot expect to hit on the most highly or most simply organized—in any rock of which we know so little as of the Devonian Flora.

Oolitic Fauna

If we have 4 species at Stonesfield of 3 genera, & of Purbeck belonging to 5 other mammiferous genera, in all the first 10 known species belonging to 8 distinct genera, this seems to imply great variety in the Fauna.[28] On the other hand their being all insectivorous, or allied to the Insectivores, may indicate a great develop.[t] of one family. Hooker says that the genera of Plants in a tropical island are numerous relatively to the number of species. But this, he observes, is no argument against the variety of the Flora of the land generally.

Extract from Charles Bunbury's Letter,[29]
November 23, 1856 [p. 46]

. . . I have carefully examined the leaves of about 75 recent species, of which nearly 30 are Oaks. Among these oaks I find 3 quite different & distinct types of venation & one of these is not to be distinguished, as far as I can see, from what I call the Laurel type, which prevails in scores of dicotyledonous families. Again, in the small & very natural genus of the Alders I find 2 very distinct types of venation—I might almost say three; & even in the Chestnuts, the venation of some of the Indian species [p. 47] is materially different from that of the European kind. All this makes me more than ever sceptical as to the referring of fossil leaves to existing genera, when they are not accompanied by any of the parts of fructification.

I cannot make out from Goeppert's work on the tertiary fossil plants of Silesia that *his* mode of proceeding is other than empirical.[30] I mean he does not say—this leaf has such & such distinctive characters—*therefore,* it belongs to such a genus—but he says—this leaf is very like such a species of such a genus, yet not exactly like, therefore it is another species of the same genus. This method does not appear to me quite secure. However, many of Goeppert's [p. 48] determinations, in the work I refer to, are supported by fruits found in company with the leaves.

February 9, 1857

Extract from Letter by Dr. Hooker[31]

. . . The fossil is the most curious thing of the kind I ever saw & far the most suggestive of Phaenogamous origin. I should like to attach a description of it—to a sketch of the real state of our knowledge as regards high & low in plants & plant development [p. 49] in geological time.

With regard to the antholite, it is hard to feel forced to say that it is an angiosperm, when I must add that I know no angiosperm very like it. On the other hand I feel that it is no gymnosperm or cryptogam.

Miss Bronte[32] on Miss Martineau[33] & Mr. Atkinson's Work,[34] *Letters on the Nature & Development of Man,* February 11, 1851

. . . Of the impression this book has made on me, I will not say [p. 50] much. It is the first exposition of avowed atheism & materialism I have ever read; the first unequivocal declaration of disbelief in the existence of a God or a future life I have ever seen. In judging of such exposition & declaration, one would wish entirely to put aside the sort of instinctive horror they awaken, & to consider them in an impartial spirit & collected mood. This I find it difficult to do. The strangest thing is, that we are called upon to rejoice over this hopeless blank—to receive this bitter bereavement as great gain—to welcome this unutterable desolation [p. 51] as a state of pleasant freedom. Who *could* do this, if he would?

Sincerely, for my own part, do I wish to find & know the Truth; but if this be Truth, well may she guard herself with mysteries, & cover herself with a veil. If this be Truth, man or woman who beholds her can but curse the day he or she was born.[35]

"Time's noblest offspring is the last." This line of Bp. Berkeley's[36] expresses the real cause of the belief in progress in the animal creation. So long as man is consid.[d] as the crowning of the brute creation, as in Hugh Miller's work, no discovery of a chimpanzee in the [p. 52] Paleozoic strata would destroy the faith in progress which we find in so many paleontologists.

Dr. Falconer thinks the assumpt.[n] of the first appearance of a highly organized being in later times so startling that it is not far short of the difficulty of the opposite or development hypothesis. It ought, therefore, to be advanced more modestly & not as an idea which can be accepted as other than a miracle, so far as we at present know the course of Nature.

If we deny that forces sh.[d] be introduced in the old world (*vorwelt*) with which we are now unacquainted, so in regard to the animate creation [p. 53] we may indeed reply that in the latter case we have no known agency which will explain the mystery, therefore we are left to conjecture.

The Bernouillis had five generations of mathematicians of eminence in hereditary succession.[37] But talent is not as a rule hereditary, but rather the exception. If genius can spring from mediocrity of parentage, what may not arise in future? Wollaston remarks that in the human race this is a difference in degree rather than of kind. If a chimpanzee became like the lowest race of Man, it w.[d] be a difference in kind.

If the negro started from a distinct stock & preceded Man, why not other & less advanced antecedent races or species?

Orangs [p. 54]

Agassiz in his letter in Gliddon & Pulszky's 4.[to] [38] says the Orang, Chimpanzee & Gorilla differ in the same degree only as certain races of Man. If they are 3 species, then so are the Negro & Mongolian, etc.

Dr. Falconer thinks the Nat. Hist. argument strong in favour of the analogy of human races & the 3 Orangs, but does not admit that the degree of difference in the 3 Orangs is comparable.

Creation

To suppose the origination of new species to have stopped with man w.[d] be the more violent hypoth.[s] if the Negro & White Man constit.[e] species.

If Man started from one stock & the different races came by constant variation, then it follows that all equal amounts of deviation from a parent stock is compatible with [p. 55] the theory of departure from one stock.

Hence we must abandon all species in the animal Kingdom which merely differ as much as the Negro & the White Man. This w.[d] reduce the present number considerably tho' not make the Gorilla & Chimpanzee one. But the Dryopithecus may have been with others a series of intermediate links.

It is not the physical conformation alone, but its union with a less degree of intelligence, which forms the strength of the case on the part of those who imagine a gradation from Man to the Orang.

Whewell denies creation since Man & says all naturalists are agreed on that.[39] Take this with the doctrine of antecedent change for millions of millions of years. What is there in Man & the various races of Man & their infirmities & mode of generation [p. 56] to make us suspect such a miracle as the abrupt stoppage of the old machinery?

Admit Whewell's propos.[n] & a great concession is made to the Progressionists—& to gradual development.

The only way of escape is to make the Negro descend from a distinctly created separate stock & to say that such acts of creation went on to the human era & then stopped.

Orangs

The absence of the intermediate gradations between the three living species make the difference, says Darwin, between this case & that of the several races of Man. Had there been no links, the Negro & White Man w.[d] have been made species. In all cases Naturalists are influenced in the [p. 57] making of species by the absence or presence of intermediate varieties or races.

Species

The ordinary naturalist is not sufficiently aware that when dogmatizing on what species are, he is grappling with the whole question of the organic world & its connection with the time past & with Man; that it involves the question of Man & his relation to the brutes, of instinct, intelligence & reason, of Creation, transmutation & progressive improvement or development. Each set of geological questions & of ethnological & zool.[l] & botan.[l] are parts of the great problem which is always assuming a new aspect.

Madeira a Great Centre of Creation

It must be supposed that (whatever hypothesis we adopt of the two, which alone I consider as feasible) there must be some proport.^{n} between the areas & the species-making power. On an island in an equal space of time there will not be as many new species introduced as on a continent. If Africa [p. 58] has lent many species to Madeira, & the climate of Africa, after being cooler has grown hot, then Madeira will appear to contain more than its share of endemic shells & insects. It will not seem to have borrowed so much as it really has. It will be a spot where many species have survived the general destruction. An island of high antiquity, which has varied in height & dimensions or has been near other islands, some of which in the same group have, like the Desertas, been half annihilated, will also appear to have more than its due proport.^{n} of species.

If we assume multiple creation & the simult.^{s} creat.^{n} of various races of the same species, all our reasoning appears to me of no avail. The proximity of species created at the same time is consistent with the doctrine of each species being necessarily fitted for all the circums.^{s} which prevail, have prevailed & are to prevail. The law which makes [p. 59] species, or governs their format.^{n}, if not prospective, w.^{d} give birth to species which w.^{d} instantly perish. On the other hand the accomm.^{n} of spec.^{s} to new circ.^{s} might perhaps be made out in like manner to require time & opport.^{s} & therefore as changes take place slowly they w.^{d} endure a long time. If the lakes of N. America & the Hot Springs of some volc.^{c} countries take long to form, so they will endure long. Time which will transmute species & fit them for freshw.^{r} under various circ.^{s} will also ensure their long duration.

"Simia quam similis turpissima bestia nobis!" Ennius? [40]

If the recognition of the near analogy of the human & the Orang type gives us pain, if the admission of it is evaded it must be owing to some defect in our education, or some overestimate or false theory of our relative position in the organic world. See Owen on [p. 60] the all-pervading similitude of structure between Man & the Pithecus "every tooth & bone homologous."—Opinion of Linnaeus & Cuvier versus Records of Creation. Linn. Proceedings. 1857, p. 20, note.[41]

Intelligence of Elephants

Dr. Falconer observes that this animal has what none of the Ape tribe or Orangs have in the same degree, the human-like faculty of fixing the attention on a given task or subject. A pitfall turfed over (near Saherampore?)[42] into which, as a trap, a female Elephant had fallen, was explored by the male Eleph.[t] who paced around it, first making a wide circuit. When he had ascertained its extent, he used his tusks ag.[t] the timbers or logs which supported the turf & worked at one end [p. 61] till he made an inclined plane up which the female ascended.

Species

Dr. Falconer remarks that no one mind can grasp the whole subject of species in the animal & vegetable world—the coexistence of Mastodon & E. Primigenius in Big Bone Lick proves the simultan.[s] existence of two extreme forms in the post-plio period.

The Mastodon & Elephants of the Sewalik prove the same for the Miocene era. These facts are against a gradation.

Creation

Falconer says that the opponents of the transmutation theory have never been able to propose another, so that altho' he himself believes in the reality of species [p. 62] he feels that there is a want of a counter hypothesis.

Uniformity in Creation

The doctrine which I laid down in the *Principles* is that the organic world is going on now as formerly—Species dying out & others coming in. If any one wishes to witness the birth & destruction of mountain chains, islands, continents, lakes & seas, he has only to look at Nature & he sees as much as he would ever have seen going on in equal or average space of Time. If he says he sees nothing, it is because a life-time of Man is an unappreciably short portion of the whole.

So it may be in the animate World, whether the Creation or the Transmutation theory be true, or even the gradual development hypothesis. [p. 63] A Naturalist, if he could obtain a sight of what took place when any species or numbers of species were in the act of being created, might accord.[g] to the transmutationist gain nothing more than he enjoys in the daily & hourly study of Nature. It is only after very long periods are compared (or their results) that knowledge can be obtained which we seek. If Man has been the result of transmutation we know that 3000 or 4000 years has been as nothing in the process by which he has been evolved out of something else. If new germs come in from time to time of new genera or subclasses, they may steal in once in 100,000 years. All depends on the rate of new creations & the rate of development. But as we see powerful intellects born of parents of humble capacities so there may have been [p. 64] great inequalities in the rate of progress.

Multiple Creations

If we believe in single stocks, then a species may mean all individuals which have a blood relationship one to another. A prolific hybrid becomes then an anomaly or exceptional [case] but if it can be perpetuated even by repeated union with the perfect species on one side or the other, it destroys the definition.

If the Negro & the White Man started from independent stocks & be of one species, then half the individ.[s] or more of a species may have no blood relationship with the rest of the species, but merely a power of blending.

A great step is made when the difficulties involved in the definition & [p. 65] comprehension of a Species are admitted & fully or largely felt.

If they who dogmatise on the theological questions wish to learn humility, if presumption & pride be the besetting sin of the human Mind, the study of Natural History, Paleontology & Geology, may be recommended as the best to convince them that, while they may make daily progress, they may satisfy themselves that the day is more & more distant, or will appear so, on which any approach to a true theory can be hoped for.

If we cannot decide how many British Roses there are, or whether there are more than one, how can we hope to decide whether fossil species are in many cases one or many. If we know not the geographical limits [p. 66] of variation, how can we define the geological, for the changes in space are more cognizable than those of Time.

Species—American Uniones

Dr. Gould[43] reckons some 250 species & says that now that they are known from the young to the adult state, the reality of the distinctions are confirmed & a few only have to be suppressed. Prof. Phillips[44] had imagin.^{d} that the series would end in showing almost all to be the same, but Gould says not so.

Unity of Plan of Creation

Abortive wings, tails & other representative & useless parts [are] explicable as transitional forms, hence coccygian vertebrae in men, are more consistent with harmony of an all-perfect scheme than of a systematic classification; they are a necessary part of the variety making degradations by disuse. We ought to consider development as a mode of explaining creation, not of getting [p. 67] rid of it. *"Natura non facit saltus."*[45] All the species, like the forms of transformation of an insect, [are] foreseen, invented, planned, not evolved by a self-acting brute machinery, for Nature is God. This may be Pantheistic, but not in the sense of denying God. There would be nothing lowering in imagining the invention of new insects for the sustaining power ever present carries out the law or design—it is the same to one to whom Time has no existence or meaning, it was not in the past or present to an eternal being. It is no more trouble —a most inadequate conception of the Deity in the *Vestiges*— supposing that the instit.^{n} of General Laws is to save trouble & intervention. Let any one be called upon to make a species to last ½ a million of yrs. or a geological period & conceive the problem.

Development [p. 68]

If an unborn infant could in 30 or 40 years become author of Shakespeare's plays—the passage from the animal to the highest intellectual is rapid & from inferior to higher races of Man may be so.

There is no more difficulty in drawing the line, if the genealogy of Man is found to be traceable to a race of mere animals with brute instincts, than there is now in saying at what period the embryonic or foetal human being became a rational & responsible creature. Millions of such undeveloped new children or infants die annually—they are as the animals or extinct inferior animals. It is only adding to the mystery to discover that before us there were transitional races if such there were, not a new enigma but the old one in a new form.

[p. 69] So there is every stage between imbecility & reason, it may be impossible to draw the line. The repugnance to admit the unity of Man with the lower animals may be the effect of educat.[n], of high aspiration not sufficiently appreciating the animal side of humanity.

If a geologist should wish to see a period when the most active mountain-making process was going on, & the most rapid creation of animals & plants, & was permitted to live 1000 yrs., when the Alps were forming & tertiary animals coming into being, he might rise from the sight & declare that he had seen everything going on as now—an earthquake perhaps had occurred & some new variety of plant or animal.

Truth to Be Welcome [p. 70]

If we once question whether it were not better to remain in ignorance, we abandon what sh.[d] be the most dear to the enquirer into Nature. If we find we are arriving at results which destroy our preconceived notions, it is because we have been taught what was unsound. The only compensation for the increased identification with the mere animal world, which we can find, is in the elevation of the human intellect implied in the discovery itself, & the desire to dive into it.

If it be intended that Man sh.[d] comprehend the terrestrial system, this can only be if the course of Nature be uniform, it may be so whether progressive or not—uniformly progressive perhaps—this no doubt leaves us little or no power to speculate on the future, but still, so far as the succession of species is concerned, if Man be made one of the mammalia & begin puling in his mother's arms, a mere animal in the embryonic & even in the [p. 71] infant state, why not equally so in the ramification of the species into races—& if these races indicate a great deviation from the common type, whatever that be, or the first race as in other animals or in plants, why not consider the variety-making power in the human as in the canine or any other genus as equivalent to the species-making power? The meaning of this, if we once aspire to regard Man as spiritually & morally belonging to a distinct class of being, is this, only by being thus identified in every thing with the rest of the animal world can he understand that world.

In the Miocene time a good human anatomist might have been bred by the dissection of the Dryopithecus—the discovery of Harvey of the circulation of the blood might then have been made.[46]

There must be some reason why Man has the power of understanding the organic world.

In all questions such as the existence of Evil, we find we cannot grapple with [p. 72] many mysteries. But Time will clear up geological mysteries—it will show whether the variety-making power keeps the animal & vegetable world as full of species, notwithstanding the dying out of old types. As in a forest, if annually 10 trees die, the growth of some from the acorn & of 1000 others of all ages to be each one year's growth older, may just compensate. But if all the oaks fail does some other tree creep in.

Instead of the Creation of a new species to balance the extinction of an old one, we have many thousand species in each century varying slightly & making steps towards becoming something different & in some instances diverging from one type in more than one direction so as to produce more than one permanent [p. 73] variety in place of a single parent stock.

Time

If the time be so great to allow of a distinct species coming from the original stock, would it keep up with the rate of dying out which is so rapid? To return to the forest it is not by sowing new acorns, but by the budding or gemmation of individ.[s] that the new must arise. We see old trees fall & the young which spring from the seed are not like them & it is at first hard to believe that the axe-man may cut, & cut down, & find just as many old ones as when he began years before—tho' never any big one starts full grown into being. They steal on & by dint of numbers keep their ground. A million may be growing & he can only fell tens or hundreds in a year. [p. 74] He may at first determine that at least he will alter the aspect of the wood, but he cannot.

With C. Bunbury: Perfection in Plants

Certain organs being specialized is by some s.[d] to be the proof of advanced develop.[t], i.e. the organs being farther removed from the state of leaves which they are all modifications of.

Goethe may have invented this theory, but Linnaeus sketched it out, but was not able in the then imperfect state of Structural Botany fully to explain it & his followers did not take it up.

In a Campanula & Lobelia the petals being all in one are more unlike leaves than if they were divided, therefore it is s.[d] that the Campanulaceae are higher in grade but this is by some thought more ingenious than sound.

July 11, 1858

Inferior ovaries [are] higher [in development] than superior [p. 75] Iris & Agave (inferior) of higher rank than Lily & Aloe —Oenothera, inf.[r] ovary—Ranunculaceae consid.[d] by De Candolle[47] as most highly develop.[d] of plants. Hooker differs from this in a paper in Linn. Soc. Trans. on Balanophorae.[48]

Coniferae, Hooker (Joseph), thinks highly organized; Lindley[49] ½ way—towards cryptogamiae.

Stamens, ovaries, petals, all altered leaves—Bracts, altered stalk.

Oolitic Flora

De Zigno[50] tells Bunbury he has a fossil Yucca in Oolite. A monocotyledon—Qy? whether inf[erio].[r] to angiosperm. A general inclination to think them so among Botanists—not much doubt whether Monocotyled.[s] may not be as C. Bunbury imagines of Coordinate Rank with angiosperms.

July 11, 1858 [p. 76]

Leguminosae

Very predominant in Australia & almost suppressed in N. Zealand. Yet in same hemisphere—abound in Siberia & Thibet, but distinct genera from the Australian.

The leguminosae unfitted for conditions which are favourable to ferns.

The question therefore may be raised whether the absence of great families or orders may not depend on states of the Earth, not the age of the planet.

Unity of Creation

The assumption that Man is the crowning link in the chain of progressive development is received with more favour because as Hugh Miller contends it is believed to be in harmony with a scheme of 6 day [p. 77] creation terminating with the advent of Man.[51] Others are satisfied that a Man had no representation in Pliocene time so we have one undisputed step in deterioration & may expect others—as that the Elephants sh.[d] give place to rodentia, Gyrencephalous to Lyssencephalous mammalia & so to Monotremata—& in like manner, Mammalia to Aves, then to Reptiles, then to fish, & fish to invertebrata. The reception of Man, as an integral part of the same progressive scheme, is thus perfect.

If we then admit the indefinite variation of creatures descended from the same stock we have 1.st the doctrine that all the existing species are races of the Pliocene, then of the Miocene, & then of the Eocene; 2.[dly], that Man in like [p. 78] manner is an improved mammifer.

But if we suppose Man to be created separately, we naturally infer that all other species were so, & that as Man has races like the Negro & White European so other mammalia may have the same & yet be all one species. Too many species may have been made but the reality of species is a doctrine not thereby impaired.

Yet Agassiz thinks that the Negro & European had separate stocks & all the obscurity, which we find in other species, meeting us in Man, the same tendency to vary, to transmit acquired peculiarities & not knowing where it will end.

Creation [p. 79]

If animal species & species & genera of plants die out fast & after many thousand years the diversity of plants & animals was found to be as great as ever & yet no one was ever in at the birth, only in at the deaths of species—then Man w.[d] be brought to the conclusion that the variety-making power was the real creative power. The develop.[t] theory, however, again requires that there sh.[d] be some cause which reproduces simple organisms. Lamarck's system was perfect, monads always appearing & some beings always attaining higher perfection, so the simplest forms never disappear but Time causes more & more complicated ones. A theory of simultaneous degradation & progressive perfection seems requisite if the Darwinian theory is adopted.

Observation [p. 80]

[Observation] will throw much light—if after 10,000 years many species extirpated & yet variety as great.

Cowslip

Oxslip & primrose same, s.[d] Linnaeus. Henslow[52] told C. Bunbury[53] that a new observat.[n] has been made of the fact, long ago recorded by him, of a cowslip & a primrose being found growing from the same root—yet the cowslip is a variety which flowers later in the season, has a different scent, colour, form, station.

Vanessa atalanta—

Single Center

This theory which explains so many facts is quite impaired by the indefinite variability system for take the human race, two pairs of Europeans—one going to Asia & one to America—may produce two races [p. 81] of same species & if 4 pair go to America there may be many common starting points, which w.[d] be equivalent to the same species originating in 4 or 5 places. They may in time, from similarity of conditions produce closely allied var.[s] wh. w.[d] be called the same species.

The occurrence of many heaths in C. of G. Hope might of course be accounted for by variat.[s] from one in course of time [and] an archipelago united.

Species Outliving Others

If we find 10 species extinct & 90 living this implies that while 90 continue unaltered 1/10th of the whole fauna is changed. Why are not the other 90 modified? Alcide D'Orbigny w.[d] maintain that they are so.[54]

July 2, 1858 [p. 82]

Distinguishing Species

Prof. Van Beneden[55] thinks that naturalists will, & perhaps must, begin with a severe analysis of species-making—first too many, then a reaction & rather too few & then will come the time to decide whether there are not such things as species & what was the law of the appearance of fossils on the earth.

Heterocerq Tails

[Heterocerq tails] appear to V. Beneden a higher develop.[t] than homocerq.

Negro Race

Rather a degraded Caucasian than the latter an improved negro for, says V. Beneden, the young negro's skull & jaws more like Caucasian race than the adult. The individual retrogrades as he grows.

Progressive Development

If the step back for Man to the Orang be a step of quite a different value from that [p. 83] of the orang to a reptile & fish it may be that all the reasoning may have a wrong bias & we may start with a deceptive prepossession. No doubt the successive repair of the theory as often as the old landmarks are carried away implies that there is some constant fact the force of which is not altered, but if that fact consists in Man having appeared last & if this is not in natural history a fact of same value as the assumed preceding ones, we may be misled. The absence of Cetacea before the Eocene, of Mammalia before the Trias, of fish before the Upper Silurian, may be connected with a gradual or possibly an irregular intermittent exaltation of types—the law is evidently as yet unknown.

It seems that the Progressionists are usually violent against transmutation. Having identified Man with the same continuous series of progressive developments in Time, they shrink from admitting the steps of that develop.[t] to be effected by lineal transmission.

Popularity of Progressive Development [p. 84]

It is chiefly owing to two [three] reasons:

1.st It admits of the widest generalization from the earliest appearance of animals to Man & coupled with transmut.^{n} embraces the whole Lamarckian scheme.
2.dly It agrees with the small knowledge we possess of land animals in earlier format.^{s} & absence of Cetacea.
3.dly Accord.^{g} to Hugh Miller & others it accords with the Mosaic record.

Transmutation, Why Unpopular

1.st Against the Mosaic creation of species in modern times.
2. Supposed to be contradicted by absence of passages in geol.[1] species, but this very questionable when the fragmentary state of our knowledge is consid.^{d}.

[Transmutation] Why Popular with Some

1. Because it evades act of creation or intervention of Deity & the format.^{n} of [p. 85] a permanent variety is far nearer to the origin of a species than any other phenomenon we can witness. Therefore, if we were led to conjecture, it w.^{d} be more natural to suppose an ass to give rise to a striped offspring with the other characters of a zebra than that a Zebra sh.^{d} come into being out of nothing. For the same reason if there be an animal having an anatomical structure like that of Man it w.^{d} be more probable that it sh.^{d} produce the first Man than that the original of the human race sh.^{d} proceed from nothing.

Agassiz v. Lamarckian Hypothesis

Agassiz argues ag.^{t} transmutation as if all adaptation to new physical conditions in the external world implied the substitution of inorganic & material agency for Mind, or a Creative Intelligence.

But when it is assumed that an original pair may have given origin to all the diversities, which we find [p. 86] in the various races of Mankind, it is not meant that such variations from a fixed standard are unconnected with the prescience & design of the first Cause.

It cannot be denied that the distinct colour & other peculiarities wh. Agassiz finds in the Negro & the different amount of intelligence is such that it constitutes what some w.[d] call a specific difference. A great many species of animals & plants & scarcely more widely apart.

A negro cannot flourish in polar [regions], nor an Esquimaux in tropical Africa. It must be the work of time to enable any one race to give rise to others by slow & gradual adaptation. If it be certain that the Creative Power has so formed species, that they can accommodate themselves, [p. 87] as must be admitted, to new circumstances & transmit acquired peculiarities to descendants, is this fact any proof that Matter has usurped the place of Mind.

Agassiz, Lecture 14, Ch. 1, 50 & 63 [56]

The independence of organic types of the medium in which they live, p. 50, is well put tho' they were made to suit the element in which they live. This might be urged ag.[t] Whewell's argument in regard to other planets & their inhabitants.

Agassiz, Physical Causes, p. 55 [57]

He contrasts the Intellectual Power of Man aiding in producing Breeds of domestic animals with the mere result of the immediate action of physical causes. Why not allow changes in Time to be also controlled by a higher intellectual power than Man? [p. 88] Permanent Varieties & even specific differences may be "ordained by the action of the Supreme Intellect & not determined by Physical Causes." p. 55.

Permanence of Acquired Varieties

Agassiz *ib*. p. 56 [58]

Are not the changes of the Americas in one generat.[n] a proof ag.[t] this operation?

No Parallax

This assertion well controverted, Agassiz p. 56.[59]

Progressive Development

When it is said that Man is the crowning work of the same series of changes with that which can be traced from the fish to the reptile, & from that to the bird & mammifer, it seems to be forgotten how great an approach they make to what is least welcome in regard to [p. 89] transmutation. First it assumes a law by which it has pleased the Creator to govern the mode of his intervention in reference to the successive appearance of animals—& that the instinct of the Gyrencephala is to that of the Lyssencephala what the Archencephala is to the Gyrencephala. It is a rule analogous at least to embryonic developm.ᵗ in the individual—& in like manner connecting the foetal & irrational with the adult & rational.

Similarity of Habits in Species of the Same Family

Agassiz P.ᵗ 1, p. 59 [60]

This may be used by C. Darwin ag.ᵗ Agassiz as showing gradual passage of species in habits as well as organization—similarity of voice, etc.

Community of Nature of man & inferior animals, Agassiz *ib.* p. 60, argued from love of parents for young.[61]

[p. 90] Agassiz cannot define the difference between the mental faculties of a child & a young Chimpanzee—*ib.*[62]

Development & Progress

Agassiz, P.ᵗ 1, Ch. 1, p. 64, asks how can mere physical agents produce animals capable by generation of producing other animals more perfect, p. 63–64.[63] The birth of a man of great genius from parents of moderate capacity & who may produce other children of faculties below par is a phenomenon so astonishing that we may be prepared for any discovery, so far as mere difficulty of our conceiving it is concerned. The man of genius may marry a woman of superior faculties & their offspring may be inferior. This would lead one to expect rather that the moment of conception was the true Creation, when a greater or less power was determined for [p. 91] the future creature.

Relation of Man & Animals

No difference in kind as to passion & moral attributes—Agassiz, *ib.* 64;[64] at p. 65,[65] the psychological resemblance of Man to the Chimpanzee & anthropomorphous creatures is urged, but why not admit the Elephant to be equally near?

The connecting link between Man & the animals [has been] made too much to depend on structural similarity, *ib.*[66]

Immortality

If true of Man, equally so of other living beings—*ib.,* p. 66.

Species

To come to any definite opinion on their reality until our geological evidence is more complete is scarcely possible. The tertiary strata & the post-pliocene will best aid us as we shall have a series of land & freshwater mammalia, plants & shells, & altho' the documents will always be fragmentary, they may become sufficient to admit [p. 92] of our forming a reasonable opinion on the probabilities. If all the gradations from the highest Gyrencephala to the lowest race of Man be set before us & it be said that the supreme intelligence following a certain plan called them in succession into being, how far will this scheme differ from transmutation as affecting the dignity of human Nature, or the drawing a line of demarcation between it & the lower animals?

The distinctness of the proposition, so far as Natural History is concerned, may be very great. The non-reality of species impairs the naturalist's chance of arriving at positive knowledge, renders all his definitions inexact & the more so the farther we proceed, his prospects are hopeless—but so far as relates to the psychological questions & the theological, it matters not whether the series of the gradually ascending scale be made up by some kind of machinery of which we as yet know nothing, or by the variety-making [p. 93] power, or the indefinite deviation from parent stocks & the transmission of acquired habits, mental or corporeal, to descendants. In both cases there must be a law assumed, a plan is obvious.

In spite of the repeated modifications rendered necessary by new discoveries, we find the progressive theory as popular as ever. It cannot be said to have been opposed by the usual bigotry, which resists new ideas subversive of old notions, calculated to remove the imagined barriers between the so called irrational animals & man & to embrace him in the same gradually ascending scale. The expansion of our ideas of time, & even the fact of extinct races of animals & plants (for races have at least been extinct tho' species may not be realities), have been opposed [p. 94] by popular prejudices & against the doctrine of transmutation, something more than mere philosophical doubt has been evinced. It has been treated almost as atheistical by the critics of the *Vestiges*. Yet no doubt the popularity of that book, the *Vestiges,* may be cited to show that the whole Lamarckian hypothesis was welcomed by a large reading public. But the progress both in animals & plants from simple to complex—first from plants to animals & so onwards to Man, then of the classes in each great branch of the Animal Kingdom, then of the orders in each class according to their development, has been hailed with general satisfaction.

This is somewhat singular in the [p. 95] history of philosophy. No doubt there are countervailing inducements.

1.st The six days & Man coming last or the Hebrew notion, but this has no more weight with geologists in general than the astronomical or other doctrines of Ch. 1, Genesis, tho' Hugh Miller favoured it on that ground.

2.dly The desire to see the beginning & end, to assume that we comprehend & embrace the whole scheme, that we can say we have discovered chaos & how order rose out of it, & that the noblest work of Nature is her last. Even the uniformitarians have thus a uniformity of progress & may embrace Man in one & the same scheme. It is satisfactory to seem to know the great outline of the scheme of terrestrial creation.

Immortality, See Ante, p. 91 [67]

As Agassiz contends for a certain immaterial & independent existence & principle in every living being, the extinct individuals [p. 96] of which we have (fossil) memorials, may at least according to such a theory be as much entitled to a future state as the newly born babe that dies or the unborn which never lives save in the womb.

To regard the whole as the beginning & the end
The womb of Nature & perhaps her grave[68]

is what none would desire to do in regard to Man & Agassiz & most naturalists acquire a more kindly feeling in regard to the lower animals, such as a philanthropist cherishes towards all fellow human creatures however humble their capacities, elevating all as his equals & as heirs to immortality & believing, as the philosopher must also believe, that [p. 97] to doubt the future state in reference to any one would impair our right to hope for our own future existence.

The baptism of the unborn, for those who regard that ceremony as indispensable to the salvation of a Christian, may be charitably viewed as a determination to set no limits to the universality of the claims of every individual to an immaterial existence from the time of conception in the womb—no matter whether it ever saw the light, or thought & acted independently, no matter tho' it may be to an Elephant, dog or chimpanzee, what a polyp may be in comparison to those animals.

If the Catholic Church be the representative of 9/10ths of the Christian world & χtianity the religion of the most advanced nations in civilization, then does this ceremony prove that theologists, when [p. 98] they endeavour to settle a question of this depth, come to conclusions as remarkable as the philosopher who speculates as Agassiz has done or as Lamarck. There is nothing new in finding that no line can be found between animal (or brute) & human nature. Orthodox opinion in the church may be as visionary or the creed of different churches may be as much at variance as that of the philosopher.

After birth when does the responsible being commence, in a minute, hour, day, week, month or year, after seeing the light? The case of idiots, imbeciles & others offer the same perplexing problem, but they leave mankind in the same state of aspiration & hope, of trust in God, of yearning after something higher yet to come, of a feeling of individuality, a belief that the discarding of [p. 99] this would not only lower the hopes but deteriorate the moral standard—that a belief in immortality betters & renders happier & is therefore more probably true than a philosophy which teaches that we are bubbles reflecting for a moment the wonders of the universe & then bursting & returning to annihilation.

Whatever opinion we may derive from Zoology & Geology if we have to embrace a gradually progressive theory & unite that to transmutation of species & find that we neither know what species mean, nor can discriminate between rational & irrational, nor tell when & how species were created (if they ever were created) save some types in the beginning, yet we shall continue as much as ever to believe that we are not ephemeral creatures if we can think so now, in spite of all the difficulties we encounter, so shall we think when an extended sphere of knowledge [p. 100] has simply multiplied the number of our difficulties.

We have never made a step yet without having to part with some cherished time-honoured dogmas, without many being unsettled & made unhappy, but there is at least as much belief in a superintending omnipotence & hope in a future state & as much morality as before these new views. When there is the grossest superstition & ignorance, there is less morality.

Progressive System

The prevalence of lower types is substituted often for the first appearance. Reptiles predominated in the time of the Stonesfield & Purbeck beds, fish in the Devonian, Gymnosperms in the Carbonf.[s]. Possibly they may imply some law distinct from successive development.

Proboscidians

Mastodons, Agassiz seems to say, preceded Elephants but Falconer finds the Elephantine [p. 101] type as early as that of the Mastodonic. It was dangerous therefore to reason from America & Europe until India was investigated.

Quadrumana

Their range was once probably greater than now—as we find them fossil in France, England, Greece, they existed in Eocene & subsequent eras, & may have been 6 fold as numerous since the Eocene period as now. The gradation sh.[d] not be assumed from the living chimpanzee. The Dryopithecus may have been superior, its skull not yet known.

New Doctrines

If certain fixed dogmas are taught on subjects of the most transcendental kind, we must, after these have been embraced & inculcated in infancy, make up our minds that every new step we make will unsettle some of these notions whenever we are permitted to draw up part of the veil which hides the plan of Nature from us.

Supra p. 95, 2.[dly] [69]

The desire to embrace as large a number of facts as possible in our great [p. 102] generalization is a legitimate bias. The presumption of imagining we are to see the whole scheme from beginning to end is a dangerous temptation, not making allowance eno' for future discoveries & present ignorance.

What then has been established? A certain amount of presumptive evidence in favour of progression—but the way in which paleontologists have regarded the tertiary as the commencement of Mammalia like the present, & other similar limitations, implies an imperfect acquaintance with the past history & present state of Geology.

Reasoning in a Circle

If Gymnosperms be deemed less perfect because they came first & then are appealed to as proving progression because being less perfect they preceded Angiospermous [plants] because having naked [p. 103] seeds they were created before plants having their seeds in seed-vessels, we evidently reason in a vicious circle. Yet the general rule being once established the geological age of a peculiar type may be used.

Simple Forms

Mosses, lichens, Marchantia, fungi, Jungermanniae like butterflies & moths, not found in the Coal, nor snakes, the absence of many of the lowest types in old terrestrial formations sh.[d] be considered when we miss Mammalia. If whole classes or orders of inferior development have yet to be found some of the higher may come to light. The genus Helix, Cyclostoma, pupa (same one) Clausilia wanting in Coal.

Progressive System

The satisfaction felt by those who believe in this system resembles that of certain persons who, having a set of articles in theology to which they have sworn, look almost with commiseration on such as have by enquiry [p. 104] seen so many flaws in the system as to be unable to subscribe to the same formula—& to doubt its pretensions to infallibility. Convinced of the general truth of their generalization, they are impatient of doubts as to some points as invalidating the truth of the whole. Yet they are forced to amend & amend to reject & invent new causes—the chain is lengthened, the links are more numerous, the beginning more remote, but still it is seen—just spared so as to be visible by Man, who beholds both extremes of the great whole—the alpha & omega. In spite of the annual destruction of monuments now in progress, & which ever has been in progress, still the earliest records of organic life have escaped annihilation. One generation has died in the blessed belief that it saw the beginning in what is now the penultimate,

& sh.[d] the present most ancient become in its turn penultimate, the newly found will be hailed as the first ever produced.

Species [p. 105]

The admission by Agassiz of the dog being possibly, as well as some other domestic animals, a mixture of several species, goes very far towards Darwin's & Morton's views for, in the course of time, natural circ.[s] & chance w.[d] bring about such mixtures.[70] However long the time required, geograph.[l] changes are so great within the life time of a species that it w.[d] be thrown into posit.[s] where union w.[d] take place, & if the hybrid race be more fitted for maintaining itself, it w.[d] survive the pure breed or original.

A great blow is therefore struck at the reality of species by this concession.

A second blow is the assumption of the original creation of a species, or races of a species, over a wide area, for if so, why not races differing so far as to raise doubts in regard to specific identity especially when fertile hybrids are admitted as founding permanent mixed races & species.

Transmutation & Progressive Development [p. 106]

We seem to be drifting towards the Lamarckian theory by two independent currents—one consisting of the arguments in favor of a successive chronological elevation in the scale of being, the advocates of which protest against the transmutation of species; the other made up of those arguments which use the proximate structure & affinities of fossils in the most nearly coeval strata as ground for inferring that such fossils imply a derivation by lineal descent—those who favour such views not insisting on progressive development.

The progressive system alone is calculated to overthrow so many preconceived notions of the independence of Man of the rest of the Zoological world, & to lead to his identification with the inferior animals so far, that its more than welcome reception is remarkable in the history of Science.

Usually truths, which steadily make [p. 107] their way against preconceived opinions & which are held by the few & are embraced reluctantly by the many, are the most sure to prevail in the long run. The eagerness with which the progressive system has been embraced, the generalizations in its favour always outrunning the reasonable induction from facts, is the most suspicious feature in it.

The doctrine in ——[?] was that Mammalia were not created till long in the tertiary period, after the beginning of the Eocene—that birds are as recent, reptiles not before the Permian, fish not anterior to the Devonian, & very rare in that —that plants of simple organization came first—sea weeds, then simple animals, then vertebrates of the lowest class, & so on to Man. That the lowest beds of the Silurian, then known, were the first indication of life on this planet & now that others are known of older date, they are pointed to with quite as much [p. 108] confidence as the first. All this anxiety to be able to point to a beginning has the appearance of a more than willingness to find facts in confirmat.[n] of a foregone conclusion. If every one of the first appearances are removed a step lower, no one who reads the history of geology can doubt that the newly found first appearance, whether of organic beings or of each class of beings, will be pointed out as the periods of the origin of things or of such classes.

The prevalence of entire-mouthed gasteropods in the lower Silurian or even Cambrian & the rarity of Siphonobranchiata may point to the earlier appearance of the former, even if we sh.[d] hereafter find a few in the Cambrian [p. 109] of the canaliculated univalves. But we must state the problem more fairly. Knowing so small a part of the strata & of the globe we must use more caution.

They who imagine that creation has ceased, because Man has entered, sh.[d] be prepared to regard Man as something apart, but the more he is confounded with the rest, the less we sh.[d] expect any cessation of laws which have been in force for millions of years. We must expect that plants having risen from the cryptogamous to the phanerogamous & from gymnospermous to angiospermous dicotyledon, will hereafter become still more developed & that Man will not be the end of the series.

Improved Habitation

There are deserts now & frozen regions, more snow, probably, than at almost any former period save the glacial. The extent of animal life, when icthyosaurs peopled the polar seas, must have been great.

August 2, 1858 [p. 110]

A Progressive Being

The length of time that some savage races have remained nearly or quite stationary, building their canoes the same way as those who hundreds of years before built them, similarly seems to indicate that it w.[d] not require us to descend many steps in the scale of races to arrive at a non-progressive species.

Progressive Development, Why Popular

What is there in the past or future which renders such a theory attractive—in the past an intimate linking together of the human race with the inferior animal,—in the future the advent of a new species to which Man may stand in the same relation as the inferior animals do now to Man. There must be some other source of satisfaction which has [p. 111] disposed the multitude to look with favour on the system & it is because it is the only one which can unite the whole into one generalization. This is a philosophical feeling, but the common herd can scarcely be suspected of willingly abandoning cherished hopes & dreams & dogmas for such a pleasure. In truth they see not clearly to what it leads, nor the illogical links of the chain of reasoning, nor how little we can expect to grapple with so vast a problem.

The future more perfect & more intellectual beings will inhabit a more paradisiacal earth in which the Lily & the Rose will be as much surpassed in beauty & complexity of their organization as is the humble lichen & sea weed by the [p. 112] most highly developed flowers which now adorn the earth—a future paradox from which our race shall be excluded. It has been the tendency to dream of a lost paradise instead of a future one to be inhabited by some superior race or species.

Mollusca in Oldest Rocks

Prof. Troeschel[71] thinks Pander's fish teeth or Conodonts are mollusca, if so, either Gastropods or Cephalopods of Lower Silurian—Agassiz reasons on absence of higher mollusca in oldest rocks, but this is dangerous & the Conodonts may be opposed to it. May there not be soft mollusca both Cephalop.s & Gastropodous with conodontic teeth. Have all the cuttle fish sepia-like or belemnite-like bones? Are not the highest without osseous appendages?

The whole period of development from the earliest period has only in the mollusca produced the difference between the [p. 113] shell-bearing Cephalopoda of the rank of the Nautilus & the Nautilus—between Orthoceras & Nautilus, whereas in the Vertebrata it has progressed from fish to Man—in plants from Coniferae to Oaks or roses—in Radiata from ——[?] in Insects or Articulata from Trilobites to Crabs.

The great prevalence of cephalopods in earlier times, wh. perhaps performed much of the function of the canaliculated univalves, is against the superior develop.t of post-oolitic times —Diluga or cephalopods in oolitic—very few canaliculated.

Our Ignorance

In our part of the world ants perform the work of carabi, in an [earlier geological] age, reptiles of Cetacea—but we may find that while in one sea cephalopods acted the part of canaliculated or carnivorous univalves, in another these last flourished contemporaneously. If marsupials can prevail in Australia & mammalia placentata in other [p. 114] regions, so may there be coeval representatives in the older times.

Aix Plants[72]

[These plants] prepare us for Mammalia mixed with Marsupials in same cretaceous land. The Purbeck Mammalia have the Wealden between them & the Aix flora.

Negative Evidence

A single bird in the chalk—coeval with pterodactyls destroys at once the theory of the gradual develop.[t] of an advanced pterodactyl into a bird—all the birds must be derived from the type already established—so must the mammalia from a triassic marsupial.

The modern Elephant must come from the Himalayan Elephant not [p. 115] from the Mastodon. The Nautilus is not derived from the Orthoceras, but from the oldest (paleozoic?) Nautilus.

A single triassic angiospermous dicotyledon in Canada found by Logan at once makes it necessary to derive all subsequent ones as variations & developments from that stock, not from circuitous genealogies from gymnosperms. The genus Araucaria existed in the oolite, proved by its fruit. It has not varied ever since. It has not diverged into an angiosperm but remained true to its type.

Monotremes

If we have nothing so low in the higher Vertebrata as Monotremes, i.e. in Mammalia—nor in reptiles, as in snakes as a class, it does not prove that as a whole the simplest forms were wanting in earlier times.

Negative Evidence [p. 116]

Had Monotremes among Mammalia, snakes among terrestrial reptiles, mosses & Jungermanniae among paleozoic plants abounded instead of being as yet undiscovered in all format.[s]; from the oldest to the white chalk inclusive we sh.[d] have heard much of the proof of the inferior grades having been the chief inhabitants of the planet before higher ones entered.

Man

Few would be willing to admit how much of the whole weight of the argument rests on the assumption of Man, who came last, being a step in the same progressive scale, only differing in degree. A wide gap would exist, as between the Purbeck animals, if all were marsupial, & the tertiary Monodelphi.

If we exclude Man, we have superior fish & more highly developed [p. 117] Reptiles in older times.

As to Birds we cannot fairly reason on them, the documents are so scanty.

Transmutation

The smallest step in its favour is more encouraging to its advocates than the greatest in favour of Progression—because the one is ag.[t] all preconceived & popular notions, the other in support of them.

In the case of develop.[t] there is the anxious desire to find a beginning, to go back to zero—to chaos—to see all, to embrace the whole, this is Genesis, the Mosaic last crowning creation of Man, there is the desire to prove unity & uniformity. Man is an interruption in the scheme of the Uniformitarian, a confirmat.[n] in that of the progressionist. It is true that the develop.[t] scheme will, if established, overturn more or at least as many received notions as the other, [p. 118] but this is evidently not foreseen & apprehended, chiefly because so many believe in catastrophes & make so little allowance for gaps to be filled up, or if not filled logically to be assumed.

That which is obvious, the coming in of Man, after all the rest & the form.[tn] of the inferior animals first, & the rising or ascending from the lowest to the highest, & the having a beginning to point to, & the frequent appeal to the intervention of a first Cause, are popular tendencies.

Therefore this theory is embraced while the other is condemned—altho' it must be acknowledged that in the *Vestiges* where no difficulties are seen, the whole of Lamarck's hypothesis was welcomed by a larger public than ever, in an equally short time, bought a scientific book.

If progress be true we must look [p. 119] the whole prospect in the face. There will be found a gradation from the Quadrumana to Man. Inferior Races of Man will have preceded—Superior will follow. If we once imagine all the 4 types of Creation to stop because Man has appeared, we make his advent so great an event that the uniformity of the system is broken. It is then like an avatar.[73]

Mollusca

From a Bryozoan to a Cuttlefish, from a creature with little more than a stomach, without a head, or eyes or locomotion, to one as active & sharp sighted as a cuttlefish, there is indeed progress. But this is not the passage, the Bryozoa existed in the earliest times & now, & so did the Cephalopoda. All that can be advanced in the way of progress is the difference between a Nautilus & a Cuttlefish. In the Vertebrata, some proceed from zero to Man, others from a fish to Man, according as they commence [p. 120] Cambrian or Silurian.

In the Mollusca, which we know best from earliest to latest times, we have the worst evidence—the least progress.

Between a Bryozoan & a Cuttlefish the distance may be as great as between a fish & an Elephant—so between a mushroom & a rose, between a moss & a lily, but a conifer is traced back as far as any cryptogam & between a conifer & an oak, no one can pretend that the distance is equivalent to the Time expended. By this time "the loves of plants" ought to have been more than a mere metaphor.[74]

Uniformity

If we are bribed by a wish to prove this, we must at least protest against stopping short here, higher mollusca, radiata, articulata, vertebrata, & plants will [p. 121] succeed the present races, new species will replace the present. Why should the plan stop here?

It is but justice to say that some steps have been gained by the transmutationist, partly the mere absence of any fact in favour of fresh creations & the continued formation of races—& the extinction of certain species & races or permanent varieties. Many living races of Man are undergoing extermination & new ones may be forming, but no new starting point is ascertained.

Beginning

Did the first egg & the first seed contain all that was to follow —was it a fecundated or impregnated egg? How completely here we are beyond our depth. It is fortunate for the progressionist that there is no possibility of tracing back the system to a beginning.

Progress

The culminating point of Reptiles, is past, of Cycads & Gymnosperms, [and] of fish. There are ups & downs.

Progressive Development [p. 122]

"It is true in the main," [75] says Owen in his Brit. Asso.ⁿ speech, i.e. Man is the most perfect? But if man, as head of the Mammalia, be the proof & with many even who do not avow it, it is the most cogent, then the progress made by this division of the Vertebrate[s] is wholly incommensurate & unlike that evinced in any other.

It seems now very generally admitted that the Reptile & Fish are not at the present epoch in the highest state of develop.ᵗ, that they are, in this, passed their culminating point, so far as their past history is comprehended. As to the Mollusca of which the history is by far the most complete, it w.ᵈ be no mere pretension to contrast the difference between the Cephalopoda of the Silurian & those of the present period, representing their progress in the whole long series of geological epochs as comparable to the supposed progress from the earliest known Mammalia to [p. 123] Man. And even the progress from the Silurian Cephalopoda to the living Cuttlefish labours under the disadvantage that we know so much less of the whole fauna of any remote period than of the present & cannot be sure that some shell-less species did not exist wh. have not left their relics behind.

Generalizations, such as the progressionists indulge in, are useful but the temptation to embrace the whole cycle of facts in one whole has often to be resisted.

Steamer Carnel, October 19, 1858 [76]

Man, a Real Saltus

In geological time the apparition of Man with his present power will always have to be admitted to be a wholly unexampled step in advance both in quantity & in kind.

It is the extreme shortness of geolog.[l] time which must always cause this. Grant that ethnologists may require, & geologists hereafter prove, 50 or 100 thousand years still it will always be a mere fragment of [p. 124] a geological period measured paleontologically, less perhaps than it has taken to establish the German fauna in Europe, that the whole develop.[t] of Man has been displayed in. Suppose the advocates of progress to show that there have been different races of Man, each of which has had their distinct & independent starting point, the least perfect preceding those of higher capabilities as they are bound in consistency to assume, for they have no right to suppose all to have been contemporaneous, still from the first to the last is but an instant of geological time. In no former period of similar duration can we see any similar jump.

We cannot find proof that any proboscidia[n] or quadrumanous animal made even a nearer approach in reasoning power to Man, than the living Elephant or Orang, tho' after ages have rolled away new Elephants & new Orangs did no doubt [p. 125] succeed each other in successive tertiary periods.

The Unity of the Creation therefore may be admitted so far as the animal nature of Man is concerned & the Moral & Intellectual & Progressive part regarded as an avatar.

To assume it to be part of the progress indicated in any of the Branches of the Animal Kingdom or by the Mammalia since the Oolitic period seems to be to compare things wholly incommensurable.

Nothing would be easier than to make out a plausible gradation between Man & the inferior animals by showing how low Man can sink & how the races may be lower & some of the lowest lost before the times of history but the Elephant & dog seem given expressly to man as against already too much importance to the anthropomorphous animals in a psychological relationship with Man.

[p. 126] It is not because it would introduce any new difficulty that we sh.[d] shrink from such a gradation, whether brought about by a series of Creations, or a graduated scale, or by generation & the indefinite variability of parent stocks, accompanied by the principles of progressive develop.[t]. This w.[d] merely be to add to an enigma of which we already feel the perplexity—where does the embryo begin to rank other than as a mere animal, where does it ascend to anything like the dignity of an Elep.[t], Dog or Chimpanzee & yet millions of them perish every century & millions of idiots, infants & others.

We sh.[d] not add difficulties of a new kind & having millions of millions at present we need not object to new ones. But the case of the human race is so different as not to be assimilated to the kind of progress which is recognized, tho' obscurely, [p. 127] in tracing the geological history of the four great Branches of the Animal Kingdom & in the progress of plants.

Serpents & Cryptogams

If there are no snakes in the older formation why is it not more dwelt upon.

Is it not an indication of the difficulty of detecting land animals of ancient date, even reptiles. Their absence, like that of Hepaticae may be because we cannot expect to find them. But if so why not apply this argument to Mammalia & if on the other hand, we really believe that snakes had not been created, nor Hepaticae, then we sh.[d] be cautious in supposing that the omission of the highest class may not depend on some rule unconnected with their rank & the age of the planet.

October 26, 1858 [p. 128]

Man w.[d] be subject to same law of origin of species as animals.

As Man is first an embryo & then born & suckled at the breast, & is subject to same diseases, epidemics, decay & dotage as other Mammalia, as he depends like them on vegetable or animal food & the population is regulated according to its abundance, as he is separated into distinct races & varieties, is limited like them to one element & to a restricted geographical distribution, so it is natural that if some general law, whether of single specific centres or of multiple creations, has prevailed in other Mammalia, it will be found to hold good also in regard to Man. Or if there has been progressive development, or gradual change by generation according to the Lamarckian system, one species being convertible into another, so will it have been with Man.

Progressive Development:
Discoveries in Autumn of 1858 [p. 129]

A triassic mammalian lower jaw probably marsupial, of a minute quadruped, found at —— recognized by Dr. Falconer. A tooth of a larger? mammal from the trias which Dr. Falconer announces as having been found by Mr. Charles Moore of Bath.[77]

Elgin, Stagonolepis Robertsoni, a large Labry[n]thodon & truly Devonian. Numerous footprints in the same format.[n] [78]

October 30 [1858]

With Wollaston: Species

80 species of land shells in the Canaries & all save Helix pisana & one or two migratory ones are distinct from those of Madeira.

As to distinct centres from wh. species have [p. 130] spread, or specific centres may require, says W.[ollaston] to comprehend the spreading from centres of Races, for if Mr. Poole, *Genesis of the Earth & Man,* be correct in supposing the Negro which is a very old race to have been started from a diff.[t] centre from the White Race, then other species of the higher animals may have done the same.[79]

November 1, 1858

If the geologist dwelling exclusively on one class of facts, which might be paralleled by the existing creation arriving at conclusions derogating from the elevated position previously assigned by him to Man, if he blends him inseparably with the inferior animals & considers him as belonging to the Earth solely, & as doomed to pass away like them & have no farther any relation to the living world, he may feel dissatisfied with his labours [p. 131] & doubt whether he would not have been happier had he never entered upon them & whether he ought to impart the result to others.

If once the question, whether the acquisition of truth is a good to himself & to mankind, whether it w.[d] not be preferable to remain in ignorance or under a pleasing delusion of a future state, it is natural since his successful scientific career implies an earnestness of character which would be without meaning if he were a being of a moment, or only stor[in]g up knowledge for a future series of ephemeral beings like himself.

The act of generalizing on geological & astronomical phenomena, or of obtaining an insight into the secrets [p. 132] of indefinite Time & Space, ought not to conduct us to such thoughts—this curiosity, given to the mind to seek a great book laid open to us of which we can decipher the records, are [*sic*] an earnest to us as some higher destiny. No doubt the humblest race of Men must be considered equal before God & if so, why infer from the intellectual advancement of a privileged few that they are superior to the lower animals in their future prospects.

Are these races to the higher ones what the child is to the Man? Can the one & the other in its more perfect state reflect dignity & rank on the embryo? The dotage [p. 133] of the old, the suspension of the intellectual powers, shows the possibility of temporary annihilation of the spiritual part of an individual. These are objections as strong as any which geology can advance. The materialist can strengthen his arguments from the facts of paleontology & on the other side we can only appeal to the knowledge of Nature, which geology affords, in order to derive from it higher hopes & aspirations for the human race.

The vastness of the plan shows how limited is the range of our survey. This is an advantage as making us humbly desist from presuming to decide.

The moral world is an addition [p. 134] to which nothing preexisting can be compared—the Free Will of Man—Memory, Pain, reasoning, instinct, sight, feeling, hearing, smelling, touch, taste, anger, rage, maternal or parental affection existed before, but not responsibility, sentiment, goodness.

The intermediate stages are the enigma. They may link Man with Higher Beings as yet unrevealed to us.

That the opinion & hope & aspiration that there is something beyond this worldly sphere in the destiny of Man has an elevating effect morally & intellectually cannot be doubted & therefore that it is better for Mankind to hope & believe in a future state.

It is not only consonant to reason, [p. 135] but conduces to the happiness & improvement of the species. If every one believed in annihilation as the meaning of Death, they could not be much in earnest nor admire earnestness in others.

By dwelling on matters beyond the reach of our comprehension, we may doubt the existence of matter & the reality of the present as well as of a future state. All may appear to be a dream—a machinery as the Fatalist assumes. Every new department of Science will enlarge the sphere of their doubts & perplexities.

But the discoverer of new scientific truths refutes, by his eagerness to interpret the plan of the Universe, the notion that he is a mere bubble which is to burst & be no more.

Owen's British Association Address, 1858[80] [p. 136]

p. 10 "To tremble at the thunder as a manifest.[n] of an offended Deity is the superstit.[n] of the savage to recognize, etc. *law* is privilege of the sage."

The "reverence for lawgiver" is the result of enquiry:

The notion that people are happier by not enquiring into Nature, because they are thereby in a less degree materialists is untrue. On the whole they are elevated & less timid & have a higher concept of their position.[81]

[p.]37. Transmutation of Chimpanzee into Man rebutted by climatal limits of Orangs.[82]

[p.]41. Purbeck Hypsiprymnus, no reference to Falconer! [83]

[p.]42. Red grouse not created for Gt. Brit.ⁿ because changes in geography before Man.[84]

" "Creation" of Zoologist "a process he knows not what"—[85]

[p.]43. Species peculiar to islands, [p. 137] whether created for those islands.[86]

" "Beginnings of things" whether a province of knowledge beyond our present capacities.[87]

" Wallace on extirpation.[88]

[p.]45. Races of Man not limited, as Agassiz has argued, to zoological Provinces of Animals.[89]

[p.]47. Antiquity of Man—Egypt[90]

" Same proved by languages.[91]

December 23 [1858]

With Tyndall & Zincke[92]

T. says the *Vestiges* struck him as making out a strong case ag.ᵗ perpetual interference of Creative Agency.

I objected that [the] *Vestiges* did not appreciate the grandeur of the idea of knowing whether a species would last.

T. granted that if Creation was [p. 138] defined to mean simply that the plan of introducing new species, the *modus operandi* was not yet brought within the domain of Science, it might be a legitimate mode of dealing with the difficulty or mystery.

Transmutation is a mere guess at present.

If man c.ᵈ observe for 5000 yrs. & learn more as Antiquary

of the races of men during last 5000 yrs., he might come to some knowledge of how the change is taking place.

T. Said that the microscopic world proves independence of Man.

Huxley

[Huxley] thinks something like transmut.[n] & progression must be true, tho' not as stated by *Vestiges* & others.

Transmutation [p. 139]

If true, an egg or seed of Cambrian or antecedent date contained in it the germ in wh. was folded up, all the animals or plants since created, or a few eggs & seeds, & also of all wh. are to exist in future, Man & his posterity included.

If Man be excepted we may except some others & so get into the Transmut.[n] hypothesis.

But it is true that if we reduce all mankind to a pair of ordinary capacity, they might contain the germ of a popul.[n] as advanced as the actual [population] of the globe. So also the 2 embryonic atoms, from wh. this pair began, would contain in them the whole future of mankind.

Unity of Creation [p. 140]

[1.st] Man, like plants & Brutes, has permanent varieties & races.
2.[dly] more than half of human beings cut off in childhood.
3.[dly] begins in an embryonic animal state.
4.[thly] subject to epidemics, disease, madness, dotage.
5. shorter lived than many of the higher animals as the Elephant.
6. spreads over newly created land.
7. becomes occasionally fossil.
8. cannot be traced indefinitely into the past any more than his contemporary animals & plants.
9. His early history & origin as much a mystery to the scientific enquirer as that of any other species.

10. confined to the land & therefore not so cosmopolite as some oceanic species?
[p. 141]
11. Hybrids between most distant races intellectually & physically intermediate.
12. Earthquakes & partial deluges do not spare man or cities. Man not a privileged class but when his moral & intellectual nature became incarnate he was subject to all the weaknesses that flesh is heir to—as frail, transient—weak—suffering. Therefore we may expect that if the transmut.[n] or progression be true or untrue of Animals, so of Man—no signs of being exempt because of spiritual superiority. The Laws wh. have prevailed for millions of years in the organic & inorganic world remain the same—like the motions of the heavenly bodies—species must be changing as of old, but whether by [p. 142] extinction & new beginnings or by perpetual modific.[n] is the question. The answer will be the same for Man & Animals.
13. Instincts very analogous to those of animals—& least marked as in the most intelligent of the mammalia.

Progressive Development

It was formerly supposed by some that when reptiles abounded an atmosphere unfit for birds & mammalia or quick breathing creatures envelop.[d] the earth. The Upper G. S. [Green Sand] bird & whale refute that idea. One individual is enough as it proves thousands.

Creation

A state of doubt as to mode of format.[n] of species [is] the sound state [of mind] at present.

If transmut.[n] true, then a few eggs & seeds probably ante-Cambrian, which might have gone into an orange, may [p. 143] have been the progenitor & have contained folded up in them the germs of all living forms, Man included, wh. have been since & are hereafter to be evolved or to exist—Man included. For that matter 2 microscopic homunculi may have been the progenitors of any number of millions of human beings. Nothing w.[d] be too wonderful as the original starting points, the might of little things, *macht der Kleinen*—if in one acorn all the trees of countless forests of the common oak of Europe—Quercus robur & all its varieties, why not so of the rest?

Unity of Creation [Continued] from pp. 140–42

14. Reproduction very analogous, 9 month.

Lamarckian System

Monads if admitted still require progression & where are the intermediates?

January 1, 1859

Globigerina of chalk an objection—monads imply all that they are afterwards [pp. 144–45 are blank; 146] to become & if an egg or a seed imply all the foreknowledge & power to create them, which is involved in all the generations of the species of animal or plant which are to come in thousands of years as descendants from such egg or seed, so of monads.

The perpetual renewal therefore of monads is as much a constant act of creation as any other system, a perpetual interference of the Deity to create the germs of Mammalia, or Man himself.

Impoverishment of Species

Simple forms must come from deterioration, if no new creations are admitted, & progressive development assumed.

For the abundance of Bryozoa is as great now as in Lower Silurian & of infusoria & foraminifera.

[p. 147] But if these simple forms are sustained & kept up in number by the deterioration of higher organisms, a very complicated hypothesis is required.

Which species or classes are rendered simpler—Shetland ponies—inferior races—Negros? Australian savages? The simplest hypothesis is to imagine the higher & lower in each great division to be constantly renovated by the introduction by some unknown, undiscovered & totally inconceivable law, & machinery of new species from time to time.

As to Man being an integral part of the same progression, the saltus is too great, all may be constant & the moral & intellectual & rational superadded as new, just as when life first introduc.[d] into the Earth.

Depauperation [p. 148]

The perfectness of the simpler forms is such for the conditions & purposes for wh. they are created, that the convertibility of the higher into the lower classes w.[d] require as great power & intelligence as the develop.[t] of the simple into the complex. Therefore the more modern times, exclusive of Man, imply a greater variety of beings convertible into higher & lower, a progressive & retrogressive system.

Hence if Man be subject to the same law higher & lower races will enter the stage.

But is not extermination rather than simplification the law for species wh. can no longer maintain their ground?

According to some all classes [p. 149] remain stationary, but there is from time to time the superaddition of more perfect beings—in each class.

Hypotheses

Lamarck, Transmutation & Progression, implies perpetual creation of monads, Agassiz, Progression & Creation of species, without transmutation—Man the crowning operation of the progressive system.

Creation of species without manifest progression—Man as a rational being not part of the same system of change—or development.

Progression & degradation of original types, the simplest being kept up in number by loss of complication of organs—as Bryozoa—depauperated mollusca accord.[g] to Agassiz.

[pp. 150–51 are blank.]

Helices & Ophidians [p. 152]

Land snakes are to the Reptilia what Helices are to the Mollusca & both wanting below the Eocene.

We have found Mammalia more spread than these two great divisions of air-breathers—Why? If pumonif.[s] mollusca & ophidians were as numerous in the old times as now, then mammalia may have been [present]; Tertiary Mammalia more common than Tertiary Ophidians, but not so of pulmoniferous Mollusca.

If Ophidians & Helices had been at the head of the Reptilia & Mollusca, their absence would have been cited as proofs of want of development in the older times.

INDEX [p. 1]

52. Creation, Dr. Falconer says, sh.[d] be modestly proposed.
55. (56.) Races of Man as distinct as many admitted species.
57. Centres of creation—islands.
59. Transmutation gives pain owing to preconceived opinions.
60. Falconer on intelligence of Elephant—pitfall.
61. Mastodon & Elephant coexisted in Mio- & post-Pliocene.
62. Man of genius—a leap—first one no revolution.
64. British roses of allied varieties or species.
66. Uniones of N. America, Gould says, 250 true species.
Tail in Man & rudiments explained by transmut.[n].
67. Species by special intervention, not derogatory to God.

[p. 153]

68. Millions of infants die—no new enigma raised.
69. All went on as now when men created.
70. Man how far spiritually of distinct class.
72. }
73. } Species form as trees grow in a forest.
Gamopetalous why suppos.[d] of higher grade, C. Bunbury.
75. Monocotyled.[s] Qy. if of coordinate rank with angiosperms.
76. Leguminosae depend on ext.[l] condit.[s], not age of plants.
77. Man crowning operation—varies like variable species.
Monads of Lamarck—provided for progress & simple forms
80. Cowslip & primrose from the same root.
81. Specific centres, how far compatible with Darwin?
Species persistent while 10/100 die.
82. " , analysis & synthesis of V. Beneden.
Negro when young most like Caucasian.
83. If Man coming last, prove progression, why not developm.[t].
84. Progressive theory, why popular & transmut.[n] unpopular.
85. Man from Orang as Ass from Zebra.
" (90) Agassiz on substit.[n] of materialism or physical causes.
88. Agassiz, near approach to a development hypothesis.
89. " on similarity of habits of species of same genus.
90. }
91. } " child & young chimpanzee of like capacity.
95. Immortality if true of Man so of beast, Agassiz.

92. (102) Progressive theory, causes of its popularity. *Vestiges.*
97. Baptism of the unborn.
98. Responsibility begins when?
100. Unsettling of dogmas.
101. Transcendentalism of theology.
103. Progression, lower organisms missing in old strata.
105. Pallasian theory admitted by Agassiz.

[p. 154]

106. Progressionists aid transmutationists by making man part of same chain.
107. Progressive theory has shifted its ground.
108. " " how far true.
109. Improved habitation—deserts now.
110. Progressive & stationary man.
(117) Progress, why popular past & future.
112. " from nautilus to cuttle-fish, fir to oak.
113. One class replacing another, cephalopoda of Silurian for gasteropods.
114. Aix plants imply chalk mammals.
115. Angiosperm in trias of Canada.
116. Many simple forms missing in oldest rocks.
Man coming last, key-stone of progressionists.
119. Progress not from Silur.^{n} bryozoa to cuttlefish.
122. " —reptiles once higher than now.
124. Man a saltus, in a very brief time.
125. " an avater [avatar].
126. Enigma of instinct passing to reason exists without geology.
127. Snakes absent tho' simpler—land animals?
128. Man subject to same law as animals.

[p. 155]

128. If specific centres for Man, so for all.
129. Canary land shells mostly peculiar.
130. Negro & White, distinct centres—Poole.
131. Immortality of soul—truth unwelcome.
135. " proof of from earnest characters.

136. Owen's, Brit. Assoc.[n] Address. Index.
139. If man be an except.[n] to transmut.[n] so may other species to be excepted.
 An embryo of male & female may produce all mankind.
140. Man & Unity of creation.
 Enumeration of points of analogy with animals—no exemptions.
142. Bird of upper Green Sand in same atmosphere as Pterodactyl.
146. All in one egg—foreknowledge.
147. Man distinct in kind.
148., 149. Depauperation—whether mode of sustaining the crop of simpler forms?
152. Ophidian in Coal, not more wonderful than pupa vetusta.

Notes

1. This letter was evoked by Lyell's reading Joseph D. Hooker's "Introductory Essay" to *Flora Novae Zealandicae,* vol. 2 of *The Botany of the Antarctic Voyage of H.M. Discovery Ships Erebus and Terror in the years 1839–1843,* 3 vols., London, Reeve, 1847–60. Hooker had sent Lyell two copies of the essay, and Lyell had reread and digested it during a sea voyage from London to Hamburg. The original of this letter is among the Hooker mss.

2. Ibid., p. xiii.

3. Ibid., p. xvii. Hooker discusses the differences between the cedars of Lebanon, the cedar of the Atlas mountains and that of the Himalayas, and decides that they are both simply varieties of one species that must once have been distributed continuously from the Atlantic coast of Africa to central Asia.

4. Johann Georg Christian Lehmann (1792–1860), botanist of Hamburg, Germany.

5. Karl Werner Maximilian Wibel (Wiebel), (1808–88), since 1837 was professor of chemistry and physics in the Gymnasium at Hamburg. He wrote on the geology of Heligoland and the rapid wasting of this island.

6. From Hamburg, Lyell went to Berlin, Warmbrunn, Dresden, Prague, Vienna. At Vienna, at the beginning of September 1856, he fell ill, but on recovering made a tour in the Styrian Alps. Thence he went to Salzburg, Reichenhall, Munich, Stuttgart, Strasbourg, and Paris, where he was on October 23. By October 30, he was again in London. He did not meet Hooker in the Alps.

7. Hooker, "Introductory Essay," pp. xi–xii.

8. Sir William Lawrence (1783–1867), *Lectures on Comparative Anatomy, Physiology, Zoology, and the Natural History of Man,* 7th ed., London, Taylor, 1838, 396 pp. Lawrence cites (p. 235) the following passage from Georges Cuvier, "Extrait d'observations faites sur le Cadavre d'une femme connue à Paris et à Londres sous le nom de Vénus Hottentotte," *Mém. Mus. Hist. Nat.,* 1817, *3,* 259–74, p. 273:

> Now that we distinguish the several human races by the bones of the head, and that we possess so many of the ancient Egyptian embalmed bodies, it is easy to prove that, whatever may have been the hue of their skin, they belonged to the same race with ourselves; that their cranium, and brain were equally voluminous; in a word that they formed no exception to that cruel law which seems to have doomed to eternal inferiority all the tribes of our species which are unfortunate enough to have a depressed and compressed cranium.

9. Joachim Barrande (1799–1883).

10. This refers to a statement by William Whewell, *On the Plurality of Worlds,* 3d ed., London, Parker, 1854, 407 pp., pp. 175–76:

> But in conceiving that all the myriads of successive species, which we find in the earth's strata, have come into being by a law which is now operating, we have *nothing* to start from. We have seen, and know of, no such change; all sober and skillful naturalists reject it, as a fact not belonging to our time.

11. Étienne Jules Adolphe Desmier, Vicomte d'Archiac (1802–68), French geologist.

12. Étienne Desmier, Vicomte d'Archiac, *Histoire des progrès de la géologie de 1834 à 1846,* 9 vols., Paris, Société Géologique de France, 1847–60. Volume 6 deals with the Jurassic.

13. Éduard Armand Isidore Hippolyte Lartet (1801–71), French geologist. His memoir describing the Dryopithecus was "Note sur un grand singe fossile qui se rattache au groupe des singes supérieurs," *Comptes Rendus Acad. Sci.,* 1856, 43, 219–23.

14. Searles Valentine Wood (1798–1880), geologist, possessed a large collection of the fossil shells of the Crag formation of Norfolk and published a number of papers about them.

15. This is evidently a reference to Searles Wood's work on the bivalves of the Crag. Searles V. Wood, *A Monograph of the Crag Mollusca, with Descriptions of Shells from the Upper Tertiaries of the British Isles,* 4 vols., London, Palaeontographical Society, 1848–82, vol. 2, Bivalves. His treatise on the univalves was vol. 1, *Univalves,* 1848, 208 pp. At p. 199, Wood gave "A Synoptical Table of Mollusca from the Crag," giving the age of the strata, Miocene, Pliocene, etc., from which each species had been collected.

16. The French translation of Lyell's *Manual* was: Sir Charles

Lyell, *Manuel de géologie élémentaire,* M. Hugard, trans., 5th ed., Paris, Langlois et Leclercq, 1856–57.

17. Lyell, *Manual,* 5th ed., pp. 169–70:

For a large collection of the fish, echinoderms, shells, bryozoa, and corals of the deposits in Suffolk, we are indebted to the labours of Mr. Searles Wood. Of testacea alone he has obtained 230 species from the Red, and 345 from the Coralline Crag about 150 being common to each. The proportion of recent species in the new group is considered by Mr. Wood to be about 70 per cent and that in the older or Coralline about 60.

18. Wood, *Crag Mollusca,* vol. 2, *Univalves,* 1848, p. 199.

19. Wood, *Crag Mollusca,* vol. 2, *Bivalves,* 1856, p. 306.

20. Wood, *Crag Mollusca,* vol. 1, *Univalves,* 1848, pp. v–vi.

21. "Grays, on the river Thames, twenty-one miles from London, in the county of Essex," ibid., p. v.

22. "Clacton, in Essex, on the coast, seven miles southwest of Walton Naze," ibid.

23. "Stutton, in Suffolk, on the banks of the river Stour, six miles south of Ipswich," ibid.

24. Gerard Paul Deshayes (1795–1875), *Description des Coquilles Fossiles des environs de Paris,* 2 vols. atlas, Paris, 1824–37.

25. Wood, *Crag Mollusca,* vol. 2, *Bivalves,* 1856.

26. In 1842, Hugh Falconer (1808–65) returned to England from India, bringing with him his collections of dried plants and fossils, including his fossils from the Siwalik Hills. Between 1843 and 1847, while he remained in England, Falconer published several memoirs on the geology and fossils of the Siwalik Hills. With respect to his collections of fossil shells, he undoubtedly consulted Edward Forbes who in 1842 was appointed curator of the collections of the Geological Society of London and professor of botany, King's College, London. Forbes was expert in the study of marine mollusca, both fossil and living.

27. Crozier, a botanical term for any structure ending in a coiled or circinate form (like a shepherd's crook), as the young frond of a fern.

28. In 1842, Richard Owen had settled once and for all the long-disputed question of the Mammalian character of certain fossils from the slate of Stonesfield, a village lying about eighteen miles

to the northwest of Oxford, England. The Stonesfield Slate is a series of beds near the bottom of the Great Oolite formation of early Jurassic age. The Stonesfield mammals therefore preceded, by the long period of time represented by the Jurassic and Cretaceous formations, the previously assumed introduction of the Mammals at the beginning of the Tertiary period. In 1842, Owen distinguished three genera of fossil mammals from Stonesfield, *Amphitherium, Phascolotherium,* and *Amphilestes.* See Richard Owen, "Observations on the Fossils representing the Thylacotherium Prevostii, Valenciennes, with reference to the Doubts of its Mammalian and Marsupial Nature recently promulgated; and on the Phascolotherium Bucklandi," *Trans. Geol. Soc. London,* 1842, ser. 2, *6,* 47–65.

In 1854, the Rev. W. R. Brodie found in Middle Purbeck beds at Durdlestone Bay near Swanage on the Isle of Purbeck, Dorsetshire, England, parts of several small jaws with teeth, which he sent to Richard Owen. On clearing away the matrix, Owen recognized that, although found together with a number of jaws of lizards, these specimens were in fact mammalian, and he established a new genus *Spalacotherium.* See Richard Owen, "On the Fossil Reptilian and Mammalian Remains from the Purbecks," *Quart. J. Geol. Soc. London,* 1854, *10,* 420–33. The discovery of *Spalacotherium* stimulated S. H. Beckles to make a thorough exploration of the cliffs at Durdlestone Bay.

29. Charles James Fox Bunbury (1809–86), palaeobotanist, was Lyell's brother-in-law.

30. Heinrich Robert Goeppert (1800–84), *Die tertiare Flora von Schossnitz in Schlesien,* Gorlitz, 1855, 52 pp. Goeppert was professor of botany at the University of Breslau.

31. Joseph Dalton Hooker (1817–1911).

32. Charlotte Brontë (1816–55), novelist.

33. Harriet Martineau (1802–76), political and scientific writer and novelist.

34. Henry George Atkinson (1812–90), friend of Harriet Martineau and coauthor with her of the work under discussion.

35. This passage is part of a letter from Charlotte Brontë to an unknown correspondent, possibly William Thackeray. Miss Brontë was commenting on the book by Henry George Atkinson and Harriet Martineau, *Letters on the Laws of Man's Nature and Development,* London, Chapman, 1851, 390 pp., which was con-

sidered an avowal of atheism. The letter is published in E. C. Gaskell, *The Life of Charlotte Brontë,* 1st ed., London, Downey, 1901, 526 pp., pp. 432–33. How Lyell came to have access to it we cannot determine.

36. George Berkeley (1685–1753), Bishop of Cloyne, Ireland. This line is from the last stanza of Berkeley's *Verses on the Prospect of Planting Arts and Learning in America.*

> Westward the course of empire takes its way;
> The four first acts already past,
> A fifth shall close the drama with the day:
> Time's noblest offspring is the last.

37. The Bernouillis were a famous Swiss family of scientists and mathematicians, originally resident at Antwerp. The most eminent were Jacques Bernouilli (1654–1705); his brother Jean Bernouilli (1667–1748); Nicholas Bernouilli (1695–1726), Daniel Bernouilli (1700–82), and Jean Bernouilli (1710–90), sons of the first Jean Bernouilli; and Jean Bernouilli (1744–1807), son of the second Jean Bernouilli.

38. "The chimpanzee and gorilla do not differ more one from the other than the Mandigo and the Guinea Negro: they together do not differ more from the orang than the Malay or white man differs from the negro." Louis Agassiz, "Sketch of the Natural Provinces of the Animal World and their relation to the different types of Man," in Josiah C. Nott and George R. Gliddon, *Types of Mankind: or, Ethnological Researches, based upon the ancient monuments, paintings, sculptures, and crania of races, and upon their natural, geographical, philological and biblical history,* Philadelphia, Lippincott, Grambo, 1854, 738 pp., pp. 58–76, 75.

39. For William Whewell's opinion, see Scientific Journal No. II, note 14, p. 140.

40. Trans: "The ape, a very ugly animal similar to us." Quintus Ennius (239–169 B.C.), Roman poet.

41. Richard Owen, "On the characters, principles of division and primary groups of the class Mammalia," *Linn. Soc. London, J. Proc.* 1857, *2,* 1–37, p. 20 n.:

> Not being able to appreciate, or conceive of the distinction between the psychical phaenomena of a Chimpanzee and of a Boschisman, or of an Aztec with arrested brain growth, as being of a nature so essential as to preclude a comparison between them

or as being other than a difference of degree, I cannot shut my eyes to the significance of that all-pervading similitude of structure—every tooth, every bone, strictly homologous—which makes the determination of the difference between *Homo* and *Pithecus* the anatomist's difficulty.

42. Saherampore Saharanpur, a city approximately 100 miles north and slightly to the east of Delhi, India.

43. Augustus Addison Gould (1805–66), physician and conchologist of Boston, Massachusetts.

44. John Phillips (1800–74), professor of geology at Oxford University.

45. "Nature does not proceed by leaps." Carolus Linnaeus, *Philosophia Botanica in qua explicantur Fundamenta Botanica,* Vienna, Thoma, 1770, 364 pp., p. 27.

46. William Harvey (1578–1657) announced his discovery of the circulation of blood in 1628.

47. Alphonse de Candolle (1806–93), Swiss botanist.

48. Joseph D. Hooker, "On the structure and affinities of Balanophorae," *Trans. Linn. Soc. London,* 1856, *22,* 1–68.

49. John Lindley (1799–1865), botanist and horticulturalist, was secretary of The Horticultural Society and wrote on botanical subjects.

50. Achille de Zigno (1813–92), botanist of Padua, Italy. In 1856, de Zigno had published a paper on fossil plants of the Oolite, "Sulla flora fossile dell'oolite," *Mem. Ist. Veneto,* 1856, *6,* 325–39.

51. See especially Lecture IV, "The Mosaic View of Creation," and Lecture V, "Geology in its bearings on the Two Theologies," in Hugh Miller, *The Testimony of the Rocks, or Geology in its bearing on the Two Theologies, Natural and Revealed,* Edinburgh Constable, 1857, 500 pp.

52. Rev. John Stevens Henslow (1796–1861), botanist, was Rector of Hitcham, Suffolk, and from 1827 professor of botany at Cambridge University, where he exerted a profound influence on Charles Darwin while Darwin was a student at Cambridge 1828–31.

53. Charles J. F. Bunbury (1809–86). See note 29 above.

54. Alcide Dessalines d'Orbigny (1802–57), naturalist, traveler and palaeontologist, was professor of palaeontology at the Museum

of Natural History, Paris, from 1836 to 1853. Lyell is probably referring to views expressed in d'Orbigny, *Prodrome de Paléontologie stratigraphique universelle des Animaux Mollusques et Rayonnés faisant suite au Cours élémentaire de Paléontologie,* 3 vols., Paris, Masson, 1849–52.

55. Pierre Joseph van Beneden (1809–94), botanist and zoologist, was professor of zoology at the University of Louvain.

56. Louis Agassiz, "Essay on Classification," *Contributions to the Natural History of the United States of America,* Boston, Little, Brown, 1857, 1–232, p. 50:

> Animals of different types living in the same element have no sort of similarity as to size; but there is no common average for either terrestrial or aquatic animals of different classes taken together; and in this lies the evidence that organized beings are independent of the medium in which they live, as far as their origin is concerned, though it is plain that when created they were made to suit the element in which they were placed.

57. Ibid., p. 55.

58. Ibid., pp. 55–56:

> This position [i.e., the stability of the organic world] is still more strengthened when we consider that the differences which exist between different races of domesticated animals and the varieties of our cultivated plants, as well as among the races of men, are permanent under the most diversified climatic influences . . .

59. Ibid., p. 56:

> It is often stated that the ancient philosophers have solved satisfactorily all the great questions interesting to man . . . Is this true? There is no question so deeply interesting to man as that of his own origin, and the origin of all things. And yet antiquity had no knowledge concerning it; things were formerly believed either to be from eternity, or to have been created at one time. Modern science, however, can show, in the most satisfactory manner, that all finite beings have made their appearance successively and at long intervals, and that each kind of organized beings has existed for a definite period of time in past ages, and that those now living are of comparatively recent origin.

60. Ibid., p. 59:

> The more I learn about the subject, the more am I struck with the similarity in the very movements, the general habits, and even

in the intonation of the voices of animals belonging to the same family; that is to say, between animals agreeing in the main in form, size, structure, and mode of development.

61. Ibid., pp. 59–60:

Who can watch the Sunfish (Pomotis vulgaris) hovering over its eggs and protecting them for weeks, or the Catfish (Pimelodus catus) move about with its young, like a hen with her brood, without remaining satisfied that the feeling which prompts them in these acts is of the same kind as that which attaches the Cow to her suckling, or the child to its mother?

62. Ibid., p. 60:

And yet to ascertain the character of all these faculties there is but one road, the study of the habits of animals, and a comparison between them and the earlier stages of development of Man. I confess I could not say in what the mental faculties of a child differ from those of a young Chimpanzee.

63. Ibid., pp. 63–64:

The successive generations of any animal or plant cannot stand, as far as their origin is concerned, in any causal relation to physical agents, if these agents have not the power of delegating their own action to the full extent to which they have already been productive in the first appearance of these beings; for it is a physical law that the resultant is equal to the forces applied. If any new being has ever been produced by such agencies, how could the successive generations enter, at the time of their birth, into the same relations to these agents, as their ancestors, if these beings had not in themselves the faculty of sustaining their character, in spite of these agents?

64. Ibid., p. 64:

When animals fight with one another, when they associate for a common purpose, when they warn one another in danger, when they come to the rescue of one another, when they display pain or joy, they manifest impulses of the same kind as are considered among the moral attributes of man. The range of their passions is even as extensive as that of the human mind, and I am at a loss to perceive a difference of kind between them, however much they may differ in degree and in the manner in which they are expressed.

65. At p. 65 in his "Essay on Classification," Agassiz has a long

footnote on the psychological resemblances between man and the animals.

66. Ibid., p. 66: "Most of the arguments of philosophy in favor of the immortality of man apply equally to the permanency of this principle in other living beings."

67. This is a reference to p. 91 of the Journal and therefore to Agassiz, "Essay on Classification," p. 66.

68. John Milton, *Paradise Lost,* book 2, line 910:

> This wild abyss
> The womb of nature and perhaps her grave.

69. Here Lyell reverts again to the discussion of progressive development left off on p. 95 of the Journal. See p. 180.

70. Lyell here seems to be referring to a statement by Agassiz, Sketch of the "Natural Provinces of the Animal World . . .":

> Whether the dog be a species by itself, its varieties derived from several species which have completely amalgamated, or be it descended from the wolf, the fox, or the jackal, every theory must limit its natural range to the European world.

Samuel George Morton (1799–1851), physician and naturalist of Philadelphia, studied the natural history of man, particularly from the standpoint of the shape of the skull in different races.

71. Franz Herman Troeschel (1810–82), German zoologist.

72. In 1848, many beautifully preserved fossil plants had been found in beds of white sand of upper Cretaceous age, near Aix-la-Châpelle in France. They then represented the only land flora, older than the Eocene, yet known. See Charles Lyell, *Elements of Geology,* 6th ed., London, Murray, 794 pp., 1865, pp. 330–33, and Matthias Hubert Debey, "Übersicht der urweltlichen Pflanzen des Kreidegebirges überhaupt und der Aachener Kreideschichten insbesondere," *Rheinl. Westphal. Verhand.,* 1848, pp. 113–25.

73. *Avatar,* a term arising from a Hindu myth concerning the descent from heaven of a deity to the earth in an incarnate form.

74. "The loves of the plants" is from Erasmus Darwin (1731–1802), *The botanic garden; a poem in two parts,* vol. 1. *The economy of vegetation,* vol. 2. *The loves of the plants,* 2 vols., London, Johnson, 1790–91.

75. ". . . as regards the varying forms of life which this planet has witnessed, there has been 'an advance and progress in the

main,'" Richard Owen, "Address" [to the meeting of the British Association at Leeds, September 1858], *Brit. Assoc. Repts.,* 1858, p. 1.

76. In September 1858, Lyell had traveled overland to Italy and thence to Sicily, where he spent some time studying the geology of Etna. He was now returning to England by sea.

77. Charles Moore (1815–81), formerly a bookseller, at this time (1858) an amateur geologist of Bath.

78. In 1844, the first casts of scales of an unknown fossil were found in a quarry at Lossiemouth near Elgin in Morayshire, Scotland. Specimens were sent by Patrick Duff of Elgin to Alexander Robertson of Edinburgh. Robertson had drawings made and sent them to Louis Agassiz, who considered the remains those of a fossil fish and named it *Stagonolepis Robertsoni.* In 1851, another fossil found at Elgin was identified by Gideon Mantell as that of a reptile, and since the fossil was found in Old Red Sandstone strata of Devonian age, it constituted the earliest remains of a reptile yet discovered (Gideon Mantell, "Description of the Telerpeton Elginense, a fossil reptile recently discovered in the Old Red Sandstone of Moray, with observations on supposed fossil ova of Batrachians in the lower Devonian strata of Forfarshire," *Quart. J. Geol. Soc.,* 1852, *8,* 100–9). Later, additional specimens of the first fossil, *Stagonolepis,* were sent to London and were identified in the autumn of 1858 by Thomas H. Huxley as also being the remains of a reptile (Thomas H. Huxley, "On the *Stagonolepis Robertsoni* of the Elgin Sandstone; and on the Footmarks in the Sandstones of Cummingstone," *Phil. Mag.,* 1858, ser. 4., *17,* 75–77).

79. Reginald Stuart Poole, ed., *The Genesis of the Earth and of Man: a critical examination of passages in the Hebrew and Greek Scriptures chiefly with a view to the solution of the question whether the varieties of the human species be of more than one origin. With a supplementary Compendium of Physical, Chronological, Historical and Philosophical Observations relating to Ethnology.* Edinburgh, Black, 1856, 234 pp.

80. Richard Owen, "Address," *British Assoc. Rept.,* 1858, pp. xlix–cx.

81. Ibid., p. lviii:

> To hide from the lightning and tremble at the thunder, as the immediate manifestation of offended Deity, is the superstition of the savage: to recognize that both phenomena are under the con-

trol of a law, and operating to beneficial ends, is the privilege of the sage. This it is which begets a true and worthy feeling of reverence for the Lawgiver.

82. Ibid., pp. lxxxiv–lxxxv:

The two species of Orang (*Pithecus*) are confined to Borneo and Sumatra; the two species of Chimpanzee (*Troglodytes*) are limited to an intertropical tract of the western part of Africa. They appear to be inexorably bound by climatal influences regulating the assemblage of certain trees and the production of certain fruits . . . By what metamorphoses, we may ask, has the alleged humanized Chimpanzee or Orang been brought to endure all climates? The advocates of 'transmutation' have failed to explain them.

83. Ibid., p. lxxxix:

In the miocene and eocene tertiary deposits, marsupial fossils of the American genus *Didelphys* have been found, both in France and England; and they are associated with Tapirs like that of America. In a more ancient geological period, remains of marsupials, some insectivorous, as *Spalacotherium* and *Triconodon,* others with teeth like the peculiar premolars in the Australian genus *Hypsiprymnus* have been found in the upper oolite of the Isle Purbeck.*

The fossil bones had been collected initially by the Rev. W. R. Brodie, but in larger numbers over a prolonged period of time by Samuel H. Beckles with Lyell's personal encouragement. The specimens were sent to London where they were described by Dr. Hugh Falconer and assigned to a new genus *Plagiaulax.* Falconer also drew the comparison between the *Plagiaulax* teeth and those of the *Hypsiprymnus* or kangaroo rats of Australia (Hugh Falconer, "Description of Two Species of the Fossil Mammalian Genus Plagiaulax from Purbeck," *Quart. J. Geol. Soc.,* 1857, *13,* 261–82, p. 263). It was perhaps natural, therefore, that Lyell should be surprised that Owen failed to mention Falconer's work in this connection.

84. Owen argued that geology has revealed many changes in the distribution of land and sea during the past history of the earth. Therefore, areas that are now islands may previously have been connected to continents. "These phenomena," he wrote, "shake our confidence in the conclusion that the Apteryx of New Zealand and

* These fossils are due to the researches of Messrs. Brodie and Beckles.

the Red-grouse of England were distinct creations in and for those islands respectively" (ibid., p. xc).

85. "Always, also, it may be well to bear in mind that by the word 'creation' the zoologist means 'a process he knows not what' " (ibid.).

86. Ibid.:

> When, therefore, the present peculiar relation of the Red-grouse (*Tetrao scoticus*) to Britain and Ireland . . . is enumerated by the zoologist as evidence of a distinct creation of the bird in and for such islands, he chiefly expresses that he knows not how the Red-grouse came to be there, and there exclusively.

87. Ibid.:

> And this analysis of the real meaning of the phrase 'distinct creation' has led me to suggest whether, in aiming to define the primary zoological provinces of the globe, we may not be trenching upon a province of knowledge beyond our present capacities.

88. Ibid., p. xci.

89. Owen argued that the geographical distribution of races of men does not correspond to zoological provinces as Louis Agassiz had said they did (ibid., p. xciii).

90. Owen referred to Leonard Horner's use of the rate of deposition of mud in the valley of the Nile as a measure of past time (ibid., p. xciv). Cf. Leonard Horner, "An account of some recent researches near Cairo, undertaken with the view of throwing light upon the Geological History of the alluvial land of Egypt," *Phil. Trans. Roy. Soc. London,* 1855, *145,* 105–38; 1858, *148,* 53–92.

91. Owen here referred to Max Müller's comparative studies of languages. Cf. Max Müller, "Comparative Mythology," *Oxford Essays, contributed by Members of the University,* London, Parker, 1856, 311 pp., pp. 1–87.

92. John Tyndall (1820–93), physicist, had been appointed professor of natural philosophy at the Royal Institution. In 1867, he succeeded Michael Faraday as superintendent of the Royal Institution. The Rev. Foster Barham Zincke (1817–93), antiquary, a native of Jamaica, had been educated at Oxford and was at this time Vicar of Wherstead.

Scientific Journal
No. IV

February 8, 1859

Falconer on Tertiary Proboscideans

The Elephant type goes back as far as the Mastodon for Euelephas hysudricus of Sewalik accompanies the old form Chalicotherium & its cranium is constructed like that of existing Indian Elephant.

The Crag Elephants associated with Hippopot.^s major inhabited, on the land, a warm climate altho' the shell evidence implies that the seas of the same era were getting colder.

[p. 2] No reason to conclude that the brain or intelligence of E. hysudricus was inferior to that of living Asiatic Elephant, so no progress since Miocene era.

Nor ought we on negative evidence to conclude that Proboscideans went back no farther than the Miocene Period.

"Same Elephants and proboscidea in Older & Newer Pliocene strata the change only coming in the Glacial or post-pliocene or Quaternary Beds."

See abstract made by Dr. F[alconer] for Portlock's President's speech for 1858.[1]

Lower Silurian Fish

See Falconer's letter; M. S. Silurian in, qy. if Pander's are teeth of mollusca.[2]

Feby 8, 1858.
Drawer of Secretary[3]

Antiquity of Birds [p. 3]

Mr. Barrett[4] tells me that lately (May 12, 1858) the tarsal & other wing & leg bones of a bird about the size of a raven were found in coprolite in the upper Green Sand of Cambridgeshire. The first instance this of birds (proved by their bones) as old as icthyosaurs & pterodactyls—see B's letter, May 18/58

Carboniferous Brachiopods

126 British species (May 12, 1858) reduced by Davidson[5] to 25 sp. & none common to Devonian & Carbonif.[s] whereas McCoy[6] thought that about 1/3.[d] were so. Spines rubbed off & other imperfections cause species to be multiplied.

May 12, 1858

Brighton Old Beach

The old beach if base of Eleph.[t] bed traced by Prestwich to entrance of Goodwood Park 100 f.[t] above the sea level & 7 m. inland.

Silurian Dicotyledonous Wood [p. 4]

J. Hooker finds that some wood sent by Sir W. Logan[7] from the Silurian strata of Canada is dicotyledonous & not gymnospermous.

Kew, May 19, 1858

Permian

[Permian] of U.S., Hayden & Meek's papers—[8]
All the European format.[s] now found in N. American continent: to be inserted in new Edit.[n]

Hippopotamus

[Hippopotamus] in Brighton, Elephant bed with E. primigenius! Falconer suggests that this being a tumultuous deposit the hippopotamus may have been washed in from older deposits like the Eocene & Miocene bones in the Crag—but Prestwich suggests that the mammoth and hippopot. of Brighton coexisted.

January 23, 1859 [p. 5]

Progressive Development

Huxley, *Builder*[9]

Sir H. Holland,[10] *Edinb. Review.*

If in future we think new races of man & animals will appear gradually so in the past they may have done so without separate creations.

Monads

Are the germs of all that is to be developed from them. Their creation like that of the first acorn, the progenitor of forest of oaks.

Baden Powell,[11] *Unity of Worlds.*

February 2, 1859

Men of genius may be born of parents who tho' unknown may have had their soul's genial current frozen by chill penury & the page of knowledge never opened to them.

February 4, [1859]

Glacial Period

Bronn[12] infers constant improvement of world—but Glacial Period scarcely reconcilable with this & the more strange as [p. 6] the anthropomorphous class inhabiting the tropics & a cosmopolite species might rather have been expected from any other class as the dominant one.

February 5, 1859

Progressive Development

The results must be looked in the face if the Lamarckian system be adopted.

1.st The reality of species must be abandoned.
2.dly No creation for many millions of years has taken place on the earth.
3.dly There is no human species historically considered, as like all others it must blend insensibly, if genealogically followed up retrospectively, into some preexisting inferior creature, its lineal progenitor.
4.^{thly} Of the races of Man the inferior must have preceded the higher & more intelligent.
5.th No line of demarcation between [p. 7] rational & irrational, responsible & irresponsible.
6. Unless all creatures are to have a future state, where are we to draw the line? This may be only the same difficulty as between born idiots & sane children, but it vastly enlarges the circle of difficulties.

Popularity of Development

1. The only generalisation wh. can embrace Man as last formed & preexisting animals.
2. Genesis—Man came last.
3. The facts, especially Man coming last, always assuming him as an integral part of the same chain or cycle of changes—& metamorphoses.

Development

If admitted, the lowest races, equally capable as they often are to maintain their ground, must have come first [p. 8] & it is not likely that the most degraded of these survive at present also other anthropomorphs of higher organiz.^{n}—than any now living w.^{d} according to the doctrine of chances have flourished & become extinct. Hence there is but little difference between the out-&-out progressionist & Lamarck, for in the one case some unknown *modus operandi* called creation is introd.[d] & admitted to be governed by a law causing progressive develop.[t] & by the other an extension or multiplication by Time of the variety-making power is adopted instead of the unknown process called Creation.

[p. 9] It is the theory of a regular series of progressively improved beings ending with Man as part of the same, which is the truly startling conclusion destined, if established, to overturn & subvert receiv.[d] theolog.[l] dogmas & philosoph.[l] reveries quite as much as Transmutation.

Progressionists

The result to which they are drifting is one too serious to all to be lightly treated. Had what is called in derision the uniformitarian view proved true, the coming in of Man might have been viewed apart so far as his Intellectual & Moral Nature is concerned, but [p. 10] when (with or without transmutation which is simply a guess at the definition of a mystery called by us Creation) it is stated that Man is last of a series of Developments from the fish to the anthropoid apes & when some consider that separate races of Man were separately created, when it is held to be the crowning operation—when the intellect and nature of Man is said only to differ in degree from that of the higher Brutes & consequently less, the lower the race compared —we embrace opinions so closely allied to those which call down unpopularity on the Transmutationist that there [p. 11] seems less to choose between the rival hypotheses than is usually imagined.

The inferiority of the skull of the Baboon to the Chimpanzee and Gorilla, with superior intelligence in the latter, favours the idea that the negro is next in the scale with many lost links. The highest of them w.[d] first be persecuted & expelled.

Dryopithecus

Owen regards it as Man to the long-armed ape, Hylobates (or gibbon?) & ergo less advanc.[d] than the Gorilla—only in stature large.

But why assume that this creature represents the Miocene Quadrumana?

February 5, 1859 [p. 12]

With Huxley: Agassiz on Classification[13]

Prophetic types—Pterodactyls are more proph[eti].[c] of bats than of Birds.[14]

Ichthyosaurus not proph[eti].[c] of dolphins.

Sauroid fish now represented by gar fish or —— & therefore an unfulfilled prophecy.

Echinoderms

[Echinoderms] not true radiata, nor in same Nat. division with corals.

Orthocerata

[Orthocerata] more complex than nautilus.

Persistent Type

The crocodiles among them.

Embryonic

Lingula Davisii not an embryonic Brachiopod

Monomyeria

[Monomyeria] not present in oldest rocks says [p. 13] Agassiz, ergo, says Huxley, the *less* perfect were wanting.

Dryopithecus

It is true says Huxley that the angle of the jaw is like (Hylobates?)

Dicynodon

New spec.[s] & many in Africa

Batrachia

What Agassiz says of progression in the class from Andreae of Miocene to present time is contradicted by the highly develop.[d] —Labyrinthodon being so ancient.

Specific Centres

[. . .] denied by Agassiz, & E. Forbes never cited on this part.

Destruction

[. . .] & new creation of whole sets of species by cosmical catastrophe in Agassiz.

Fish

[. . .]s.[d] to be in all parts of the Silurian.

[p. 14] Notes on Grove's[15] L. C. on *Progressive Development*

Continued from M.S.

p. 10. Question of miracles we are not to assume abnormal or exceptional causes when the normal will account for phenomena of Creation of organized beings from amorphous matter, we know nothing nor any creature of inorganic matter.

11. Hybrids. If the greyhound & bulldog interbreed, they are both called of the same species—why so, if the horse & the ass are not so, for the mule will interbreed with a half mule (Qy.).

[p. 15]

12. If a mule can be merged by interbreeding with one of the perfect stocks, it goes far to disprove species.
13. Darwin's & Wallace's theory of selection of varieties best fitted for new circumstances.
 " Elimination of weaker races.
[13.] Acquired habits transmissible—the pointer. If so much in one life what in a series of generations.
14. N. American Indian more acute sight & hearing than European notwithstanding inferior mental powers. Why then sh.[d] not Giraffe obtain longer neck.
 " Cultivated plants.
 " Cuvier's observation that each organ of an individual bears a relation to a whole. This supports view of indefinite variation [p. 16] as a tree-climbing creature will acquire all the necessary changes.
15. Phases of resemblance to inferior animals, which man passes through, brain of fish, reptile, etc. & so of plants.
16. If a Plesiosaur appeared in your fishpond you w.[d] wonder.
 " How were first beings created? & if a primeval creation of one germ or one pair of beings from which all others spring by progression or retrogression, why not other creations & continual intervention.
17. Metamorphic rocks may have contained fossils.
 " Owen has shown that no one attribute can be abstractedly predicated of the inorganic that is not applicable to the organic & [p. 17] vice versa.
17. Whether a certain number of original types first formed, or in the divine mind from eternity, & then stamped on matter from which there may be variations preserving the generic character of the type.
19. No types really exist, but abstractions of our own; when we classify, we cannot refer our abstract ideas to the Creator. The doctrine of types [is] a concession to that of indefinite variability.
20. Changes greater at periods when the earth altered more & natural groups then formed.

February 12, 1859 [p. 18]

With Sir B. Brodie: Creation[16]

Three improbable modes:

1.st Adam & Eve
2.dly Lamarckian, from Gorilla
3.dly Sudden change of offspring of anthropoid species at time of conception.

Chief objection to this hypothesis [is] that it requires coincidence of both sexes being thus suddenly & as by accident born of unlike parents.

Gorilla

At College of Surgeons the keel on the skull for attachm.t of temporal muscles makes cranium appear very much more different from Man than the Chimpanzee.

Retrogression

The degradation of reptiles since the time of the Deinosaurians may imply that so also Man may not be succeeded by higher races or species.

[p. 19] If a 2.d glacial period causing Europe to Lat. 50°, & N. America to Lat. 40° N., to have a sparse population of animals and plants sh.d recur or if the heat to become unfavourable to the best develop.t of Man, he might be reduced to an inferior state physically.

Cephalopods

We should compare Silurian with Cretaceous & then the progress will be in 6 or more great Geol.l epochs from an Orthoceras to a Belemnite—no soft animals without shells ascertainable in either period.

Human Period

If we take the time from the Trias to the Pliocene Period for the progress from a marsupial to an elephant, & then the passage from the Eleph.[t] [p. 20] or gorilla to Man as 50,000 years, it would be a fraction of a period which has been sufficient for the greater [advance] & 5 or 6 geol.[1] epochs for the less advance. The quantity of progress is as disproportionate to the time as its kind is different.

Moreover it resembles in this respect the total want of correspondence between the relative mental powers of parent & child so that 3 or 4 generations may be conceived to form a passage from the humblest intellect to the greatest genius. This is the greatest saltus, the most decided known breach of continuity in Nature.

For this reason in the *Principles of Geology* I proposed to consider the advent of a moral & intellectual being as belonging to a distinct order of [p. 21] things, but the animal nature to be subject to all the same laws as previously so that it might be like a spectator & not derange the system.

The idea of an improved habitation need not be taken up. The glacial period may have been inferior to the Miocene. The superiority of Man as an animal, take away his intellect, may be doubted.

If Man be part of one continuous system of improvement then whether we adopt the Lamarckian develop.[t] or not we must take all the laws wh. regulate the lower animals as strictly applicable to man, such as inferior races coming first—a gradation in physical organization.

February 15, 1859 [p. 22]

Permian Palm

Geinitz[17] in *Die Leitpflanzen des Permischen* Pl. II, fig. 6, describes & figures a palm altered to Gulielmia of Martius wh. he calls Gulielmites—an example of Monocotyledon of old date.

Creation

Par.[adise] Lost. B. viii (p. 55, vol. 1)

The rib he formed & fashioned with his hands.[18]

The philosopher treats the creation of an adult man from nothing & of a woman from the rib as a tale fit for the credulous multitude akin to their faith who still believe that there is one rib less on the left side owing to the loss sustained by their progenitor in the Garden of Eden when a rib was fashioned into a helpmate of the other sex.

March 7, 1859 [p. 23]

Is Not the Earth[19]

With various living creatures, & the air
replenished—Knowest thou not
Their language & their ways, They also know
and reason not contemptibly—*ib.*

Progressive Development

The same indisposition to believe in the extent of gaps which will be filled up hereafter in the fossil series which causes the Progressionist to see no danger in his creed, makes him shrink from the Transmutationists as one who is running into materialism, while the simple Progressionist is safe.

But both tend to the same goal tho' one of them is not aware of the direction in which he is drifting.

Spirit

What is Spirit, see Macaulay on Milton[20] in *Edinb.*[g] *Rev.*[w]— All spirit of one essence, Channing,[21] cited by Taylor[22] in a sermon—If spiritual & animal instinct are, as Agassiz says, then talent, after [p. 24] skipping a generation in a family may reappear in a grandchild, just as acquired peculiarities in domestic animals & cultivated plants. The inseparability of the human & animal may be established by many arguments.

Proboscidea

Who can doubt that if it had pleased the Creator, a new species of Elephant might have been gifted with a proboscis even more fitted to pick up a needle & pull down a tree than the Elephant now living. The approach in intelligence wh. this last makes may be meant to show that it is not the form wh. is of any essential meaning, but all depends on the union of spirit & the animal nature.

The longevity of the Elephant very great? ask Falconer.

March 13, 1859 [p. 25]

Anthropomorphism

There must be depth of meaning which no one has yet fathomed in the close analogy of the anatomical structure of the Gorilla & other quadrumana & Man. There were in the Eocene Period (Macacus Eocenus) such forms on the Earth before the Alps were. Millions of years ago the circulation of the blood & the skeleton like that of Man [were present] on this planet. The coexistence now of such forms with man & that too in animals, not superior in intelligence to the Elephant or Dog, may mean that the anthropomorphic structure is not of itself significant of any intellectual or moral or spiritual superiority, but that we are united to such a form as a link with the pre-existent & long established physical systems of the planet.

If every act of thinking even in a [p. 26] reverie be so far material that one is the worse for it, if there be a loss of oxygen, if as Owen says (lecture on fishes, Jermyn S.[t]) there is something burnt, & repaired by food & the digestive organs be required for the waste caused by so exercizing the brain, there is a mysteriously inseparable connexion between the soul & body & the question is how & when it began on the earth?

Anthropomorphs

If we had known no living quadrumana, or none of the tail-less apes, or of the Orangs, the discovery of the first fossil Chimpanzee or Gorilla or Gibbon would have created a profound sensation among naturalists.

They would have inferred in all likelihood that the creature bore the same relation to an extinct fauna that Man bears [p. 27] now to the living one, or if not so high that at least the fossil indicated by its affinity to Man in stature & structure, in intelligence far beyond that of any Proboscidean. The absence of fabricated articles & tools or buried towns in the same strata might be advanced to prove that the extinct creature had remained in a much ruder state than even the least civilized savages whose axes, or canoes or arrow-heads we find in strata. But a creature with precisely the same number of bones in the skeleton, just 2 neither more nor less, w.[d] naturally be supposed to approximate [more] nearly to Homo sapiens L.—than a dog or an Elephant. But if in some unexplored part of Africa or Borneo, a Chimpanzee or Gorilla was then discov.[d] living, an entire change w.[d] be produced in opinion & it w.[d] be admitted that the near proximity [p. 28] of osteological structure was no test of a corresponding intelligence or moral nature or spiritual. That there is some deep meaning in the similarity may be granted but it may be not that wh. the progressionists desire but the opposite, it may be to show that the structure does not determine the amount of brain.

Instincts

If it be so difficult to draw the line & say where instinct ends & reason begins whether in the passage from the infant to childhood, or from the lower animals to Man, we must expect to see these perplexities reappear when we enlarge our view of Nature. Many have been driven to the alternative that brutes have souls, rather than pretend to be able to draw a line of demarcation. Then comes the passage from Gyrencephalous to Lyencephalous & thence to the Cephalopod & coral.

Negative Evidence [p. 29]

If we find many Icthyosaurs in Lias or Proboscideans in Eocene (Paleotheria) there is always some species of these genera in each format.[n] only just known or of which a small portion of one individual alone has been discov.[d] amidst heaps of the commoner kinds. Scarcity like this in older rocks w.[d] imply impossibility of proving their former existence. Yet comparative scarcity may be argued provided we do not overlook cases like the Flysch. It is when we have no tertiary stratum as full of shells, crustacea & invertebrata, as the Lower Silurian & yet no vertebrata that we get a strong case in favour of scarcity at least of Low.[r] Silurian fish.

Belgium supplies a case of Lower Miocene & Eocene formations rich in invertebrata & without a single Cetacean or mammifer as yet found fossil—But there are fish. It is a strong case in favour of future possibility of carbonif.[s] mammals.

Owen's Lecture on Fish, March 17, 1859 [23] [p. 30]

Silurian Onchus

Sharks are the most perfect of fish. Therefore Nature began with the creation of as highly develop.[d] fishes in Upper Silurian, as now live. See Murchison new (3.[d]) Ed. of *Siluria.*[24]

How is this reconcilable with progressive develop.[t]

It is indeed said that no vertebrata in Older Siluria.

Conodonts

. . . may have been teeth of naked mollusca. Qy. why not of cuttlefish which left no other memorials?

Owen said that the defensive teeth of Onchus seem to imply that they had some other bigger enemies of the tribe of fish [p. 31] Mr. Perley of N. Brunswick told me that other fish w.[d] fly or be eaten up bones & all by the Onchus.[25]

Cestraceonts of chalk very big, more so than living Australian —wh. nearest representative—as usual antipodal species most like sharks.

Hybodus, a shark—no airbladder in sharks.

[May 18, 1859]

With Mr. Taylor[26]

The religious instinct or sentiment exhibited in the forms & dogmas & creeds of all nations, the spiritual affections apart from the mere reasoning of the intellect, are truths & deep realities of the soul which science can neither create nor destroy. They are as much facts, as positive [p. 32] as any material phenomena. The abstract being, to whom merely negative attributes can be given, does not satisfy this instinct & cannot be the whole of the immaterial world with which these instincts, these mysterious feelings, are connected.

The finite intellect cannot discover the relations, nor expect to understand them, between the moral & reasoning or purely intellectual part of our Nature. But no future progress of science will lessen this however much it may disturb old forms of belief. If Man c.[d] be shown to come gradually out of the almost endless or beginningless past, it w.[d] not destroy those inner convictions of the heart & soul.

If there were an insensible passage from a being without to one with a soul, it w.[d] still be the same—merely a mystery added to many before existing.

Creation [p. 33]

Whatever be its laws or modes of operation, the truth if known, must be as marvellous as any one of the 2 or 3 hypotheses hitherto made. If being with a soul was produced by one without a soul—if a rational soul came out of nothing—or if one individual being made, a rib was fashioned into a helpmate of the other sex, whatever conjecture be made, the real truth, if different, must be at least as marvellous. It has been developed in a moment of geological time.

They who shrink with horror from the consequences of transmutation assume as astonishing a link between a series of successive appearances leading from mere instinct to a semblance of reasoning & from that to a rational & responsible being. They allow the fact of Man coming last to prepossess them [p. 34] in favour of receiving the progressive system in geology & that too in spite of having frequently to correct & alter their inferences. If true, it leaves little to choose between them & the transmutationist so far as contradicting established opinions.

Owen, Lecture, March 25 [1859] [27]

The Ray family

But for a single ichthyodorulite in the Devonian the Ray w.d not be known before the Lias.

This shows the danger of negative evidence.

The sharks were Cestraceonts (Pleuracanths) in the early periods when they had ganoids to crush with their pounding teeth, but when (after the Lias?) the cycloid fish & unarmed or those without dermal armour came in, the sharks with sharp [p. 35] teeth for tearing came in & the Cestraceonts were reduced to the small Australian living species.

No progression, says Owen, indicated by the Sharks but succession of very distinct types.

Now that the ganoids are few the Cestraceonts are few. [See Fig. 8.]

Fig. 8

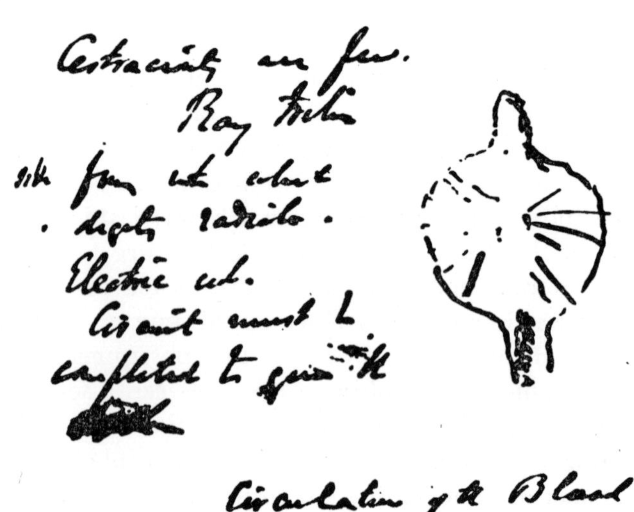

Huxley says is true of all mammalia not merely of Quadrumana, Bell's (Sir C.) discovery of the two kinds of nerves of motion & sensation—also true of all animals (vertebrate) except a [p. 36] few fishes.

An intelligent being therefore might in the oldest geol.[1] periods have made Harvey's & Bell's &, in short, all anatomical discoveries.

Quadrumana

Dryopithecus argument not good as far as angle of jaw goes. Same number of bones in orangs as in Man & of teeth.

Specialized Forms

The Pterodactyl, says Huxley, commenting on this doctrine of Owen's is of all creatures, tho' long extinct & going back to the Trias, an example of the most special adaptations of certain reptilian organs & departing in this way from the archetype.

Owen, on Fish, March 31, 1859 [28] [p. 37]

Myxine & Lamprey, Cyclostoma, w.[d] leave no fossil remains because cartilag.[s] but might leave horny teeth possibly.

In Paleozoic rocks therefore the highest organiz.[d] without the lowest.

So the lower might in mollusca be without the highest cephalopods if the latter were without hard parts.

Pterichthys

—is to supersede Asterolepis—the pectoral fins are armed. [See Fig. 9.]

These Devonian fish, Cephalaspis, Caecosteus, Pterichthys, all of them [p. 38] with dermal skeletons & internally cartilaginous.

Chaenidra

—allied to Cestraceonts in many points. Australian species now called Call [?]—passage of teeth into bones of jaws.

Fig. 9

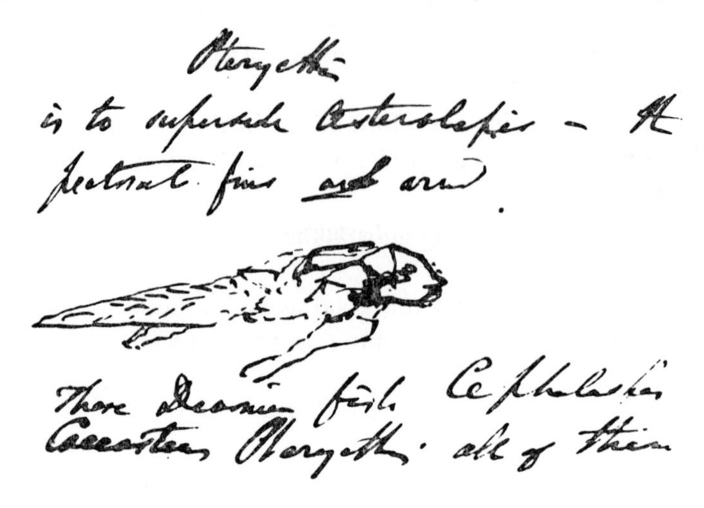

Owen's Lecture, April 17, 1859 [29]

Some of the Old Red fish which have such strong external skeletons had also a bony internal [skeleton]. The Coelacanths had a bony coat to an internal cylinder of gristle.

Nile Deposits

Mr. S. Sharpe objects to Mr. Horner that when the Statue of Rameses was created at Memphis, the city was surrounded by a wall & protected [p. 39] by embankments & that when New Cairo was erected 1000 years after Christ the floods first covered the site of Memphis.[30]

No deposit took place therefore till 850 years ago. Instead of 8 foot in 4,000 yrs. or 3½ inches in a century, the deposit has been much faster.

[S] Sharpe takes 2000 instead of 3000 yrs. B.C. for age of Memphis. The measure therefore is altogether fallacious.

So at Heliopolis the obelisk was opposite a temple wh. must have been protected from the inundation by embankments so long as the city was inhabited. Cities inundated w.[d] be unhealthy.

New Orleans [p. 40]

Horner has assumed that from the moment when the old monument was built the deposit went on as before the building when the country was open & not built upon.

This is an error.

Moreover the local currents by which the deposit is rendered irregular w.[d] vary after artificial embankments & after buildings —a new state of things & of disturbing causes are introduced.

Plato

So far as the soul of Man is connected with perception, it is mortal—it died with the body of the percipient [being]. But as the individual body after death unites itself with the great body [p. 41] of the Universe from which it sprung, so also the soul, so far as it is not represented by the bodily perceptions, returns to the great soul of the world of which it is an emanation & remains undestroyed & indestructible.

Plato's views of the immortality of the soul developed in the Phaedo. *Penny Cyclop.*, Plato, vol. 18, p. 241, column 1.[31]

Metempsychosis

The passage or transmigration of souls. It was a Hindoo doctrine that the souls of men pass after death into different bodies either of man or animals unless an individual has lived a holy life in which case his soul is absorbed into the Divine Essence. In this case souls pass like rivers into the great Ocean being reunited with the Supreme Being. 'His vital faculties [p. 42] and the elements of which his body consist are absorbed completely & absolutely; both [name] & form cease & he becomes immortal without parts or members'—Colebroke's extracts from the Brahma-sutras, Trans. of Roy. Asi. Soc.[y] vol. II—[32]

Egyptian metempsychoses Herod.[otus]. vol. 2—123.[33]

Progress

The D. of Argyll says, lecture p. 28, 1859, *Geology its past & present*,[34] "I own that I regard the universally-admitted fact that the creation of Man was the last & latest work, as one wh. almost leads us to expect as probable, an analogous rule in the preceding history of the world."

[p. 43] ib. p. 32. Before the Human Period the agencies of change differed in degree from those since exerted.[35]

p. 27. Man being latest a "telling fact." [36]

p. 35. Climate was warmer.[37]

[p.]36. Glacial Period.[38]

[p.]39. Herculaneum buried by lava.[39]

If Man being the latest creation is the last of a connected series of progressive changes, beginning with the Invertebrata & leading thence to the lowest of the Vertebrata, & from there to the gyrencephalous mammalia from the most intelligent & last created of which as a crowning operation the step was made to Man—if all these were links in one & the same chain of analogous events then indeed our old opinion of the relations [p. 44] of Man to the inferior animals must undergo a total reformation. Our anciently cherished ideas are subverted much in the same way as by the transmutationists or Lamarckians. In both systems the difference of brute & human intelligence, that of the Elephant & of the Australian savage, are mere questions of degree not of kind—the mode by which the succession is brought about is different nominally, rather than really, in the one case by a conjectural exaggeration or multiplication in time, of the law which admits of variation in the offspring of the same parents, an indefinite [p. 45] & almost insensible deviation in the course of successive generations from the parent stock, in the more orthodox system by successive 'creation,' the species next in the order of Time resembling each other more nearly than those farther remote. The modus operandi of creation being in the theory confessedly unknown, mysterious unsearchable perhaps, but at any rate hitherto undiscovered—yet being governed by laws & one law being that of continual progress towards more complex organization & higher intelligence—In its consequences, so far as they are subversive of old received doctrines of the wide distinction between rational & irrational brutes [p. 46] & human, these two theories of transmutation & progression approach far more near than the progressionists are willing to confess, or desirous of stating.

Malta

Cave with Rhinoceros bones in it. This confirms the Gozo ? elephant of Smith of Jordanhill.[40] Lord Ducie[41] told S.[mith] of the cave.

Prestwich on Boucher[42]

[Prestwich] saw B. de Perthes' axes & flint knives in gravel which has furnished El.[ephas] primigenius & Rh.[inoceros] tichorinus. In Brit. Mus.[m] similar tools f.[d] 100 years ago with Eleph.[t] bones.

Manual, New Edition[43]

Excavation of Valleys—subaerial chiefly seen when rising out of the sea & then the sea acts [p. 47] but submarine currents level, as they operate on highest portion with most power. The quantity of sediment in deltas is part only of what is borne down & yet if restored w.[d] fill up great valleys.

Progression

Assuming the intelligence of the inferior animals to differ only in degree, there is no assignable reason why the Elephantine family sh.[d] not have become the dominant one or given rise to a ruling & rational species. The gap between Man & the Gorilla is as great as between Man & the Elephant in all that regards the intellectual & the moral world. See Kirby & Spence on instinct.[44]

Agassiz

—has admitted frankly that 1.st the different races differ in degree of intelligence & were created from [p. 48] independent original stocks.

May 20, 1859

With Huxley

Cestraceonts new in Australia—do not feed, he thinks, on fish—no ganoids there, perhaps they feed on crustacea.

No proof that the old icthyodorulitic fish fed on ganoids—as likely that the ganoids fed on them. Perhaps the icthyodorulites do not imply such big fish. Lateral or medial line [is] a sensorial not a lubricating cord, which runs thru the medial line (or lateral) in [p. 49] fish, a German, Leidig[45] discov.[d] this some 10 years ago—Agassiz & Owen mistaken.

Progressive Development

—whether with or without transmutation [is] the same thing. Huxley believes in it but suppose that most of the progress was made in times anterior to our oldest fossilif.[s] rocks.

Archegosaurus

The proof given by H. v. Meyer in favour of absence of an ossified spinal vertebral column is strong in favour of the small develop.[t] of Carbonif.[s] reptiles.[46]

May 21, 1859 [p. 50]

Syngnathus Brevicollis

4 specimens in Aquarium, Zool. Gard.[s]

Prehensile tails—after swimming by aid of pectoral fins which progressed upwards beyond the neck (being as normal behind the gill opening) & by aid of the dorsal fin, which quivers rapidly, they are arriving at an upright stick coil, their tails round it & support the heads upright.

If only fossil, how little idea of their singular habits, could one have—the pipe-fish family.

Transmutation

—may agree with retrogression as well as progressive developm.^{t} & in that has the advantage of the progressionist.

Grove[47] thinks the cerebral developm.^{t} may have been accompan.^{d} by muscular & other degradation.

Rudimentary Organs

All of these, says Grove, may perform [p. 51] some function tho' not that wh., when fully developed, they were intended or used to perform.

No scientific mind can imagine an Eleph.^{t} adult coming at once out of nothing.

Archetype

That of Owen implies development & transmutation—not in the divine mind & disembodied but in a corporeal state.

Transmutation

An individual has had thousands, perhaps millions of ancestors; out of these may come the peculiarities of children born of [the] same two parents. Each child may represent some one of the ancestors. So if species crosses have taken place & old characters come out [p. 52] again. A constant adaptation of parts to circumst.^{s}. The arms of a blacksmith get develop.^{d} but at the expense of some other organ, so intellectual develop.^{t} is at cost of physical.

The equality of the Greeks & moderns is because 3000 yrs. [are] nothing but there may have been inferior races before.

The scheme [is] too vast for us to know or take in all the details.

Retrogression

The Reptiles indicate retrogression from the time of their greatest or culminating development. Fish, ganoid & sauroidal, may be adduced to prove that the common type of reptile & fish were more united & that the osseous fish on the one hand & the reptiles on the other were more specialized afterwards.

The Labyrinthodon [p. 53]

The archegosaurus, as the oldest known reptile, implies a simpler type than any known creature of that class for none living have the continuous notochord—more embryonic than any living. H. v. Meyer to be cited for this discovery.[48]

Man

If the antiquity of Man be as great as the glacial period, still it remains but an insignificant part of our & the last geological epoch. Great changes of geography have occurred with the glacial period, but it has been justly considered as a mere episode in the Pliocene tertiary era. Hence if we assign to Man the last step in a progressive system, we must admit that no similar jump can be found in antecedent eras. We have not yet discovered any fossil species which can be consid.[d] as in any way intermediate between Man [p. 54] & the Orang family. On the other hand the intelligence of the Elephant or Dog may well be consid.[d] as equal to any known anthropomorph, & therefore some extinct proboscidean of Older Pliocene or Miocene age may & probably did come as near to man as the Gorilla. The transmutationist ought to have been able to prove that in proportion as the anthropomorphous genera approximate physically to Man they are more intellectual & that the inferior Races of Man come nearer to Orangs physically.

The progressionist has no right to object to the transmutationist that the notion of advance by generation has any peculiar difficulty to contend with in the case of cranial development or mental power if it be once conceded, as by Agassiz, that the distinction [p. 55] between Man & the Brutes is one of degree merely. For it so happens that the inferiority or superiority of Parent & Child in mental gifts may be such that, if it c.[d] be expressed in figures, it w.[d] far exceed any contrast in stature, strength, agility or other corporeal gifts which c.[d] be found between parents & their offspring. The difference between a Man of the highest genius & his parents is probably the greatest saltus in Nature—go back at this rate of diminution a few steps & you may arrive at a race as low as to compare with the irrational & soulless beings, which might be self supporting nevertheless & strong. Which ever way endeavour to conceive the passage from brute to Man, whether by some unknown law such as [p. 56] we express by the term creation or by generation, it may be supposed even by the theory of the transmutationist to have been sufficiently sudden as to amount to a transition at once from irresponsible to responsible, from ignorance of the difference of right & wrong (if dogs have not that already) to a knowledge of it.

May 27 [1859]

With Huxley

No ganoids in the Southern Hemisphere—Qy. as to Amia.

8,000 fish now known, suppose half in S. Hemisphere & no ganoids, this w.d resemble no cycloid or ctenoid in N. Hemisph. in palaeozoic & old mesozoic ages.

Amia (see Pictet)[49] is ganoid in affinities but without ganoid scales & w.d not be a ganoid of Agassiz.

Wealden [p. 57]

In Morris' Catalogue[50] only 19 Wealden reptiles & a footprint. Say that there may be in N. Germany & England 25 Wealden reptiles exclusive of the Purbeck. Probably there were more species then in the world than there are now. But we know ______ reptiles living.

It was not till about 35 species of Reptiles had been found in the upper green sand that we obtained a knowledge of the existence of one Cetacean.

We must therefore be cautious remembering that not far above is the placental & next below swarms of marsupial mammalia.

Species

The varieties of Helix polymorpha, says Mr. Lowe, in the several isles of the Madeira group may well be called semi-species.

Flysch [p. 58]

Azoic Period—Subazoic era—Alpine scarcity of org.[c] remains in Eocene times—probably period of great richness in animal & vegetable life.

Owen on geograph.[c] Distrib.[n] of Quadrupeds, see Art. Paleontology—p. 172–5, gives abstract of it.[51]

Plagiaulax

—[is] carnivorous like Thylacoleo—Owen, *Ibid.* [p.]175. Ask Falconer—qy. if at all grooved in teeth.[52]

Quadrumana & Proboscidea

Why sh.[d] not the Elephants have culminated in some being higher than Man. Unless some species are yet undeveloped they have been arrested & a saltus occurred unless some species are yet undiscovered in a fossil state between the Gorilla & Man.

[p. 59] It may be said that the mystery of the link between Man & the Inferior animals cannot be cleared up. But if once it becomes part of a chain of analogous changes & is reasoned upon as a step in the order of development we can no longer treat of it apart. It is an integral part of our system—as much in one scheme as in another.

There is something suspicious in our theory when we find that it is the airbreathing inhabitants of the Land which form the exception, that insects, ophidians, mammalia, birds are wanting in the oldest rocks. Cetacea, it is true, have not been traced back so far as marsupials as yet. If we adopt the same method in Vertebrates as in Invertebrates we must take Mammalia, Aves, etc. separately.

Progression [p. 60]

The argument of Owen is this:

> There are great gaps—& new & higher organizations coming in.
>
> These gaps indicate leaps in Nature—or jumps.
>
> It is so improbable that they will ever be filled up that we may suppose Man to have had no intermediate between him & lower animals.
>
> It is more probable to assume that each type came in in the same manner as the first did in the origin of things than to suppose an indefinite departure from a parent stock.
>
> The progress of geology will always be lessening the distances but they will remain.

We may assume some other law of development, similar to that, says Agassiz, of embryonic develop.[t] not unlike it in having breaks by which the independence & reality of species are caused.

[p. 61] Thus a law of progressive development is believed in leading up from the inferior animals to man, & accord.[g] to some high authorities very analogous to that of embryonic development. Man is the culminating point reached by this mechanism or law, but there are too many breaks such as can never be expected to be filled up to allow us to suppose that even in indefinite time one species can pass to another. The allied species must be created by the intervention of the first Cause. Where are the passages say they? The answer is the *disputed species,* fossil & recent—Varieties—Race widely distinct—yet called of the same species.

So long as it could be supposed that the development or succession of species was sometimes progression & sometimes retrogression & not on the whole clearly advancing, we might imagine Man to stand apart from the rest, but the [p. 62] identification of him with an order of ever advancing succession implies such a unity of system as makes it too arbitrary to attempt to draw a line.

Antiquity of Man

The chief bearing on the transmutation theory of the addition of 50,000 years to the era of Man's advent would be not the geological age being changed. Man w.[d] still belong to the yesterday of the Newer Pliocene—the human period might still belong to that [period] of all the shells being of living species. But it would open a wider prospect to the preexistence of numerous extinct races—& give time for the formation of all the present races from one parent stock.

On the other hand the countless undiscov.[d] races of quadrumana may afford lost links. Moreover [p. 63] when we find that the physical geography of the globe has been so much altered since the glacial period—Malta rhinoceros—& Elba & Puggaard,[53] Moen, & upheaval of Jura since the E.[lephas] primigenius & union of England & discovery of same & sinking of Holland & burying of E. primigenius in ice—all these may prepare us to find that Man has inhabited countries now submerged. Cretan cities submerged, Lieut. Spratt.[54] Hence in the East—if Man first inhabited the East & in regions now submerged, human remains may be treasured up.

If monuments of Grecian cities difficult to find & Egyptian, how much more so the remains of rude savages.

Unity of Creation [p. 64]

The uniformity of the law may consist in a gradual development, in the main progression, tho' at times also retrogressive.

The claim of the transmutationist to be a speculation meriting more favour than any other is that it is at present the only one which even pretends to bring the successive changes under a law or within the dominion of science—When we say that it is a mystery, that each new species may be introduced by a fiat or by some power similar to that which was exerted when life was first introduced into the planet we confess to a feeling of despair as [p. 65] to the mode of succession being ever brought within our reach as a scientific subject of investigation.

The transmutationist, if his conjecture be correct, claims to place the appearance from time to time of new kinds of beings in the earth on the same footing & as much a matter of observation as the appearance of new individuals—If we have not yet succeeded in defining what we mean by a permanent variety or race as distinguishable from a species I cannot deny that the last 30 years of study has made the Lamarckian theory more reasonable [and] less visionary, because no definition has been found of a species, except one which begs the question by assuming the invariability, [p. 66] which has to be proved & which the Transmutationist denies. Moreover, they who deny the doctrine of the latter are compelled to admit it so far as concerns the formation of races & the transmission of acquired faculties & new forms.

There are none who so much aid the transmutationist as they who push the doctrine of progressive development farthest—since their assimilation of the successive appearance of species & genera to embryonic develop.[t] in an individual is the very opposite of that arbitrary fiat which at other times they invoke, it is creation working by law, according to a prescribed pattern & by a force analogous to that [p. 67] displayed in an individual from the embryo to the adult.

Again why sh.[d] there be a series of beings most like their antecedents. This indeed might be accounted for by assuming the necessity of each type being of necessity created in harmony not only with all the surrounding conditions of organic & inorganic nature with climate, physical geography—the peculiar coexistent species & those foreknown to be about to come on, & as all these change gradually, the change of species must be equally gradual whatever conjecture or whatever one can make, but it is also an advantage which the Transmutationist may justly claim that this likeness in form & attribute of beings nearest in space—a [p. 68] nearness & resemblance always increasing the more we fill up gaps in our geological records. This is a necessary consequence of his system, a system not invented to suit that particular class of facts, but taken from the phenomena of the production of species by generation.

The deviation from a parent stock is a positive fact—its possible limits are unknown; if time be allowed they may be greater than we can verify in a thousand years. In such a period we might be unable by any climatal exposure, or feeding or breeding to obtain a black man from a white pair without any intervention of one of the negro race, or vice versa [p. 69] obtain a white from a black, but this is no reason why, if time be allowed, the conversion may not be possible. So the Transmutationist has a right to speculate on the possibility of an amount of deviation which a few thousand years cannot enable us to test by experiment. It cannot be proved or disproved experimentally, but the two hypotheses are so extremely distinct that we may discover which is the most likely to be true by the tendency of one or the other to explain facts—the drifting of men unwillingly or unconsciously towards the hypothesis they wish to avoid.

If by assuming that the tendency to depart from a type, as in the case of the white & black men, might if indefinite [p. 70] explain the largest number of facts—then this theory is to be preferred because it calls in a natural, & dispenses with a supernatural agency. But if the centres of creation imply the reality of species, & the gaps in the system are such as we cannot ascribe to mere defect of annals we may, by a different hypothesis, however arbitrary, explain more facts.

Much of the force of the objection against the transmutationists, founded on the improbability of the jump from irrational to rational by the ordinary way of generation, is removed when the doctrine of progressive development is [p. 71] adopted, as it commonly is, Man being the crowning operation of that develop.[t] For the unity of the develop.[t] being thus admitted, or even eagerly appealed to as a proof of its truth, it becomes simply a question of how it is to be brought about—& they who think one race of men, however dissimilar, can by gradual deviation give birth to another of superior faculties, moral & intellectual, may be as unable to stop as in the deviation from physical characters. Indeed the greatest jump which we know of in Nature is perhaps that of parents of ordinary powers to an offspring of great genius—other children of the same parents being perhaps [p. 72] of very ordinary capacities.

We cannot say how manyfold an increase of intellectual power may thus be witnessed in the contrast between the son & either parent, but we may say that it is so remarkable that a similar augmentation in relative stature, or muscular strength, or sight or hearing, would be such a phenomenon as would be deemed a miracle, a greater one than has been witnessed. If this be so we have only to suppose a very few such deviations from an original stock & we pass more rapidly from a race of the humblest capacity to one of the highest than from a [p. 73] white man to a black.

They at least who, tho' opposed to the transmutationist, admit that the difference between the moral & intellectual powers of Man is one of degree & not of kind, could find no such difficulty in the development of the powers of the inferior animals to the lowest grades of human capacity as they find in the development of one set of specific peculiarities to another.

June 3, 1859

Man

The appearance of Man at so modern an era as when all the shells, land & sea, were already existing as now—& the conviction we have that he does not penetrate far into the past as into the Newer Pliocene—& that there was no being holding the same relation to the remoter animal & vegetable populations of the [p. 74] globe; which Man now holds to animals & plants, over contemporaries is a discovery the most important wh. geology has brought to light. We cannot explain it by any conceivable plan of Natural Operation witnessed by us, we can only at present confess our ignorance & affirm that every hypothesis, save that of a miracle or such an intervention of the first cause as may have given origin to the first animal & vegetable life on our Planet, may have in like manner created Man. The transmutationist affirms that in like manner we must assume that every species or at least every marked type requires the same hypothesis so that we are reduced to the dilemma of introducing upon the stage a continued succession of miracles. It is also objected that the constant [p. 75] coming in of higher & higher organizations so far as that doctrine is true & the favourite doctrine that Man is the last link of the same chain, the crowning event so far as the series is already developed, implies a system, a plan, a law of development analogous to that which transmutation assumes & that as the more we fill up the gaps in time the nearer the successive creations come to each other; the progressionist is always coming nearer & nearer as if by an involuntary course to the Lamarckian hypothesis, drifting in spite of himself to a doctrine which renders the indefinite variability from a parent stock less violent.

The assertion on high authority that domesticated varieties do not [p. 76] revert to their original stock when exposed to their original circ[umstance].[s] & conditions, whether in plants wh. are cultivat.[d] or animals bred & selected, has to be consid.[d] & fairly weighed.

The Negro race not reverting to the White or some other from which in common with the Anglosaxon, it may have been derived, is an example of the unwillingness of acquired varieties to revert.

I have stated in the *Principles of Geol.*[y] my reason for believing in the reality of species wh. the transmutation theory requires as a consequence & I still feel the force of these arguments, none of which seem wholly answered & refuted, & but few requiring any modifications.[55] But I [p. 77] admit that the impossibility of defining a species after 30 years & the increased knowledge of the wide limits of variability, an increasing conviction that all our geological documents are not only fragmentary but a mere minute scrap of the records of the past, & that as yet imperfectly read & interpreted, these consider[ation].[s] make me unwilling to dogmatize with the confidence I once felt.

Any other hypothesis becomes more & more inconceivable, the greater the number of perfect & adult types required to be created anew—not by hundreds, but by millions in a manner we have never witnessed—according to a plan & yet by some machinery other than that which the scientific observer has witnessed. The naturalist who [p. 78] like Prof. Agassiz assumes the difference between Man & the brute to be one of degree only & also that distinct races of Man had distinct starting points makes admissions which greatly aid the transmutationist—for all races have not the same capability & the inferior, according to the system of a constant advance in organization, must have come first, the highest race last. Hence—we do not require a long period for a passage from races far inferior, intellectually speaking, to any known to the historian or ethnologist to the rudest of known savages. *Natura non facit saltus*[56] is a maxim which might be addressed ag.[t] the transmutationist, but if once the Lamarckian could render his case of the slow change in physical organization leading [p. 79] from one species to another by a great number of generations, the transition in regard to mental powers might present the least difficulty of all.

Elephant

No one can deny that the intelligence of an Elephant exceeds that of any marsupial quadruped known to us & that there is a resemblance in some of the faculties of the mammalia—the dog, the elephant, chimpanzee, that we cannot find in any reptile & still less in any fish. If therefore we are entitled to affirm that in the older times there were no mammalia, not even marsupials, & that the succession was from them to Gyrencephala & thence to Man, we may affirm that the passage from the less to the more perfect leads to Man. If it be said that there is an hiatus between the highest of the mammalia & Man [p. 80] even between the lowest human race & the most advanced of the Gyrencephala, an enormous saltus, this cannot be denied. The transmutationist must retreat on the undiscovered world of lost & inferior races of Man & unknown Quadrumana—& he may maintain that to suppose these is a less improbable hypothesis than to imagine the first human beings or the first of any preexisting type to have started into being out of no womb. Why show us a creation with just as many bones & in shape so like osteologically, if there were not in this a deep meaning; then corporeally Man belongs to the existing world which has been going on for millions of years. The discov.[s] of Harvey, etc. c.[d] all have been made before the Pliocene Period & of the nerves. The passage from an unprogressive, stationary, canoe-building, flint arrow- [p. 81] making savage, & a rapidly progressive Man, may require time, but [not] (saltus).

Geographical Distribution of Species

Lowe's transl.[n] of Mousson's Canary shells. *Annals & Mag. Nat. Hist.* 3[d] Ser., Feb. 1859, p. 81.[57]

June 5, 1859

Letter of C.L. to Lowe

Differences [illegible] between land shells & insects, the latter show greater resemblance of Canaries & Madeira [insects] than [of the] land shells! Is it owing to greater powers of migration —or because insects more constant to specific types?

Atlantis Theory

M.[a] & P.[o] S.[o] land shells so different—yet the plants so like in same M.[a] & P.[o] S.[o]

Transmutation Theory

If we shrink from supposing a new instinct, such as that of birds to migrate springing up in a new generation & becoming transmitted, or a more intelligent animal descending from a less one, if the difference between marsupial & placental [p. 82] seem too great, what shall we say when the same mechanism produces a remarkable genius. In every rude tribe some one has sprung up & has been deified—the habit of ascribing inspiration to such remarkable gifts, as if they came directly from heaven, implies the same unwillingness to suppose that the less c.[d] give rise to the greater—the inferior to the higher "cui mens divinior." [58] The idea that this is created in the individual at the moment that the embryo, from which the individual is formed, came into being is not unnatural. We know not how long Man may have inhabited the earth as an almost non-progressive being before a start was given by some more highly gifted individual. But that he may advance his contemporaries, the race must have become capable of profiting if we believe in the distinction between races.

June 5, 1859 [p. 83]

Absence of Mammalia

Between 30 & 40 sp. of reptiles in the one foot ? bed of phosphate nodules of upper G.[reen] S. [and] at Cambridge & not one Cetacean.

Successive Creations

It w.[d] be a mistake to suppose this to be a theory, but an equal one to assume that because we have no theory, therefore transmutation, the only scientific hypothesis, is true.

The passage from fish to reptiles is nearly complete—from reptile to bird, & from bird to reptile not so, but as birds are so rare in a fossil state we can get no aid from palaeontology in that direction. Such gaps make the hypothesis of transmutation hard to believe. If the geolog.[l] hiatus between Permian & Trias & between Chalk & Tertiary were filled up, still birds being deficient, we sh.[d] have but little help.

The Dinornis family shows what revelations might be made in that class.

Transmutation [p. 84]

The discovery of links between Secondary & Primary by the identification[59] of St. Cassian with Keuper, links wanting in Muschelkalk & Bunter, is no anomaly. The Muschelkalk is a poor & limited marine fauna, the Bunter still more barren. The gradation w.[d] be doubtless almost complete had we the filling up of these gaps.[60]

Reptiles

Only one fragment of Nothosaur[us] in marine S. Cassian beds among 800 sp.[s] & 100 miles in length.

At least we might have thought that reptiles were rare in Keuper Period, but no, they were almost at their point of culmination. Mammalia perhaps were rare because the 80 triassic reptiles were [p. 85] near land & part of the coal period of Upper Trias freshwater almost, tho' not lacustrine.

Species

By a kind of innate perception, an intuitive sense, an instinct, even children recognize species; to abandon this creed is like giving up a belief in the existence of the material world because we cannot disprove Berkeley's argument.[61]

A great naturalist like Agassiz is aware that he cannot always decide the demarcation between some species & others, still he believes in them.

All our knowledge seems in danger if we give way to this scepticism. We must strive to efface all the boundaries we have discov.[d]—we must labour towards realizing chaos—think [that] order, & the possibility of definition [are] an earnest of [p. 86] the small progress we have made towards complete knowledge.

This faith in generalization is then a delusion proving our ignorance.

The same may be said of all our Geological lines of division. They disappear with time & as we advance; but our science does not vanish with them. They will never all be filled up, but if they c.[d] be, the fullness of our knowledge w.[d] be augmented.

These gaps are real helps & the stationary period in which Man lives because of the shortness of his life is to him a reality.

The geography of the earth is not changed in a few thousand years, tho' ever changing. We may navigate the seas, & cultivate the land & [p. 87] plan our docks & political schemes on the assumption that the land is motionless. England is disjoined from France tho' it was not always so.

We may speculate on the constellations—changing—some stars opening, others closing behind—this great movement may be true & yet the constell.[s] of the era of Hipparchus[62] may serve the astronomers now. They may be the objects of scientific precise measurement & yet not be real if Time enough be allowed. In speculations these forms of the stars are the accident of the point in space to which the earth's orbit is confined, or the solar system for a period astronomically brief, tho' it last possibly several millions of years, very long in reference to Man, & by him the change may be disregarded except when speculating on the state of [p. 88] things in great astronomical periods. A book publ.[d] several thousand years ago may be accurate now & define the constellat.[s] well,—we might sail by them if we had no compass, measure our longitude by them—yet believe them to be fluctuating & unreal, even as objects seen from the earth except for a limited period. If we name them, class them accord.[g] to their magnitude & shapes, our science will be available now & for tens or hundreds of thousands of years or ages & yet they may be unreal.

Monotremata

If hereafter some terrestrial vertebrate fauna be discov.[d] in the Coal Period it might consistently, with the gradual development of forms of higher & higher orders or the lowering of the grade [p. 89] as we recede from modern times, be made up of monotremes or some class below marsupials. But in that case the absence of such fossils in the beds of Coal seems to imply rarity of airbreathers in old deltas. We seem to have most hold of the land in the Coal Period & yet grant even salt-marshes—least hold of an airbreathing fauna.

June 8, 1859

With Huxley

Take of each period the number of the percentage of extinct genera, i.e. the number not passing from the preceding to the present Period—say 20, then take the number of new species which appears in the present period for the first time, say 40, & this percentage will express the difference of the Period.

Ganoidei

<table>
<tr><td>Lepidosteus
Polypterus (Egypt)
Amia (N. America)</td><td>Holostei</td><td rowspan="2">all from the northern Hemisphere</td></tr>
<tr><td>Accipensers
Scaphyrinchus
Spatularia</td><td>Chondrostei</td></tr>
</table>

[p. 90] The Lepidosteus has perfect exo—& endoskeleton—vertebrae opistocoelian. The other three genera have persistent notochord gristle not bone. All the genera except Accipenser are freshwater, the sturgeon (Accipenser) is marine, but like the Salmon goes up rivers.

The lower jaw of Lepidosteus divided into six pieces as in batrachan reptiles.
The air bladder in Polypterus & in Lepidosteus. Polypterus [is] heterocercal.
Spiral intestine causing spiral coprolite.
Amia no exoskeleton or only thin flexible scales.
Pharyngobranchii
Marsupibranchii
Ganoidei
Elasmobranchii
Teleostei

Amphioxus

[p. 91]—A Pharyngobranchian, pointed at both ends, hence the name—transparent—no eyes nor teeth, nor heart, nor brain, white blood—the notochord persistent & undivided into vertebrae—very rudimentary as a vertebrate.

None of these known in oldest rocks, but not preservable.

Lamprey & Myxine, Marsupibranchiate. The nine eyes, on lamprey so called, because the pouches called eyes but which originally look like eyes. Myxines (form?) parasitic, no exoskeleton, persistent notochord, no limbs, no pelvic bones, but they have teeth.

Progressive Development

The evolving of the lesser out of the greater, it may be said *omnia rerum in pejus ruere,*[63] degradation, degeneracy—all this is intelligible, but [p. 92] the higher out of the lower grade, intelligence out of mere instinct, or genius out of mere intelligence can that be conceived. Certainly not if the inferior is regarded as more than the mere instrument in the hands of a higher power. If a series of generations of savages continue & then a superior Mind arises, one who is obeyed during life & deified after death it is not the _____

Pteraspis

—of Lower Ludlow found May 10, 1859 by Mr. Lee of Caerleon; the fossil was found at Leintwardine (near Ludlow).[64]

Pliocene & Post-Pliocene Rhinoceroses [p. 93]

Glacial

1. Rhinoc. tichorhinus Rh. antiquitatis Blum	complete nasal septum
2. Rhinoceros hemitoechus (Fal.) (Syn.) Rhin. leptorhinus of Owen, partial bony septum	Grays—Northampton & Gowes Cave
3. Rhin. Etruscus (Fal.) Syn. Rhin. Leptorhinus, Cuvier, *pro patre* Oss[emens] Foss[iles][65] figs. 1–20, cf. pl. 49 partial bony septum & tall slender limbed species	crag & bottom of Val d'Arno
4. Rhinoc. leptorhinus Cuvier pro patre (Rhin. megarhinus, de Christol)	

Cuv. Oss. p. 47, Fig. 7, exterior skull; no bony septum.[66]

[p. 94] Hippopot.

Wollaston's Letter Received June 10, 1859: Species

The conviction in W.'s mind after the study of 50,000 specimens of Madeira & Canarian insects of the reality of species & their fixity of character, nothwithstanding that so many of the insects are common to Mad.a & Canaries, as many as w.d be or might be looked for if the whole tract from Mad.a to Teneriffe had once been [p. 95] continuous land. This belief in species which grows in the minds of so many honest working naturalists is perhaps the best argument in favour of their reality. It is like the real existence of matter—& one's instinctive belief in it. Wollaston remarks that we must be modest & make allowance for our ignorance, our limited knowledge of all the phases of a species wh. requires time. A working naturalist who devotes his life to seeing things as they are is worth more then than the most brilliant speculations.

Wollaston doubts whether sluggish nature (Helices?) are acted upon from without. He scarcely allows enough for permanent varieties—insular ones.

The Helices of Madeira, he observes, have not altered so much as a granule or punctus since they became fossil. [p. 96] The Tarphii (Coleoptera) like the Helices are sluggish creatures & in their case the species are as restricted as those of the family of Helix.

Mousson & Lowe, however, speculate on permanent varieties of Helix Pisana & whether some of them are, or are not, true species.[67] It is doubted whether Helix ustulata of the Salvages is an aberrant var. of H. Pisana (qu. does not Wollaston doubt it?)

Mousson, p. 84 *Annals of Phil. 1859.*[68] doubts about H. gemminata whether entitled (as being so very aberrant from the typical H. Pisana) to rank as a species!

Men of Genius [p. 97]

Out of a population of ordinary commonplace people in the course of time a man of great genius arises. It may be according to strict law, but so capricious appears the apparition that it is regarded as an emanation from heaven—*mens divinior*. No one is willing to attribute the new power which may exceed many times what had been exhibited by any individual & is so different in kind that no combination of many individuals can be conceived capable of doing what it achieves by its imagination. Yet the mode in which this new power is ushered into the planet is by ordinary generation.

It may be the nearest phenomenon we ever witness to that 'jump' which Nature is said 'never to take'. A few such leaps in a backward direction & to what a humble grade of humanity might we not descend. We cannot estimate how far the intellectual power in such an individual is multiplied [p. 98] but it would seem not rash to affirm that such a sudden augmentation in stature, muscular power, sight or other more material & physical quality, is never observed & w.[d] therefore be regarded as a greater marvel, a more unexampled departure from the parent type.

If this be true it does not seem that the greatest knot the transmutationist has to untie is the rising up of a rational & responsible being.

Heterocercal

The paleozoic fish accord.[g] to Bronn's Table II, p. 19. *"Entwickelung's geschichte"* [*sic*][69] comprise only 437 species (in 1858) suppose they are not 450—& the existing fish 8000, it might well happen that if Homocercal were [p. 99] as rare in Paleozoic Times as are now the Heterocercal, they w.[d] never yet have been found. Assuming that fish in Paleozoic Times were as numerous as now or that the proportion was similar of Homocercal to Hetero—as it is now inversely. The Heterocercal ganoids with scales are now (25!) 25/8000 or 1 in 320, it might well happen that the first 437 or 450 might not yield one. But the paleozoic placoids being separated the number of true heterocercal ganoids became still less.

Placoid	Ganoid	[Total]	Bronn
238	199	437	Tabl. IV, p. 21 [70]

The Paleozoic Ganoids were less than half of the then known fish, therefore as there are so many placoids now, the exceptional nature of the ganoids is not quite so great. The Teleostei with bony spinal chord is certainly a remarkable change, & the absence of bone in both placoids & ganoids indicates a resemblance to the embryonic type.

Paleozoic Fish, Bronn, Table IV, p. 21 [p. 100]

Placoids already known		Ganoids	Recent
Genera	64	4	492
Species	238	30	8000
Recent Placoids			Paleo.
Genera	68	57	101
Species	230	199	437

What is remarkable is this that tho' we know so few Paleozoic fish (only 437 in all) we know as many paleozoic placoids, genera & species, as we know living ones.

Therefore the Placoids predominated in a vast degree—and they are not imperfect forms. The variety in them compensated for want of variety in Teleostei. Sir P. Egerton[71] tells me of a Triassic Homocercal fish.

June 1859 [p. 101]

Microlestes

Mr. Etheridge tells me that Mr. Moore[73] says he found the Microlestes tooth with Saurichthys in the yellow Magnesian conglomerate of Triassic age near Bristol (Mendip?).

Etheridge also says that the Paleosaurus & Thecodontosaurus were in triassic, not Permian, conglomerate.

Mammalia in Limburg beds at length (1858? or before) Phoca ambigua found at Elsloo near Maestricht is the first instance of a mammifer in the Belgian or Limburg beds—see vol. 3. Tracts. Bosquet's Catalogue.[74]

June 17 [1859] [p. 102]

Extract of Letter to Huxley

Species

Linnaeus thought genera as real as species. If we find in Geology or among our fossils or in the living world no transitional forms between Kingdoms, Sub-Kingdoms, or even as you hint between Classes, Orders, Families & Genera, if there are certain fixed & absolutely limited groups, can we reconcile the want of such evidence with the theory of Transmutation?

Must not such groups have come first into the world by virtue of some *modus operandi* different from gradual Development not in the way of indefinite variation from a parent stock in the course [p. 103] of thousands of generations?

Each newly created type (meaning by creation some act or law not yet within the domain of science) may have been, as a general rule, in advance of the highest of the preexisting ones, Man being the last of the series—a race of savages having at first but small cranial development, & out of this the Negro & White Races & other extinct varieties may have been, or may be hereafter, evolved in the same way as all permanent varieties.

Primrose & Cowslip

If you saw a gorilla delivered of some new species of the Order Primates, genus Homo & superior to the species Kentuck[75] [p. 104] you would trust the evidence of your senses tho' the phenomenon might be without precedent in your experience, a fact independent of the vast chain of causes & effects in the Universe.

If all the Kingdoms, Classes, etc. & Genera were in the oldest Periods, we might by develop.[t] get in millions of ages all our species, but if Placental mammalia are wanting in Lower Silurian & we require such an event as the first appearance at some period long subsequent of a Monodelph, we might compare this with the coming in of any other new type—ending with Man. It becomes difficult to know where to stop, why not new specific types in like manner?

[p. 105] The greater the number of these phenomena the less miraculous do they become. They may be conceived to be in obedience to some Law unknown to us, but analogous to that manifestation of power which gave origin to the first link of the chain.

Admit the higher groups alone to have come in, in this way yet to be discov.[d], & the event is rare & exceptional, embrace the genera in the same category & we shall be led on to sub-genera if not to species till such creative acts or first appearances cease to be praeternatural. But why then do we not witness them at least as often as extirpation?

Shall we in time find fossil men intermediate between some extinct Quadrumana & the Bimana? If we can come to no conclusion in regard to the first coming [p. 106] in of our own species, an event comparatively speaking so modern, how can we hope to bring to light the coming in of any antecedent type. In going back from the known to the less known, from the modern to the more remote periods, at the first step in our retrospect, the greatest of all difficulties, the absence of any representative of Man in the Miocene, no signs of any dominant & implement-making brute of the order Primates or Archencephala, or in other words with negative evidence of the fact that the advent of Man was post-Miocene & if he belong to a new type, then the creation or development of a new type is a thing or event of yesterday, geologically speaking.

[p. 107] Shall we then modestly say that the true philosophy in these matters is the Philosophy of Silence. But the Transmutationist cannot be pooh-poohed. Their hypothesis is better than none. They are making some progress by awaking attention to the variability of species, by challenging their opponents to prove that it is not, or may not be indefinite.

They who think they came from a chain of developm.[t] beginning with the invertebrate & ending with Man & who, admitting some undiscov.[d] Law which they often liken to embryonic development, make the human species the last link of the chain & regard it as the crowning operation have also an hypothesis which is as speculative & perhaps as liable to the charge of being materialistic as that of the Transmutationist.[76]

[p. 108] Haldemann paper.[77]

Did Geoffroy St. Hilaire publish on "Progressive Developm.t"? [78]

Permanent Types

Davidson[79] says that Lingula, Crania, Rhynconella & (Discina?) are genera of Brachyopods which do not vary from Cambrian & Silurian to the recent Period. Even the old *species* nearly the same as the modern. Such has been the harmony of all the condit.s of the planet throughout, to these genera.

Brachyopods at all depths from shallow sea to great depths, see Woodward Conchol.y [80] & in all climates d.o—so in time 70 sp. known in living state more than in any Secondary format.n

Origin of Man

Once admit an antiquity of some geological importance, or implying the occurrence [p. 109] since the spread of Man of considerable changes in Physical Geogr.y—& all idea of any other origin, but that of a long rude stationary condition of the first settlers must be abandoned. A struggle with the powerful carnivora—subsistence on fruits. A frugivorous state inferior to that of the Hunter State as requiring no weapons or stratagems except those of cunning & agility to escape the more powerful beasts of prey.

The Garden of Eden, Milton's Paradise, the Golden Age, all vanish. In its place we have an ancestry for tens of thousands of years as unprogressive as the Elephants or Orang tribes. The Progressionists are bound to place the lowest they can discover any monuments of, as the first in time of these rude ancestors or progenitors. They prepare the way therefore for the transmutationists. Why the Malays of Mr. Crawford [81] have not become quadrumana is a difficult [p. 110] question for the Lamarckians.

One of the most remarkable facts is the coexistence for thousands of years of the rudest savages with the civilization of Asia. Retrogression after war & conquest may often have taken place, a relapse into a primitive condition.

How could Man exist for tens of thousands of years & not advance beyond the arrow & flint hatchet-making condition?

An indigenous tribe [exists] in the Malayan peninsula with quickened sight & hearing, frugivorous, much dreading the tigers, climbing trees to escape from them.

Empires in Asia may have risen slowly & fallen while these savages made no progress. How much more then, if an aboriginal race existed before [p. 111] any civilization.

The progressionists must, if consistent, assume that the race which possesses the humblest intellectual power came first, but it may have had other attributes giving it advantages. One w.[d] suppose that the strength & tree-climbing power of the gorilla w.[d] have given it a decided advantage over the Man, but the latter might have had in all states, more cunning.

Truth to Be Sought at All Hazards

In Faraday's lecture on "the education of the Judgment" or against Table Turning he says, "The point of self-education wh. consists in *teaching the mind to resist its desires & inclinations, until they are proved to be right,* is the most important of all, not only in things of Natural Philosophy, but in every department of daily life." [82]

[p. 112] The temptation, says T. Spedding,[83] to mental unfairness which beset us on every side are such, so many & so subtle, that Faraday lay down this rigid and even ascetic rule of intellectual discipline.

To give up a high & godlike ancestry from which we have degenerated is one of those inclinations which beset us & it seems very unlikely to be proveable by Science.

Unity of Original Stocks

If at a certain period certain individuals of a species formed a new race, other individuals of the older species might give birth to the same or a closely allied race, & these by intermarriage to a hybrid species, & in this way there might be several starting points.

The theory of Agassiz of the separate [p. 113] origin of the Negro & White Man would therefore agree so far with the transmutation theory, for if a Gorilla gave rise in one climate to a negro, it might in another to a copper col.[d] Indian or Red Man.

Can the doctrine of specific centres be maintained in its integrity consistently with the theory of Transmutation?

Its abandonment by Agassiz is a great weakening of his system of "Independent acts of Creation" for they w.[d] be best reconcilable with specific centres—& the facts of Geographical Distribut.[n] would follow as a matter of course. All migratory powers become unmeaning according to Agassiz's view. Why sh.[d] we explain the Salvages' Helix Pisana by migration? Why not, according to Agassiz, creation?

June 19, 1859 [p. 114]

Species—Astronomy

In ten millions of years when our astral system shall have accomplished half its revolution round the Pleiades (Index B.[k] 91) all the constellations will have vanished & have given place to others.[84] But the fact of the instability in the forms of the constellations which are varying constantly, annually, hourly, does not prevent them from remaining for thousands, nay for hundreds of thousands of years, so unchanged that the Navigator may overlook such trifling modifications arising from alteration in their relative situations which the delicate instruments of the astronomer alone can detect—changes which it requires millions of years to bring about may be overlooked—appearances which last for thousands of generations of Man are to them at least realities—they [p. 115] partake of the fixed & the immutable, yet they are the merest accidents of a unit in the indefinite cycle. In the organic world there is no similar recurrence to the old forms as in astronomy. A map of the heavens may be useful in spite of such changes.

Creation Miracles

The independent acts of creation have been becoming always more & more miraculous as time advances. The complex according to the natural system is evolved out of the more simple. The progressive development theory assumes that Nature began with simpler forms & evolved higher ones by degrees as Time passed on—not by transmutat.^{n} but, says Agassiz, by creation (give his definit.^{n} of creation).[85] It follows therefore that the miraculous power which called new species into being was more & more preternatural as time advanced. A power always giving birth to more complex organisms not preceded by simpler ones.

[p. 116] On one side Agassiz gives a quasi-embryonic development & sees analogy in the successive coming in of higher & higher forms—on the other he introduces a power which violates more & more all modes of operation which lead from embryo to adult. The creation of an Eleph.^{t} is a greater miracle than of a marsupial, that of Man the last & greatest.

One might imagine a tendency to a more complex organization if an embryonic analogy c.^{d} be established but not so it be a series of independent acts.

If however it be said that these acts, tho' each is independent, may be govern.^{d} by fixed laws, as each individual may owe its peculiarities directly from God tho' produced by generation, we may then ask why the law of transmutation will not suffice. [p. 117] The answer is because of the absence of transitional forms in the great Kingdoms of Nature.

If as Pictet hints, there may be a law like that of alternate generation yet undiscovered, then we may say that it is a mystery not yet solved.

Unless we can believe that some approach will be made towards solving the mystery by the transmut.^{n} hypothesis we sh.^{d} not adopt it. At present it is the only scientific hypothesis.

That new forms still rise like exhalations from the earth is no scientific theory.

Catastrophes & Miracles

It has been said that the orderly course of Nature is uniform, but it is liable to be broken by occasional catastrophes both in the Animate & Inanimate Worlds—that Mountain Chains [p. 118] have resulted from the recurring catastrophes, & the greatest of all these convulsions, or that which produced the Alps, was the last, or if not, one of the latest. So also in the organic world the violations of the regular course of events have been rare, but have gone on increasing in magnitude & the creation of Man, like the rise of the Alps, is the last & greatest miracle.

Transmutation

—is a mere hypothesis, but may be useful as such—& lead to the establishment of a Law. If it be admitted that the gradual change of an inferior to a superior race of Man is not a violation of law, then the change from a species of Orang between the Gorilla & Man may not be so. If every species enters the world by same law, if each is not a miracle like the first creation, then there must be some undiscovered mechanism which if known w.[d] perhaps most nearly resemble the variety-making or monstrosity [p. 119] making power. There may be laws of structure, but this is pure conjecture, causing passages suddenly to new varieties. If as we close up the series of geolog.[1] documents we find the leaps less great, we are approaching a time, still very remote, when the series will more & more resemble the effects of an ordinary cause.

Savage Tribes

—have but a dim perception of Natural Laws nor of God & are therefore timid & superstitious. If we desire that Geology w.[d] not unveil a primeval state, when Man was without knowledge —with quick senses & cunning, but scarcely any conception of the existence of God, scarce rational or responsible, we sh.[d] not thereby escape the consciousness of the fellow-beings who may be numbered by thousands, who are alike ignorant & almost imbecile & who will amount to more millions, as Time flows on, than all who once peopled [p. 120] this planet in the infancy of our Race if it began thousands of centuries ago.

Transmutation

The passage of one order of animals into another ought to be more easily provable now than 30 years ago, instead of which it seems less so. As many good naturalists believe in species as ever. But perhaps they have not reflected much on geology. If there are new proofs are there not also confirmat.[s] of old ones, successive coming in of new forms—Orders, Classes, a strengthening of the old proofs as well as of the extension of the variety-making power.

Imagination fails us to conceive the shortness of the time & the magnitude of the whole operation. But if we nowhere can detect a beginning & yet do see varieties & races forming, may this not be the most [p. 121] probable conjecture, until some positive act of spontaneous generation occurs. If a new species appeared it w.[d] probably steal in, not accomp.[d] by thunder & lightning & a burning bush—perhaps as a monstrosity—a giant or a dwarf, a six-fingered child—an individual of great genius, a calculating idiot.

Catastrophes

More progress has been made in the abandonment of sudden & violent convulsions breaking the Order of Nature than in the belief of sudden interferences in the organic world—acts of production corresponding to those of extirpation or extinction.

Inspiration, Genius

Knowledge derived from a supernatural & divine source, directly to the individual, not developed materially & by generation—like the physical attributes of certain tribes of Man—muscular strength, ordinary intelligence.

Antiquity of Man [p. 122]

Had men soon reached such a state of civilization as to build cities & construct roads, aqueducts, railways, docks, pyramids, we sh.[d] have expected to detect them more easily than the monuments of men who merely made bows & arrows & inhabited caves.

But frugivorous savages might long inhabit the earth & leave no signs of their existence, but their bones & skeletons. To bring these to light is the result of much labour & fortunate chance.

Inspiration

I see no objection to assigning the genius of a prophet or his moral power to direct communication with God, provided other instincts & new ones fitted for new circums.[s] & showing themselves at times, tho' normally dormant, be ascribed in like manner to some power independent of generation transmission.

Antiquity of Man [p. 123]

So long as we merely ascribe ten or 20 thousand years to the sojourn of Man on the globe, the format.[n] of the existing races by derivation from a single stock seemed impossible—but grant not only that Man, but Man in a rude state & therefore divided into separate communities having no intercourse for thousands of years, inhabited the planet & the difficulties disappear. In this respect the origin of many species from one parent stock is rendered easier the more we enlarge our ideas of time.

Progressive Development

There are two schemes of Progressive Development—one by the way of transmutation or generation, the other supposed to be effected equally according to a plan of the Almighty's by means of a succession of independent Acts of Creation—accord.[g] to which new species are introduced or called into being by a power like that to which the origin [p. 124] of life in the planet was due.

By both, the evolution of Man is made part of one & the same system of successive changes—if it be a miracle, a direct interference, so are all the antecedent interpositions of the Creative Power. It comes after the higher mammalia, the Placental for example, in obedience to the same law of chronological development which caused the Placental to succeed the Marsupial & these to make their appearance on the stage after the Birds & Reptiles had flourished & so on till we reach the humblest of the Vertebrates or Fish life, but the invertebrate had long been established—so far as our present evidence gives us the means of knowing.

[p. 125] In 183[6?] I endeavoured to imagine that tho' Man as a species of animal might succeed other mammalia according to the fixed antecedent laws, yet the addition of his moral & intellectual Nature in this planet was an innovation in kind as well as in degree & ought not to be made a part of the same series of events as the developm.[t] of the inferior animals but exceptional.[86]

But if we make the admission, such as Prof. Agassiz makes, that instinct differs only in degree from reason & that different races of Man started from distinct origins, we then bring Man into the same category, or if allowing all the 5 races, one remote starting point if we assume a low [p. 126] cranial development for the earliest created, with perhaps as perfect a physical develop.[t] as the finest of the hunter tribes & if we only assume that the intellectual & moral nature of Man differs not in kind but degree, then we encounter the stumbling block of the soul-less Man & that is simply what we equally meet with in the millions of savages & low intellectually organized individuals which coexist with civilization.

These must already have exceeded the early inhabitants of the planet belonging to the human family. If we cannot explain why God permits them, any more than the origin of evil, we are not [p. 127] prevented from believing in a future state on their account, or if perplexed no new perplexity will trouble a philosophical spirit by discovering that at a remote period of the history of Man a difficulty existed, which now continues & seems destined forever to endure, in consequence of some general laws of physiology & psychology which are beyond our comprehension.

If the human race was develop.[d] to a civilized from the rude state, whether that rude stock develop.[d] by transmutation—or creation, it was doubtless in obedience to a plan & another part of that plan causes so many of the existing individ.[s] of the species to have their moral & intellectual faculties arrested at a point short of their becoming responsible & rational beings.

[p. 128] The evidence of the geologist in favour of a low

primaeval moral condit.[n] & mental [condition] of Man may become strong, but can never equal the proof of what a witness of many fellow beings, whose notions of a God & whose powers of generalization are so weak, as to make us pronounce them irrational.

It will not be to Ethnology & Geology that we shall owe this psychological enigma—Sedg. Discourse p. 128. 129 app.[x] [87]

Creation—Genius

What can be more natural than that the sudden appearance of extraordinary talents or goodness or inventive powers sh.[d] be regarded with adoring gratitude, sh.[d] be looked to as direct emanations of the Divine Spirit, the first Cause—that the origination [p. 129] of such ideas sh.[d] be referred to a direct interference of the Great Spirit of the Universe. A new race we may imagine to arise gradually, more distinct than the negro from the European by some law of gradual change, far more than after the earth had been inhabit.[d] for ages by an unprogressive tribe of savages, if the era at length arrived for the advent of such moral or intellectual phenomena suddenly developed in some superior individual. Would not this be the nearest to Creation—to a 'leap' that we can witness or have witnessed. Are the fixed varieties of domesticated races comparable to this? If they do not amount to the origination of new species they may still less be compared to such a sudden developm.[t] as this.

Imbecility [p. 130]

There are difficulties like the origin of evil, mysteries like the gradual passage from rational to irrational surrounding us & ever near to which, perhaps wisely we habitually shut our eyes. We cannot grapple with them. The more we extend our knowledge the more we shall find them because the works of the Creator will bear throughout the impress of the same scheme. But if in spite of knowing that men without a soul exist now, we believe in the soul's immortality or in the future development of such unborn souls or prematurely dying infants—how much less reason have we to fear that any revelation of the former existence of millions of such beings sh.[d] shake our faith. It may be that additional enigmas will be [p. 131] added. As when we reflect on the number of microscopic beings or others inhabiting the depth of the sea, we may ask what relation have they to Man. We get reconciled to this puzzle & perhaps reflect that the creation has some other end besides its subserviency to Man, perhaps far higher ends of wh. this is one only—Geol.[y] then unfolds to us a wider extension of the same independence of Man, of beautiful & grand manifestations of creative power. Harvey might have discovered the circul.[n], Bell the nerves, before Man's [appearance]. This new discovery of the past world adds to the perplexity as if made for the first time.

Ideal World

To retreat into an ideal spiritual world "unfettered by one gross companion's fall,"[88] to close our eyes to the material actualities of our [p. 132] condition as one of the animals, to live in an imaginary superiority to that intimate relation of the physical part of our being which clogs the intellectual part & the soul, to cherish aspirations beyond this narrow & sensual existence, this ambition causes the identification of our mortal state with that of the lower mammalia to be unwelcome & when we find our genealogy less dignified it is the same inclinat.[n] w.[d] lead us to wish that the Fuegians or Negroes had some different or independent origin.

Insects, a million & ½ of species, according to Schauw's estimate, Harting p. ——[89]. There are now living this prodigious number of species. Yet entomologists believe in the reality of species. No class of naturalists has upon the whole more faith in [p. 133] the unalterableness of species. How many acts therefore of creation are here implied. Yet amidst her leaps, how much order has Nature preserved. The insects in amber are not the same. In each period therefore of Geology the miracles amount to millions, the special interference becomes not exceptional but a rule. If we allow a million of years for a geological period or ten millions these interferences will become more than annual.

But if to escape from this conclusion we take refuge in the format.[n] of types alone as due to creation, we ask whether sub-generic types are to be included—also are races as well as species? If varieties corresponding in value to those of Man for which Prof. Agassiz w.[d] have distinct original creations, we then may double or triple the number. [p. 134] Changes are now continually in progress—species are dying out, undergoing extermination as of old—this has gone for many successive periods, for 30? according to a plan—it must be enduring—to suppose the variety-making power to be at rest is impossible, can it be invisible or is it only seen in some small movements corresponding to earthquakes & volc.[c] eruptions & floods in the inorganic world?

Is it as insensible as the growth of trees in the forest or is it by starts? intermittent or perpetual—

Constancy of Species

In 9 millions of years all the constellations may disappear & give place to a distinct set, but in 18 millions they will again be as now—

Not so the species, once changed we can never expect them to return [p. 135] for there is no cycle, dependent as each are upon the ever varying condit.[s], past as well as present, of the earth & of organic Nature.

First Man

If we suppose a start or leap, we may imagine one of the other sex of superior endowments—a rational fellow being & from them other rational offspring.

The coexistence of irrational adults is a corresponding phenomenon not more astounding.

Species

Linnaeus' definit.[s] may be useful a million or ½ a million of years hence, or most of them, as may—list of stars—or the photographic chart of the heavens.

Time

If geologists sh.[d] ever obtain reasonable grounds for estimating the time of our geological period or a nearly entire revolution in species & sh.[d] find it to be as long as the greatest astronomical [p. 136] cycle or the revolut.[n] of an astral system round the Pleiades. If instead of one million years it may require about 20 million for bringing about the entire change, we sh.[d] immediately take a new view of the amount of lost forms & unknown transitions of which no evidence survives—the existence & disappearance of islands & continents in part of a period, just as the map of Europe (& of N. America) has changed enormously since the glacial period, a brief episode in one revolut.[n] of organic life. It may not improbably have taken several 100 thousands years for this episode to have been accomplished—all our ideas of denudat.[n], of alternate upheaval & subsidence of varieties of species, will be modified by [p. 137] such (perhaps more just estimate) of the duration of Time.

We cannot take a country like Sweden or Greenland as tests of rate of rising or sinking unless we allow for vast intervals of rest. If one area of 1000 miles diameter be in appreciable motion—upwards or downwards & another area of ten times the size be apparently stationary, we must strike one average with reference to the whole—and if we thus obtained a few inches or the fraction of an inch for a century of rising we must allow also for movements in opposite direction before we can arrive at a probable mean.

As to unusual disturbances we have no more right to call them in without proof than unusual periods (or exceptional) of absolute immobility.

Man of Genius [p. 138]

This phenomenon, the birth of a man of remarkable power from ordinary parents is the most remarkable of all—& the time may come when the transmutationist may even find it harder to explain, anatomical, physical, corporeal, physiological jumps than psychological by actual appeal to facts in Nature. The passage from one race to another may be harder to imagine than from a creature too humble in mental power to know right from wrong & be responsible for his acts, etc., to a rational & responsible agent.

The elevation of faculties moreover is here exemplified. It may be attended with no moral improvement—it may be with depravity.

June 29, 1859 [p. 139]

Chat with Mr. Atkinson[90]

Ancient miners of Siberia of silver used flint instruments. Shafts of galleries still left leaving the harder rock & cut away softer parts. Skeletons of the old miners [were] found in these mines & flint instruments—Barnaoul mining district of Albi, central administration of mines. On the Obb Mr. Atkinson obtained in the auriferous alluvium a casting in brass, 14 f.^{t} deep in undisturbed beds, with forest over the alluv.^{m} in which trees of Cedar 7 f.^{t} in diameter—supposed [the brass casting] to be harness of horses.

Progressive Development

If fish & reptiles most perfect in oldest beds or where they first appear & now less perfect, what [p. 140] assurance have we that there were not, in older strata than the Cambrian, higher forms of mollusca. Perhaps less perfect fish & reptiles, lower than Archegosaurus may turn up in Cambrian or Laurentian, but they who think [that] the oldest now found ought to be considered the first created, until prior ones are detected, must admit a contradict.^{n} in shells having been created less perfect. The history of fish is at variance with Crustacea & Mollusca—which are s.^{d} to have been less develop.^{d} at first than afterwards.

Man the Crowning Operation

Harting after describing the gradual development in terms of Plants & Animals p. —— observes that man may be cited as the [p. 141] exemplification of the crowning operation.[91]

Ascending from invertebrate through the vertebrate to Man by the gradual development of the Brain.

What matters it whether this development is effected in millions of years by a series of 30 millions of special or miraculous interferences worked according to a plan—each million producing species on the whole resembling most those who preceded & follow, & that which followed being always nearer the existing & more remote from the older set. The type always approaching nearer the present & departing from the remotest always adapted to the changes in physical geography & climate.

Grant that species are realities & even genera, as Linnaeus thought, [p. 142] that some undiscovered machinery, different from that of generation or the mere variety-making power, has been always at work—still we see a law & this the progressionists admit, a law of progress & a law of passage or resemblance according to nearness of time & chronological proximity.

In all its consequences of a moral & theological nature there is so near & continually a nearer & nearer approach, that to draw the line is not easy. If the same develop.[t] which made an Elephant appear later than a marsupial & this later than a bird, this than a reptile, this than a fish, & a fish after a cuttlefish or nautilus, also governed the time of Man's birth, we need not regard the transmutationist [p. 143] as degrading human nature especially if we adopt the idea of some naturalists that sudden passages without all the intermediate links are possible in the physical, as they seem to be in the intellectual succession by generation.

If there had not been an idea that the proof of Man coming last harmonized with the develop.[t] from plants to animals in the 6 days of creation, this identification of Man with the succession of events w.[d] have met with the same opposition that the vastness of Time & the number of former creations have had to encounter, sublime as are the views of Creation which in truth they reveal to us. If true this developm.[t] ending with Man will be the first example of a great truth, really subversive of old & [p. 144] time-honored creeds, having been welcomed theologically. Unless indeed it sh.[d] prove true that the transitional version of the gradual development is correct & then it will have had the usual sanction of theological repugnance & opposition.

Whichever we adopt, progression with or without transmutation—beginning with the invertebrate & ending with Man, there is really no cause for alarm, as if any new views overturning our high estimate of the position of Man in the scale, or his relation to the lower animals had been brought to light.

The relation of Man corporeally to the Orang is not altered or increased by these discoveries. If there be a deep meaning in this relation it might be [p. 145] discovered independently of geology—(Harvey & Bell). Man without a soul.

Valley of Thames Drift

If the Grays beds are post glacial with Cyrena & E.[lephas] antiquus then there were two cold Periods.

For the Happisboro[92] forest, with E. antiquus & Cyrena & Hippopot.[s], was older than the till & Scandinavian erratics of Norfolk—wh. we may compare in age with Muswell Hill. If after the latter came the Grays beds, we have then the Cyrena & E. antiquus & Rh. leptorhinus a 2.[d] time after extreme cold & followed by mammoth tichorhine Rhinos.[s], Bos moschatus & Reindeer implying a return of colder climate tho' not as cold perhaps as that of the Boulder Clay.

Fisherton, Salisbury [p. 146]

The Elephant brick earth of Fisherton Anger, near Salisbury,[93] is, says Prestwich, most like the Menchecourt, near Abbeville, drift with flint knives, except that there are no marine shells at Salisbury.

Creation

The turn of mind as well as the amount of intellectual power may be exceedingly different in the child from its immediate & still more from its remote progenitor. I once regarded the difficulty of a sudden passage from irresponsible & irrational to rational as the great difficulty, it now appears the same, only perhaps a less difficulty than the physical change.

To what conclusion then do I arrive? Is it to that state of

doubt, which however philosoph.[1], is of [p. 147] all states the least satisfactory? Is it to abandon the views advocated in the successive Edit.^{s} of the *Principles* in regard to the reality & constancy of species. Are all the arguments adduced in the 2.^{d} book answered & overcome—I think not. But I then regarded the constancy of species, & their sudden coming in, as certain as the extinction of some is sudden. The transmut.^{n} hypothesis as bordering on the visionary & as a somewhat mischievous speculation. I now regard it as admissible & as having made real tho' small progress—no longer to be treated as chimerical.

The admission of Man as the crowning operation of one & the same series of changes & progressive develop.^{t} compared at least analogically to embryonic development, & tho' pushed too far, having [p. 148] a real foundation, allowing for retrogression occasionally & the future change of all the limits of the extension of families—the successive exaltation of the sum of all the living types, their resemblance to the antecedent & next sequent set of animals & other generally acknowledged systems of those who oppose transmutation, all imply a drifting in one direction—viz. towards a theory nearer to that of the transmutationists than to that of creation by arbitrary or independent or special intervention & volition.

The agreement with Genesis, in the fact of Man having appeared last [p. 149] in the series, may have led some like Hugh Miller to favour this doctrine & others to take advantage of this coincidence in popularizing what they believed to be a scientific truth, but the greater number has arrived at the conclusion by identifying Man with animals as part of the same system, physiologically & psychologically identical, only differing in degree—& when this development by the successive coming in of new & higher types is embraced—when the races of Man are duly taken into account—when the formation of permanent varieties is granted & recognized—when the so called creative acts are counted by millions in each Period—when the adaptation of species to climate & new [p. 150] physical condit.^{s} is granted to resemble the accommodat.^{n} of acclimatized races to those conditions, we are getting nearer & nearer to the identif.^{n} of the two theories.

Birth of Man

What should we expect to see if admitted to behold the birth of a new species, Man for instance. Should we expect that at his nativity the "front of heav'n was full of fiery shapes," [94] that a meteoric storm brought down the germ of the new creature or that it stole in with as little disturbance as when an infant destined to grow up & change the destiny of a race or of mankind [p. 151] is born without any extraordinary attendant phenomena stealing quietly into the world. Would the first jump from irrational to the humblest of the rational individuals, which ever existed, imply so great a revolution as we might witness in our time & yet not deem it a revolution—or miracle.

Is it that we consider the machinery of generation, the mere earthly, corporeal material mechanism, that which is common to Man & the lower animals, capable of itself of evolving this intellectual & moral power—as the cause of the phenomenon. Is it not regarded as simply passive, as inert as matter [p. 152] so far as causation or creation is concerned? Genius has been often deified—regarded as an emanation from the Mind of the Universe—inspired. The term creation of an individual may be more appropriate than of a species. But the abortions are the stumbling block.

Geological Inferences from the High Antiquity of Man

1.st Long in a rude & stationary condition.
2.dly Coexisted with quadrupeds, now extinct.
3.d Survived a change of climate & geography.
4. May have lived wherever existing shells are found in a deposit, & therefore in the glacial period south of the icy regions.
5. May have been for ages only slightly superior in power to one of the lower animals.
6. Development in our time may show a [p. 153] geometrical ratio of progress of a higher race.
7. Birth of Man of Genius greater miracle & the appearance of the first rudest Man.

8. All the races may have sprung from one stock—& new races higher & lower may hereafter arise.
9. However much we may explain the coming in of Man as no revolution, still if it took a great part of a geolog.[1] epoch, it is so great a change that it disturbs our belief in the Uniformity of the Course of Nature. We cannot reason from the present to the past, nor from the present to the future, except when we exclude disturbing causes arising from the interference of the spiritual, moral & intellectual worlds with the material.

Progressionists

They have laboured to make Man the last & crowning operation of a system of development from the simple to the complex, from the invertebrate to the vertebrate, from fish to marsupial—from marsupial to Man—from the intelligence of the lower animals [p. 154] carried to its utmost point & thence to the human species.

Even before we had proofs of man having existed for myriads of years, before he attained a civilized state, it was a consistent inference that the least advanced races came first in the order of Time.

30 years ago when I publ.[d] the 1st Ed.[n] of this work I endeavoured to show that Man, as a rational & progressive being, stood in so exceptional a position & the proof of his forming an integral link of the same chain was so defective except as an animal, & then his superiority so doubtful, that I hesitated to regard the uniformity of the sequence of changes if Man was to be included.

The chief objection to the hypothesis [p. 155] of transmutation was naturally the inseparable connexion which it established between Man & the lower animals. But when once the human race is recognized as only the last of a series of developm.[t] embracing not only the physical structures, but the instincts and that intelligence, which is considered to differ from mere instinct, & when once a variety of races of man are acknowledged, differing in their capacity, all develop.[d] chronologically, it was clear that the two theories were approximating so nearly, that it w.[d] not matter greatly whether the progression so gradual was effected by the variety making power or by some hitherto undiscovered process.

We behold how the passage [p. 156] from a very humble to an ordinary, & from an ordinary to an extraordinary, mental or moral capacity may be made. It can take place without disturbing what we call the everyday course of Nature.

Can we explain it?—not in the least. Is it not a miracle? What in the fall of a meteoric storm, if we witness it for the first time, can compare to this? Suppose the first instance to occur of a man of very superior intellectual powers? What w.[d] the sudden appearance of a new plant or bird or quadruped be in comparison? The one they would follow & obey & continue to worship after death—they w.[d] regard his inventions as sent from heaven, his instructions as emanating from some higher power. Before experience [p. 157] how much more easy w.[d] it be to imagine that a dog sh.[d] give birth to a lion than that in the course of many generations such a difference sh.[d] arise as that between the humblest & most intellectual of human beings.

If there is a case where we sh.[d] have a right to call in the direct intervention of the Almighty it is here rather than in the first introduction, by whatever machinery, of a new species. And yet the phenomenon occurs & at his "nativity" we do not find—"the front of heaven, etc." [95] there is no more sensation than if any one of the lower animals had produced & man had never been born.

Development

—is said by Von Bär,[96] cited by Huxley, to be "the gradual differentiation of the simple into the complex." [97] "Development," says Huxley," not only begins in an ovum but ends in the [p. 158] production of one.[98]

"The vital changes are cyclical; beginning with an ovum, they end in an ovum. Tho' there is a physical continuity between the progenetrix of a race & her latest progeny yet the material substratum presents a series of distinct tho' constantly recurring cycles of form."

How can the egg be simple if it is the last term of the cyclical series? It contains within it all that is to be evolved between the egg & the adult—& the capability of future perpetuation—variety & racemaking—sets of different "recurring cycles of form"—cowslips & primroses all in one seed. If a seed or egg tho' simple [p. 159] may contain in itself the germ of all the future adults of distinct permanent varieties, it is then, however simple, the germ of things complex which Time has to develop.

Some of these future varieties can only come into play when states of the globe, not yet evolved in geology & physical geography, are accomplished. Whether progressive or retrogressive there is a chronological order predestined for them. Yet the egg or seed in all its apparent simplicity contains within it the latent germs of all this prospective organic, instinctive, intellectual & spiritual & moral world to come—a microscopic spermatozoon.

Omne vivum ex ovo. Every living thing proceeds from an egg was said by Harvey.[99] It is a maxim usually [p. 160] opposed to 'spontaneous or equivocal generation.'

That creatures organic may arise spontaneously, or 'not from preexisting forms of life', but make their appearance otherwise, is in truth the theory of creation by the direct volition or interference of the supreme being. It w.[d] be unjust to suppose the advocates of spontaneous generation to mean that the origination of living beings c.[d] take place independently of the will of God.

Spontaneous Generation

The whole course of modern investigation tends to show that the probabilities ag.[t] it are enormous. It was a refuge of ignorance. May not the assumption of it in geology be equally so? At least [p. 161] it is safe to assume that the new form came from preexisting forms of life rather than spontaneously, equivocally, independently—they are governed in their entrance into the world by some law—not by 3 million of miracles in the course of each period.

Species (Huxley)

Naturalists assume that species have physiological characters by wh. they can be distinguished from all other kinds.

1. The individ.[s] of a species are supposed to be descended from a single parent or pair of parents originating independently of all other living beings.

2. Their modifiability is supposed to be limited.

No evidence of unity of parentage of any one species, how did the first origin [occur]—[was it] created?—the question then passes out of the domain of science. [p. 162] In fact we know that individuals are modifiable by conditions parental, climatal, educational; any one may be made to deviate & it will tend to transmit the deviation to its offspring.

No evidence of the existence of a limit to the modification by climate & training—very slow process.

Mere races very persistent—

Hybrids—
Individuals
Species
Families
Orders
Classes
Subkingdoms

all these are expressions of fixed morphological laws—no real transition from group to group.

Amphioxus [is] not a link between Vertebrata & Annulosa.
Cirripedes [are] not between Annulosa & Mollusca.
Polyzoa [are] " " Mollusca & Polyps.
Echinoderms [are] " " Polyps & Annulosa.

All fixed & absolutely limited groups.
[p. 163] Laws of definite combination as of Animal Forms.

Nature Creates Only Individuals, *Indigenous Races,* p. 361 [100]

This doctrine, which w.[d] introduce the direct agency of the Creator as the immediate source of individuality, labours under the difficulty that abortion, inferior, diseased, immoral, stupid, insane creatures are not the result of laws but of special intervention—as Voltaire w.[d] say, granted, provided you allow that the Devil *a noyé les centres*—

Centres of Languages

Each organically diverse tongue had a certain centre to wh. we may trace it back, Gliddon p. ——[101] If we could not assume that it had *one* centre there w.[d] be an end of one of the most important branches of ethnology & philology. So if not only the same language but the same species c.[d] spring [p. 164] up independently in two places, there w.[d] be an end of the interest in physical geography of plants & animals in a retrospective & geological sense.

Languages are intensified by isolation & time—hybrid when they radiate & touch others—idioms & dialects wh. are contemporan.[s] resemble in space what the change of the same tongue does in time at two distant epochs.

The transmutation theory has the advantage that it will not allow of such arbitrary fiction as the simultaneous creation of the same species in distinct areas. It offers same restrictions.

The creation of as many types of a species as are required to account for distribution, when not understood, shows one of the effects of the licence to wh. the theory of miracles leads.

Unity of Creation

The imputation of being orthodox theologically [p. 165] need not attach to the monogonists, for Agassiz & others have shown that the old Testament, understood literally, implies two races with some of wh. Adam's sons intermarried.

The Astronomy, Meteorology, Geological sequence of creation of animals & plants, chronology being rejected, the agreement in regard to derivation from a single stock w.[d] cause them to suspect their scientific & philosophical acumen.

Miracle or Direct Intervention

If we call in a special act of the Creator in the case of each individual of high moral or intellectual endowment, we may find among the offspring of the same parents an idiot, an insane, an ordinary or stupid individual—& by this device to take refuge in some law, to escape a Manicheistic theory.[102]

Varieties

[p. 166] When once a plant has gone wrong, *ebranlée affolée,* there is no end to the vagaries it will play. De Candolle—vol. 2. p. 1100 [103]

Fossil Plants Monocotyledon

—in Bronn's Table II. p.[19],[104] we find 48 as the number of the Coal Monocot.[s] & all the other format.[s] much less. The Miocene 41, wh. is the highest number.

Development, Bronn's Entwickelungs Gesetz, 1858 [105]

His argument about the superabundance of the sinu-palliata above the integri-palliata—the former living in shallower water & by sea beaches is dangerous—for in Paleozoic times the deep-sea forms were most common, those which we now have extant. So of Palliobranchiata, they were deep sea forms.

[p. 167] If it be said that the abundance of deep seas (like lakes in Canada containing Uniones) gave rise to a variety of Brachiopods, we must grant that there seems some reason to admit this prevalence of sea.

Boucher de Perthes, Vol. 1 [106]

Sepulchral monuments of oldest races reveal often much of their history—show that they wished memorials & to live after their death.

The Abbeville hatchets [are] many times older than the Pyramids.

Forests [were] full of game & rivers full of fish before the first savage inhabitants of Gaul knew how to sow.

When there was a surplus population (p. 18) in Asia, Africa

& parts of Europe, antecedent to 500 B.C.—there must have been savages at least in Gaul tho' no records of them.

[p. 168] Bones, shells, stones (flints) are the materials of the first instruments, wood also, but that decays.

October 26, 1859 [107]

Selection

In order to select a certain number of properties, powers, forms, qualifications must be laid before the breeder.

The inventive power is always at work, every variety or monstrosity so called may be looked upon as offered & according to the circumstances it may or may not be chosen & accepted. The original creative power *"varietas Naturae insatiabilis"* [108] throws these out; if they are not accepted they form only one of that larger number of individuals never intended to reach even an adult state. They are no more to be regarded as factors than the majority of beings cut off in childhood or nonage. They are manifestat.[s] of the inexhaustible resources of the creative power. If they last two generations they are more [p. 169] likely to last three & to get a fixed set, [namely] a genealogical tendency to perpetuate the like. But it is not the selection which is the primary power. The law may be uniform as one which secures variation so that no two individuals of a species were ever quite alike in every physiolog.[l], physical, psychological peculiarity. Variation may be the rule admitting of no exception.

Glacier Motion

Alfred Wills, A.N., author of Ch. 1, in *Peaks, Passes & Glaciers.*[109]

Prof. J. D. Forbes[110] called the viscosity of ice, bruising & reattachment correspond.[g] to Tyndall's "fracture & regelation." [111] *National Review,* 1859,[112] Glaciers & Glacier theories (Hopkins[113] omitted?) See also *N. British* by Brewster.[114]

London, October 28, 1859 [p. 170]

[Lyell to Darwin] [115]

My dear Darwin,

Many thanks for your clear explanation of Pallas's views.[116] Carpenter[117] tells me that the variations of period of gestation in cows, rabbits & some other domesticated animals, is said by experimenters of authority, qy Gärtner? [118] to vary as much as ⅛.[h]. Nevertheless if it be true that dogs have the same period as the wolf, I sh.[d] infer that they all came from that wild species if we are to derive them from some one wild species of the canine genus rather than from another or from foxes or jackals or any other species [p. 171] differing in period of gestation. If some varieties of dog approach the fox, others the wolf, this may have to do with "atavism," as De Candolle calls it, or inheritance from their common ancestor ascending beyond any of the living wild species?

As soon as I get your book I shall attentively reread it & try to reconcile the admission with several propositions there laid down as to sterility of hybrids & laws of variation which as I read them seemed to me fundamentally opposed to the admission of our existing races of the dog coming from several wild or aboriginal canine species.

I understand you to say that selection under domestication rarely does in a shorter time what Nature might do in a longer. [p. 172] I also thought, to borrow Hooker's language, that variation was a centrifugal, & crossing or hybridism, a centripetal force. This would all accord with the doctrine that after a wolf, fox & jackal existed descending from some remote progenitor, one of them, the wolf, might produce the Esquimaux dog, the bull dog, the Italian greyhound, the shepherd's dog & so on, but if the three wild species above mentioned could produce any or all these races of dogs (or incipient species) by coalescing & interbreeding, I find it impossible to hope to trace any clue to the past transformations of species or their [p. 173] probable birthplaces.

I cannot help thinking that by taking this concession, one

which regards a variable species, about which we know most (little tho' it be) an adversary may erect a battery against several of your principal rules, & in proportion as I am perverted I shall always feel inclined to withstand so serious a wavering. I marked your passage of northern species over tropical plains, & approved of it, as you may remember I once proposed to bring certain Madeira shells over Africa during the glacial period.

C.L.

November 5, 1859 [p. 174]

Carabus of Mundesley

Weston says the elytron is Broscus cephalotes—common on most of our coasts. Jno. O. Weston, Taylor Institute Oxford.[119]

Development, Bronn, Tracts 2, xvi, p. 9 [120]

If it be the law that the less perfect or lower forms, as ferns e.g., when they had the world to themselves, & ample space & verge eno', produc.[d] varieties without number, some very high, exceeding their actual grade, but when the more exalted dicotyledons came they were reduced to smaller dimension, then the manufacturing of species of passage w.[d] be greatest.

Just as Reptiles approach nearest to Birds & Mammals, when the Pterodactyl & Iguanadon prevailed. This w.[d] agree with Owen on Wealden.[121]

Darwin's Theory

The question is not whether this theory of the variety-making, viz. the species-making, [p. 175] power is better than any other antagonistic theory for no other *vera causa* has yet been invented or suggested.

I have formerly supposed in P. of G. 18—— that some unknown power or law or Creative Force interfered—a miraculous power is simply to cut the Gordian knot.

The question therefore is whether this hypothesis accounts for many phenomena, while no other hypothesis founded on fact does so. Spontaneous generat.^{n}, if it had been establish.^{d} in any one single case, w.^{d} be no doubt a *vera causa*—but this I do not believe.

Now since the last edit.^{n} of the *Manual* was published in 185[5] many excellent public[ation].^{s} such as those of De Candolle & Hooker, of Wallace & Darwin have made it far more probable that the variety & race-making force is greater than we supposed, [p. 176] that there is a great & constant tendency to divergence from parent type, but also that the admiss.^{n} of such a tendency to be indefinite w.^{d} solve a multitude of problems in departments of Geol.^{y} & Natural History which are perfectly *independent,* geographical—geological—morphological problems in the vegetable as well as in the animal creation—& in the animal, throwing light equally on the invertebrate & vertebrate. If this be so the hypothesis becomes sufficiently strong in the total absence of any rival one whatever for a succession of miracles, of which no one has been witnessed by man, cannot be s.^{d} to be a scientific hypothesis at all strong enough to make it our duty, as we pass in review the phenomena of successive periods, to halt occasionally & consider how few the facts presented [p. 177] to us can or cannot be reconciled with it—& how far the missing links in the chain can be explained by the fragmentary nature of the evidence inferred from independent facts & reasonings.

At the same time the doctrine of progressive develop.^{t} must be taken into account & it will be shown that so far as Man is concerned many of the popular objections to it are the same as to the Darwinian hypothesis & the object.^{s} tho' often coming from those who embrace progression ought not to proceed from them.

Mill's Essay, p. 452, Vol. 2, on Whewell,
Moral Philosophy

"The person who has to think more of what an opinion leads to than of what is the evidence of it, cannot be a philosopher or a teacher of philosophers." [122]

Trees of a Forest-like Species [p. 178]

If a forest be large enough we may cut down several trees annually choosing adults and the aged ones & yet at the end of 50, 100 or any other number of years, find that the number of trees & the quantity of timber & foliage not diminished—& the variety of ages from the seedling to the oldest inhabit.[s] of the forest has not in the least diminished—we may ascertain that no fullgrown tree has started into being to replace those which have been removed & that no one has ever acquired suddenly a sensible augmentation of height or dimensions. The change has been constant & insensible & when we resolve to estimate it, being divided amongst so many thousands that it can only be measured by comparing individ.[s] from time to time & thus we discover that some are almost stationary for a time while others grew fast & that altho' the aggregate amount of individual change & growth maintains the aggregate [p. 179] diversity of forms & numbers & bulk yet some during the period of our experiments & observations are altered much less than others.

Antiquity of Man

Man's cosmopolite character will ensure him a longer endurance on earth than species which have a more limited geograph.[l] range. But to be cosmopolite implies high antiquity, the formation of many & distinct acclimatized races, some fitted for tropical, some for temperate & others for arctic regions. This fact alone therefore of so many distinct races implies a very high antiquity & the farther we trace back the history of the race the fewer will be the races till we sh.[d] end in our primordial race—according to the doctrine of inheritance from one parent stock.

Hence the discovery of greater remoteness of origin for Man & that remote ancestry lay in a savage state, ignorant of metals, favours the doctrine of the derivation of Man from one stock. The independent creation of each separate [p. 180] race is absolutely necessary for those who deny a very remote antiquity.

The same reason, which leads us to anticipate that Man will outlive the greater number of contemporary mammalia, will lead us also to expect that he may be traceable back into the past farther than the living mammals. But as man has only maintained himself as cosmopolitan by varying & diverging from a common stock & as he has already survived some changes of climate, we may safely conclude that some races have died out so that the extremes which now separate the European & Australian do not probably represent the whole amount of divergence wh. has already been witnessed on the globe.

If there be a law of progress & intelligence, the leading feature is the superiority & predominance of one race of Man over another, we must suppose aboriginal races inferior to any now existing in that particular, altho' they may be, as hunters, physically superior. This craniological develop.[t] may have been less [p. 181] than strength; sight, agility, power of endurance greater. Their senses [were] keener. Their instincts [were] more perfect —their corporeal gifts of a higher order.

Progressive Development

In the case of cephalopods we find them in all format.[s] of wh. we have a suff.[t] knowledge to enable us to reason on the general state of the mollusca—down to the Lower Silurian—we do not find first the cephalopods, then the gastropods & so in succession the other acephala disappear as we look back into remoter times according to the old-received notion & the way in wh. the simplest form precedes the more complex, for we find the highest order in all periods. But what do we find, the dibranchiata, as displayed by the Belemnites, in full force as far back as the Lias accomp.[d] by numerous shells, Nautilus-like forms, Ammonites & others & the dibranchiates also were persistent retrospectively as far even as the middle Trias. As to the soft cuttle-fish without internal bones like the —— we cannot discover them. At a time [p. 182]—Nov. 18, 1859—when the Nautilus (of 2 species) of the present day was represented by [the] Ammonite of innumerable forms & when the Sepia is represen.[d] by so many Belemnites, it wd. be a violent hypothesis that none of the soft forms existed. But when we examine the paleozoic strata we find only the tetrabranchiate & it is inferred that here at last we arrive at indications of a less complex develop.[t] of cephalopoda. The inference is only made by assuming that none of the soft bodied cephalopods like the —— then existed. It is therefore founded on negative evidence, but here comes in the general leaning towards a belief that as we recede in time we reach periods of less perfect development. The progressionist falls back upon the fact that tho' the reptiles were once more develop.[d] than now, yet the earliest known reptiles were less so, that mammalia cannot be [p. 183] traced back to the remoter periods, & which enters, much oftener than it is avowed, into the argument, 'that Man came last.'

Man Came Last

It is the only theory calculated to alter & subvert old receiv.[d] doctrines (viz, the hypothesis of a successive develop.[t] by some as yet undiscov.[d] law, or creation according to laws) which has ever been received with popular favour, other new doctrines tending greatly to enlarge our conception of the Universe & its Maker in time & in the extent & variety of his work, having been opposed with bigoted hostility, the only one which has been defended even beyond the fair & legitimate inference & which tho' often having to be corrected has stood its ground. It tends so to identify [p. 184] Man with a series of improvements, gradual & by steps, both in intelligence & structure with the preceding series of animals & the inferior races of Men with the higher, as to leave it, so far as consequences are concerned, no great choice which doctrine we adopt as to the mode in wh. it has pleased the Creator to introduce each new & advanced form into the organic system of the earth.

But they who have most steadily advocated this view are least charitable to those who conceive that hereditary descent or generation can have been the mode in which the changes have been brought about. They have prepared the way for the adoption of that view by showing that without it they must multiply miraculous acts, each race being called separately into being & independently—but they have been drifting towards it reluctantly & perhaps that [p. 185] implies that both are true. They approach what they desire to avoid, but this may show that the current of facts is too strong & that the wind which fills their sails, the gale of their wishes & preconceived desires to isolate man, is too feeble to enable them to keep clear of the rocks on which their old theology will be wrecked, tho' doubtless to create a new & improved one.

Indeed, the difficulties are not new for he who already knows that he himself has in a few years grown from an embryo animal to an adult state, that so may the majority perhaps never reach beyond the embryo, that so many inferior races with small range of abstract ideas coexist, to say nothing of those hundreds or thousands of the most favoured races who[se] reason is so feebly ripened that we may doubt if they are responsible beings, when it is considered that these facts were already known, we may simply consider the doctrine of transmutation, if establish.[d] for all species [p. 186] including the human, as only showing that in the past scheme, as compared to the present, there are the same difficulties & that there is a unity—that had we better included in our estimate of Man's position on the earth, his relation to the rest of animate Nature, we sh.[d] have found less wh. appeared anomalous in his relation with antecedent parts of the creation.

Progress Overstated

We must beware not to underrate the importance or truth of progress because it is so often overstated, founded on mere negative evidence as that the mammalian came in with the tertiary strata or the oldest member of the Eocene in wh. they have as yet been found.

Development

They who go farthest in support of progress are often most opposed to transmutation. So far therefore as they are in truth drifting towards the same [p. 187] goal their evidence is the more weighty in favour of transmutation or the mode of progression being by generation or the law of divergence by hereditary succession.

The repugnance of the defenders of progress to transmutat.[n] may be the stronger because they have already so far embraced the conclusion of a progressive series, not only physically but psychologically from the simpler forms up to Man, that it must follow, if transmut.^{n} be proved, as a general law governing the changes of one set of species to another in the next geological epoch, that Man has also come into the planet in the same way. If the embryology be assumed to point to developm.^{t} it will be immediately applicable to transmutation.

Transmutation [p. 188]

It is enough that there is somewhat better evidence in favour of the law of variation, that it should explain more phenomena than that of independent creation. If it be the only *vera causa* known, if it be the only hypothesis which we can advance & it explains more phenomena, than even an arbitrary theory expressly invented or imagined can explain, we may accept it for the present. It is entitled to preference.

For altho' it may be inadequate without assuming what we cannot for want of Time prove at present, it is only Time that may be required.

The laws of variation still acting around us, Darwin, p. 344, 345.[123]

Species

If Natural Selection & the principle of variation be true, the formation of New Species becomes a series of Acts of Prescience, not of after-thoughts. A self-repairing ever renovated [p. 189] system instead of the restoration of a decaying building—not new gains to make up for losses, but a perpetual supply of equivalent products & if we couple this with a progressive tendency, it is a still more elevated manifestation of increasing dignity, not of power as that from the first is omnipotent but as displayed on earth.

Monads are not required because higher organizations are not a necessary part of the hypothesis, provided there is a tendency in plants & animals to occasionally produce higher grades of organization. Whenever a simpler form is better fitted it may succeed both—suppose two [p. 190] varieties—one having greater fecundity of offspring & another some more specialized organ to compete, the former as simple as the parent & to whose variation the greater prolific power had gone & exhausted its strength, it may happen that the prolific variety may succeed & the other die out. Just as the increase of a bird may cause the extinction of a mammal or that of an insect the disappearance of cattle as in S. America. It is not always the highest in rank among species or genera wh. give way. So may it be with varieties —hence Prof. Huxley's permanent or persistent forms.

If the Teleosaurs of the Lias are now represented by the [p. 191] Gangetic Gavial, it does not follow that some ramification of that same type may not have led to the evolution of higher forms.

Development

Mr. Darwin has given a scientific & philosophical explanation of the Lamarckian system—not requiring the constant creation of monads, nor volition, but simply to grant what cannot be denied, that at a remote period certain types existed.

We have nothing to do with the beginning of life. As in regard to the inorganic world we may discuss whether the existing causes have given rise to the changes on the earth's surface in physical geography [p. 192] without being required to say whether we believe in an original fluid & incandescent nucleus or in the nebular hypothesis, so in regard to the origin of life we may start from the first period in which we find the several existing invertebrate types already in being & the fish as representing the vertebrata.

We may leave it to future geologists to decide whether the several types of vertebrate did or did not penetrate farther into the past—whether plants or animals were first introduced or all simultaneously, whether one, ten or a hundred original [p. 193] types were formed & whether as seeds & eggs or as adult beings or simple cells. These may become subjects of future science but at present we may be content to see whether in the many millions of years which have elapsed since the upper Silurian strata were formed it is more reasonsable to imagine the appearance of successive forms to have been brought about by ordinary generation, the principle of divergence, natural selection, the endurance of the stronger & most improved types, rather than by a series of acts resembling or analogous to the unknown exertion of power by which the first living beings or the planet itself was called into existence.

Astronomical Cycles [p. 194]

The revolution of the sun round a centre of gravity in the Pleiades requiring more than eighteen millions of y.[rs] for its completion may, as some think, comprise a change of climate (an approach to some stars wh. may modify the distrib.[n] of light & heat or the inclin.[n] of the earth's axis. Suppose such a period to be included in a part of the tertiary system. Half a century ago such an hypothesis w.[d] have been impossible.

But other & greater cycles may be imagined. Centres of gravity not within our astral system, but to which our own & other countless systems like the milky way may be comprehended & which may be to the revolut.[n] calculat.[d] by ——, what this 18 millions of years is to the time required for the Precession of the Equinoxes & even this revolut.[n] of only —— years must produce some slight periodical effect on organic life—when the extremes of [p. 195] declination of the earth's axis are considered. Plants approaching so near the pole in the carbonif.[s] era for example.

Large Genera

Hooker considers genera like the Hieracium, Rubus, Rosa & Salix in England as bramble, rose & willow, as newer & therefore not divided into distinct species—not yet time for isolation or extinction, variation unchecked.

Pallas' theory of several original species of Dogs.

Letter to C. Darwin, November 21, 1859 [124]

The admission which I least like among your familiar illustrations is that while the various pigeons have descended from one stock, the dogs have come from two or more species. It seems to me that an occasional cross with some one of the several wolves, the grey & the [p. 196] prairie wolf, or even perhaps some foxes & more distant species of the canine genus would have served all the purposes you require. Let these have been more frequent than in the case of most domestic animals, let the hybrid have always interbred with one of the pure parent stock, but the infusion of the foreign blood into the domesticated wolf, or whatever single species the dog came from, may have given rise to a tendency to reversion to some of the ancestral crosses, why should [p. 197] not they satisfy all your well-ascertained facts, leaving your canons of variation & hybridity unshaken. It is for the followers of Pallas to prove their opinion by some facts. The period of gestation of the dog & the wolf, if as constant as Bell believes, is a strong argument. As Bransby Cooper was a six months' child,[125] I doubt not you can find great occasional irregularity in the gestation of domestic animals & of Man, but still it is surely worth a good deal.

Life on Earth Not All for Man [p. 198]

The existence of life millions of years before Man was the first shock to our exclusive reference of all to us, the next is the approxim.[n] of Man to the inferior animals, lowering our pretensions. The next is the antiquity of Man as helping in the belief in his long existence in a lower state—against which we can only place the elevation of Man by the discovery of his true position, consid.[d] as an inhabitant of the earth.

His instincts, the religious especially, may point to worlds beyond but superstit.[n] [is] imperfect instinct, which knowledge—

Anticipation of Men

It has often been a subject of speculation whether some being, say incorporeal, visiting the earth before Man in the Pliocene Period might have foreseen the coming of Man.

They w.[d] have supposed the Eleph.[t] with its intelligence & trunk capable of picking up a needle or pulling up [p. 199] a tree by the roots to have been perhaps with an improved proboscis rather than an ape.

But there have been so many changes in mammalian life in

the tertiary period that the quadrumana extinct may be tens of thousands, to say nothing of the wide gap between the Chalk & Eocene. We cannot speculate till we know about the last species of Orangs & of races of Man.

Harvey might have discovered the circul.[n] of the blood, Bell the nerves, Owen the knowledge of the skeleton of the Orang, before Man. Linnaeus c.[d] not distinguish Man & the higher Apes.

Rudiments

The 5 vertebra of the tail of Man same as in higher Apes.

Soul's Revelation of Itself

It is not in the lowest, but in the most intellectual, advanced & educat.[d] races, that we see most of the religious aspiration after some thing beyond the material world. Not that the difficulties & doubts do not rapidly [p. 200] augment but that with increasing psychological developm.[t] there is more revelation of the spiritual from the only source from which it can spring, from the spiritual intercourse & communion, with something internal & external & not of this earth.

If an analogy would lead us to expect there will be future progress the difficulties will augment with increasing knowledge but there will be more faith in a higher nature.

The unity of Creation will point upwards as well as downwards. We regard the humblest races as of more dignity because they belong to the same species as the highest & so, if there be future development, we shall have to modify, not to abandon our theological speculation.

INDEX [p. 201]

“ Brighton Elephant bed.
4. Dicotyledonous wood, not gymnosperm, J. Hooker.
“ Silurian of Canada.
“ Permian of U.S. found.
“ Hippopot. in Brighton Eleph.^{t} bed. (glacial?)
5. Progressive develop.^{t}, if future races will be created, so past.
“ Genius by generation.
“ Bronn, earth improved habit.^{n} (but glacial eras).
6. Anthropomorphs tropical, yet Man cosmopolitan.
“ Effects of Lamarckianism, rational & irrational blend.
7. Develop.^{t} popular—Man last & lowest races of Man earliest.
33. 9. 11. 23. } Transmutation & Progressive system essentially the same.
9. Uniformitarian view less unorthodox
12. Pterodactyls prophetic of bats not birds.
“ Orthoceras more complex than Nautilus.
“ Embryonic form wanting in oldest rocks, no monomyeria.
13. Labyrinthodon, ancient batrachia, not embryonic.
“ Specific centres denied by Agassiz.

[p. 202]

14. Grove on progressive develop.^{t} & creation.
15. N. American Indians had quicker sight & hearing.
16. If one creation, why not many?
17. Generic types in Divine Mind.
“ This concedes indefinite variability.
18. Retrogression, Dinosaurs—glacial cold.
19. Cephalopoda—small progress since Silurian period.
“ Human Period—rapid progress.
20. This is consistent with Nature having made the greatest saltus in the way of generation in individual men of genius.
22. Monocotyledons ancient. Permian Palm.
“ “The rib he fashioned” Milton.
24. Elephant, superiority of—47, 58.
25. Anthropomorphs, deep meaning in the analogy of structure.

26. If no living orang known, a fossil one w.[d] have caused profound sensation.
" Instincts & reason, souls of brutes.
29. Negative evidence may often prove scarcity rather than total absence.
30.–34. Fish-sharks most perfect, came first, no progress in fish. Owen.
31. Spiritual things as much realities as material phenomena.

[p. 203]

35. Circulation of blood, etc., before Man.
36. Specialisation of parts in Pterodactyl very ancient.
38. Fish of old Red with dermal skeleton & bony internal one —Owen.
39. Horner's Egyptian, 1300 yrs. & Sharpe's objection.
40. Plato on immortality of soul.
" Metempsychosis—of Brahmin.
42. Argyll, D. of—Man coming last implies analogous rule of progression in the past.
" Malta—cave with hippopot.[s]
" Valleys might be filled up if delta mud replaced.
48. Ganoids might eat icthyodorulite fish?
" Lateral line in fish not for lubricating.
49. Archegosaurus embryonic.
50. Syngnathus in Zool. gardens.
" Transmutation, Grove on.
52. Equality of Greeks only proves 4000 yrs.
53. Embryonic labyrinthodont.
55. Advance of Man by generation affords no peculiar difficulty.
56. Ganoids now in S. hemisphere.
57. Wealden reptiles only 19, ergo, a mammal may be found.
" Semi-species of Helix, Lowe.

[p. 204]

57. Land animals fail soon as we recede.
58. Azoic period of Flysch.

60.–61. Progression of Owen & Agassiz—embryonic development of Owen
62. Antiquity of Man—still a newer Pliocene animal.
" Lost quadrumanous races coexistent with Man.
64.–65. Species, usual definition begs the question of invariability.
66. Creation by embryonic laws comes near to transmutation.
68. Transmutation or genealogical system why superior.
69. " accounts for more facts.
71. Progressionists nearly concede all.
72. Saltus of Man of genius (79), 82.
73. Man's late appearance.
76. Non-reversion of domesticated races.
77. Miracles, millions of, required.
79. Elephant, scale of intelligence.
80. Passage from brute to Man.
81. Madeira insects & Plants more like the Canary I. ones, than shells are like the shells.
83. No mammalia tho' 30 to 40 reptiles in coprolitic bed of Upper Green Sand, Cambridge.

[p. 205]

83. Gaps between reptiles & birds, etc.
85. Gaps in Geology filled up—St. Cassian.
" Species, not to believe in them, seems to some to abandon all knowledge.
86. But in Geology, divisions arbitrary & gaps fill.
87. Sun moves towards Hercules, causes all the constellations slowly to change.
88. Monotremata to be found in Coal.
89. Periods, length & difference of, to be estimated by proportion of genera which pass.
" Ganoidians, range of, Huxley.
91. Amphioxus.
92. Evolution of higher and of lower intelligences out of mere instinct.
" Pteraspis, Lower Ludlow.
93. Rhinoceros, species of, as newly named by Falconer.

94. Wollaston on constancy of specific character in Madeira insects & shells.
96. Helix Pisana, vars. of, Salvages species.
97. Man of Genius sudden leap (121).
98. Ganoids not more dominant perhaps in paleozoic times than are now the Teleosteans.
" Bony skeleton absent on embryonic type.
100. Placoids, vast predominance of, in paleozoic times.
" Homocercal fish in Trias—Egerton.

[p. 206]

101. Microlestes, in English Trias.
" Thecodonts of Bristol in Trias, not Permian, Etheridge.
" Phoca in Limburg beds (Lower Mio.), Belgium.
" Species—to Huxley—leaps from class to class—coming in of Man—progression without transmutation equally materialistic.
108. Persistent types in the Brachiopoda.
" Antiquity of Man, coexistence of barbarism & civilization, 110 (123).
111. Faraday, on education of mind.
112. Ancestry of gods aimed at & degeneracy coveted.
" Single or several starting points of Man?
113. Agassiz, by abandoning specific centres, loses much.
114. Orbit of sun supp.[d] 18 millions of years (134).
115. Miracles always increasing by calling more & more complex organisms into being.
116. Embryonic theory of Agassiz.
" Creation governed by fixed laws—alternate generation—Pictet.
118. Man last & greatest miracle as the Alps last & greatest catastrophe.
" The undiscovered law of Creation may be the variety-making power, coupled with inheritance to fix it & Selection to direct it.
134.
119. } If for ages there were only savages so now how many contemporary savages, imbeciles, idiots, monstrosities.
120.

[p. 207]

123. Progressive develop.[t] by creation or by transmutation. Man being last.
125. Lyell's hypothesis in 1830 of Man being exception.
128. 130. 138. 146. } Genius—the great leap—Savages like imbeciles.
132. Aspirations for world beyond—Negroes.
" Millions of Species.
136. Sweden, movement of, cannot be supposed at Amiens.
138. Psychological leap may be the greatest after all.
139. Progressive developm.[t] greatly shaken by reptiles having been higher & fish as high in older times.
153. 140. } Man, crowning work, Harting.
141. 149. } Developm.[t] theory stated & shown to imply nearly as much as transmutation.
143. The stand sh.[d] have been made ag.[t] progress.
145. Glacial Period & qy. warmer period with hippopot.[s] after it? Thames drift.
149. Hugh Miller regarded Man coming last as supporting Moses.
150. "The front of heaven was full of fiery shapes."
152. Antiquity of Man, inferences from.
154. My original objection of Man being last link.
157. Man of Genius greatest miracle.
158. The egg or seed tho' simple contains all that is to follow of physical or spiritual.
161. Species, says Huxley, if we depend on their origin or first pair not within domain of Science.
162. Groups, says Huxley, do not run into each other.
163. 165. { Individuals if created, why often abortive, mad?
163. Centres of languages as of Species.

[p. 208]

164. Creation independ.[t] acts—abuse of the theory by Agassiz —same species twice created.

166. Vagaries played by a plant wh. has once gone wrong.
" Monocotyledons in Coal as many as Miocene, Bronn.
166. Bronn on early mollusca being simplest.
167. Savages in Europe, B. de Perthes, in Gaul.
168. Power in creation above selection.
170. Pallas's views. Letter to Darwin.
" Hooker, variety making centrifugal, hybridiz[ation] centripetal.
174. Weston on Mundesley Carabus.
" Diversity & improvement greatest when a class had the world to itself.
176. Indefinite power of divergence is the question.
177. Mill—a philosopher is not thinking of what an opinion leads to, but of the proofs.
178. Growth of new Species like trees in a forest.
179. Cosmopolite Man must be ancient. All the races have to be traced back to one stock.
180. Extinct races of Man probable.
181. Cephalopoda, soft ones in oldest rocks? Dibranchiata.
183. Man came last, popular tho' subversive of old ideas when linked with developm.[t]
185. }
187. } If adult Man from embryo, why not all?
192. Pregeologic beginnings of life not our business.
194. Astronomical cycles.
" Hooker thinks Rosa & Rubus modern creations.

[p. 209]

195. Pallas's theory of dogs, objections to.
" Period of gestation.
198. Man, antiquity of.
" Anticipation of Man on earth. Elephant.
199. Man's rudimentary tail.
" Soul's revelation of itself.
200. Unity of system pointing upwards & downwards.

Notes

1. In 1857–58, Joseph Ellison Portlock (1794–1864), major general in the royal engineers and a geologist, served as president of the Geological Society of London and on February 19, 1858 delivered the usual president's address at the anniversary meeting of the society. See Major-General Joseph E. Portlock, *Address delivered at the anniversary meeting of the Geological Society*, London, Taylor and Francis, 1858, 153 pp. Hugh Falconer had prepared an abstract of recent discoveries of fossil elephants in Pliocene strata for Portlock to include in his speech.

2. This note is bewildering.

3. Lyell was presumably going to look for Falconer's letter in the drawer of the secretary, a combination desk and bookcase.

4. Lucas Barrett (1837–62), geologist and naturalist. See page 74 for further detail.

5. Thomas Davidson (1817–85), geologist and palaeontologist, was persuaded by Leopold von Buch in 1837 to devote himself to the study of brachiopods. He later undertook to do a monograph of British fossil Brachiopoda for the Palaeontographical Society. The first volume appeared in 1850 and the third in 1870. With supplements, the whole work makes six volumes.

6. Sir Frederick McCoy (1823–99), naturalist and geologist, was appointed professor of natural science at the University of Melbourne, Australia, in 1854.

7. Sir William Edmund Logan (1798–1875), geologist, in 1842, was appointed the first director of the Geological Survey of Canada, a post he held until 1870.

8. F. V. Hayden and F. B. Meek, "Descriptions of new organic remains from north-eastern Kansas, indicating the existence of Permian rocks in that territory" (1858), *Albany Inst. Trans.*, 1858–64, *4*, 73–88.

Ferdinand Vandiveer Hayden, M.D. (1829–87), geologist, in 1853, went with Fielding Bradford Meek (1817–76), a palaeontologist who had previously worked with James Hall on the New York State

Geological Survey, to collect fossils in Nebraska, the Badlands of South Dakota, and other areas of the Great Plains.

9. Thomas H. Huxley, "Science and Religion," *The Builder,* 1859, *18,* 35–36.

10. Sir Henry Holland (1788–1873), physician. We have not been able to locate the article in the *Edinburgh Review* to which Lyell refers.

11. Baden Powell, *The unity of Worlds and of nature: three essays on the spirit of inductive philosophy; the plurality of worlds; and the philosophy of creation,* 2d ed., London, Longman, Brown, Green, Longmans and Roberts, 1856, 556 pp.

The Rev. Baden Powell (1796–1860) had been appointed Savilian professor of geometry in Oxford University. He wrote on theology and philosophy as well as on physical and mathematical subjects.

12. Heinrich Georg Bronn (1800–62), palaeontologist of Heidelberg, Germany.

13. Louis Agassiz, "Essay on Classification," in Agassiz, *Contributions to the Natural History of the United States of America,* Boston, Little, Brown, 1857, 452 pp., *1,* 3–232.

14. In his "Essay on Classification," Agassiz discusses the reptile-like fishes that existed in geological periods before the appearance of true reptiles. These fishes resembled reptiles and performed similar ecological roles in the natural order. Similarly, the pterodactyls had filled the place of birds before the appearance of birds in the course of geological time. Such types Agassiz called *prophetic types,* because he considered that the reptile-like fishes were prophetic of the reptiles to come and the pterodactyls were prophetic of the later appearance of the birds.

15. Sir William Robert Grove (1811–96), English barrister and later judge, who also was an eminent physical scientist. It seems to have been the custom of Grove to circulate letters on subjects in which he was interested and the letters "L.C." probably refer to "Letter Circulated," the letter in this case dealing with progressive development. See "A Letter having been circulated, addressed by Mr. Grove to Sir C. Lyell, on the subject of Mr. Grove's opinions as to the desirableness of the Physiologists of the Royal Society forming a Society for themselves," etc. (*Correspondence between W. R. Grove and J. Bishop,* London, Bowerbank, 1848; listed in British Museum Catalogue) for an example of such a circulated letter.

16. Sir Benjamin Collins Brodie, F.R.S. (1783–1862), surgeon and physiologist.

17. Hans B. Geinitz, *Die leitpflanzen des Rothliegenden und des Zechsteingebirges, oder der permischen formation in Sachen . . .* , Dresden, Teubner, 1858, 43 pp. Hans Bruno Geinitz (1814–1900), German geologist, wrote various works on the Permian and Cretaceous formations of Germany.

18. "The rib he formed and fashioned with his hands" (John Milton, *Paradise Lost,* Book 8, line 469).

19. Ibid., lines 369–74:
What call'st thou solitude? Is not the Earth
With various living creatures, and the air,
Replenished, and all these at thy command
To come and play before thee? Know'st thou not
Their language and their ways? They also know,
And reason not contemptibly . . .

20. Thomas Babington Macaulay, review of *Joannis Miltoni, Angli, de Doctrina Christiana libri duo posthumi,* Charles R. Sumner, trans., 1825, *Edinburgh Rev.,* 1825, *42,* 304–46, p. 313:

> What is spirit? What are our own minds, the portion of spirit with which we are best acquainted? We observe certain phenomena. We cannot explain them into material causes. We therefore infer that there exists something which is not material. But of this something we have no idea. We can define it only by negatives. We can reason about it only by symbols. We use the work; but we have no image of the thing; and the business of poetry is with images, and not with words.

21. William Ellery Channing (1780–1842), Unitarian clergyman, was from 1803 on minister of the Federal Street Church, Boston, Massachusetts.

22. Probably Isaac Taylor (1787–1865), artist, writer on religion, and a Unitarian.

23. This appears to be a reference to the Fullerian course of lectures on physiology given in 1859 by Richard Owen, which Lyell must have attended. See Richard Owen, *On the Classification and Geographical Distribution of the Mammalia,* London, Parker, 1859, Appendix A, "On the extinction and transmutation of species," pp. 58–63.

24. Roderick I. Murchison, *Siluria, a history of the oldest fos-*

siliferous rocks and their foundations; with a brief sketch of the distribution of gold over the earth, 3d ed., London, Murray, 1859, 592 pp.

25. Moses Henry Perley (1804–62), Canadian naturalist and businessman.

26. Isaac Taylor (1787–1865). See note 22, above.

27. See note 21, above.

28. Ibid.

29. Ibid.

30. In 1851, Leonard Horner (1785–1864), geologist and educator, had arranged to have an Armenian engineer, Mr. Hekekyan Bey, supervise the excavation of a number of pits around the obelisk of Heliopolis and later around the statue of Rameses to determine the thickness of sediment that had accumulated since these monuments had been set up. From the known age of the monuments, Horner calculated the age of the sediments accumulated in the Nile Valley as a whole. He found 9 feet 4 inches of sediment accumulated around the temple of Rameses, which he estimated to have been in place 3,215 years in 1854. The rate of accumulation was therefore about 3.5 inches per century, and the 30 feet of sediment in the Nile Valley at that point represented an age of about 13,500 years. Leonard Horner, "An account of some recent researches near Cairo undertaken with the view of throwing light upon the geological history of the alluvial land of Egypt," *Phil. Trans. Roy. Soc. London,* 1855, *145,* 105–38; 1858, *148,* 53–92.

In 1859, Samuel Sharpe (1814–82), Egyptologist, had criticized Horner's conclusions in an unsigned article in the *Quarterly Review.* Sharpe contended that inasmuch as the temple that contained the statue of Rameses was one of the buildings of Memphis and would have been protected from the annual inundation of the Nile, "We may therefore feel certain that the Nile deposit did not begin to accumulate at the base of the statue till Memphis had fallen into ruins about the fifth century of our era." The period of accumulation was thus shorter and the rate of accumulation more rapid than Horner's measurements indicate. See Samuel Sharpe, "Art. IV. [review of] 1. *Aegypten's Stelle in der Weltgeschichte—Geschichtliche Untersuchung in fünf Buchern.* Von Christian Carl Josias Bunsen. Vols. I–V, Hamburg and Gotha, 1844–1857. 2. *Egypt's Place in Universal History—An Historical Investigation, in five books.* By Christian C. J. Bunsen. Translated from the German by Charles H. Cottrell. Vols. I–III, London, 1848–1859. 3. *An Account of some recent Researches near Cairo, undertaken with the view of throwing light upon the Geological History of the Alluvial Land of Egypt.* Instituted by Leonard Horner, Esq. From the Philosophical Transactions, Parts I and II. London, 1855 and 1858," *Quart. Rev.,* 1859, *105,* 210–32, pp. 230–32.

In 1864, Sir John Lubbock pointed out that Sharpe's objection was invalid because after the embankments had broken down sediment would accumulate in the previously enclosed area more rapidly, but would do so only until the level of sediment in the enclosure equaled that outside. It would not rise above the surrounding level. See John Lubbock, "Age of the Nile Valley," *The Reader,* 1864, *3,* 399.

31. "Plato," in *Penny Cyclopedia of the Society for the Diffusion of Useful Knowledge,* 27 vols. in 14, London, Knight, 1833–43, *18,* 233–41.

32. Henry T. Colebrooke, "Essay on the Philosophy of the Hindus," *Trans. Roy. Asiat. Soc. Great Britain & Ireland,* 1830, *2,* 1–39, pp. 29–30:

> Of a dying person the speech, followed by the rest of the ten exterior faculties (not the corporal organs themselves), is absorbed into the mind, for the action of the outer organ ceases before the mind's. This, in like manner, retires into the breath, attended likewise by all the other vital functions, for they are life's companions; and the same retreat of the mind is observable, also, in profound sleep and in a swoon . . .
>
> But he who has attained the true knowledge of God does not pass through the same stages of retreat, proceeding directly to reunion with the supreme being, with which he is identified, as a river, at its confluence with the sea, merges therein altogether.

33. Herodotus, *The Persian Wars,* book 2, ch. 123, A. D. Godley, trans., 4 vols., London, Heinemann, 1920, *1,* 425:

> Moreover, the Egyptians were the first to teach that the human soul is immortal, and at the death of the body enters into some other living thing then coming to birth; and after passing through all creatures of land, sea and air (which cycle it completes in three thousand years) it enters once more into a human body at birth.

34. The Duke of Argyll, *Geology: its past and present, being a lecture delivered to the members of the Glasgow Athenaeum,*

January 13, 1859. Glasgow and London, Griffin, 1859, 50 pp., p. 28. George Douglas Campbell (1823–1900), eighth Duke of Argyll and Marquis of Lorne, was an ardent naturalist and geologist. He was interested in evolution and took a position strongly opposed to Darwin's after the publication of the *Origin of Species.*

35. Ibid., p. 32.

36. Ibid., p. 27.

37. Ibid., p. 35.

38. Ibid., p. 36.

39. Ibid., p. 39.

40. James Smith (1782–1867) of Jordanhill, geologist and historian.

41. Henry George Francis Moreton (1802–53), second Earl of Ducie, agriculturalist.

42. Joseph Prestwich, "On the occurrence of Flint-implements, associated with the remains of animals of extinct species in beds of a late geological period, in France at Amiens and Abbeville, and in England at Hoxne," *Phil. Trans. Roy. Soc. London,* 1860, *150,* 277–317. Prestwich's discussion of Boucher de Perthes is on p. 279. He read this paper to the Royal Society on May 26, 1859. Lyell had evidently attended and taken notes.

43. Charles Lyell, *Elements of Geology,* 6th ed., London, Murray, 1865, 794 pp.

44. William Kirby and William Spence, "On the instinct of insects," *An Introduction to Entomology: or Elements of the Natural History of Insects,* 4 vols., London, 1815–26, *2,* 466–529. In his *Principles* (11th ed., 1872, *2,* 327, Lyell refers to Kirby and Spence's description of thirty distinct instincts in insects, citing p. 504.

45. Franz Leydig (1821–1905), German zoologist, demonstrated in 1851 that the lateral line of fishes was a sense organ rather than a mucus-secreting gland. See Franz Leydig, "Über die Nervenknöpfe in den Schleimkanalen von Lepidoleprus, Umbrina und Corvina," *Arch. Anat. & Physiol.,* 1851, pp. 235–40.

46. Hermann von Meyer, "Über den Archegosaurus der Steinkohlen-formation," *Palaeontographica,* 1851, *1,* 209–15. Lyell was also in personal communication with Meyer. See Lyell, *Elements,* 6th ed., 1865, p. 510.

47. Sir William Robert Grove (1811–96). See note 15, above.

48. Meyer, "Archegosaurus."

49. After discussing the characteristics of the subclass of Ganoid fishes, which had been established by Louis Agassiz (bony fishes or cartilaginous fishes with well-defined bones in the cranium, protected by bony scales or indurated plaques), and which had been further distinguished by Johannes Müller by their possession of multiple valvules in the *bulbus arteriosus,* says François Jules Pictet (1809–72), naturalist and palaeontologist of Geneva, "M. Vogt has since established that the remarkable genus Amia, which lives in South American rivers also has multiple valvules. It must be grouped with the Ganoids" (*Traité de Paléontologie ou histoire naturelle des animaux fossiles considérés dans leurs rapports zoologiques et géologiques,* 2d ed., 4 vols., Paris, Baillière, 1853–57, *2,* 23–24).

50. John Morris, *Catalogue of British Fossils . . . with references to their geological distribution and to the localities in which they have been found,* London, Van Voorst, 1843, 222 pp.; 2d ed., London, privately printed, 1854, 372 pp. John Morris (1810–86), pharmaceutical chemist and geologist, in 1854 was appointed professor of geology at University College, London.

51. Richard Owen, "Palaeontology," in *Encyclopedia Britannica,* 8th ed., 1858–59, vol. 17.

52. In 1857, when Dr. Hugh Falconer described the fossil genus *Plagiaulax,* he considered these animals to have been herbivorous marsupials. Owen, on the other hand, argued from the upward curve of the incisor in *Plagiaulax* and the presence of grooves on the teeth that they resembled those of the carnivorous fossil *Thylacoleo.* Falconer denied the presence of furrows in *Plagiaulax* at all comparable to those in *Thylacoleo* and, when Lyell examined the actual fossil specimens, he agreed with Falconer. See Lyell, *Elements,* 6th ed., 1865, pp. 381–82.

53. Christopher Puggaard, "Übersicht der Geologie der Insel Möen," *Neues Jahrb. Mineral., Geogn., Geol. & Petrefakt,* 1851, pp. 791–809.

54. Captain Thomas Abel Brimage Spratt (1811–88), vice-admiral, hydrographer, author, in 1832 went as a young naval officer with Lieutenant Thomas Graves to survey in the Mediterranean, where he became interested in the geography, geology, and archaeology of Greece. See Thomas A. B. Spratt, "On Crete," *J. Roy. Geogr. Soc.,* 1854, *24,* 238–39.

55. Lyell, *Principles,* 1832, *2,* 1–65.

56. "Natura enim non facit saltus," Linnaeus, *Philosophia Botanica,* Stockholm, 1751, p. 36.

57. Joseph Rudolph Albert Mousson (1805–90), naturalist of Zurich, "On the land shells of Lanzarote and Fuerta Ventura; with observations on the Molluscan fauna of the Canary islands in general," R. T. Lowe, trans., *Ann. & Mag. Nat. Hist.,* 1859, ser. 3, *3,* 81–91.

58. "Igenium cui sit, cui mens divinior atque os Magna sonaturum des nominis huius honorem" (trans: "To him whose nature and mind are of a more divine stamp and who speaks with a magnanimous voice you may give the honor of this name"), Horace, *Satires,* book 1, sect. 4, lines 43–44.

59. The St. Cassian beds are a marine formation of Upper Triassic age outcropping on the southern edge of the Austrian Alps. The Triassic beds of northern Germany and England had been deposited for the most part in brackish or freshwater so that, until the St. Cassian beds were studied, very little was known of the marine fauna of the Triassic period. Previously, there had seemed to be a sharp difference between the marine fauna of the Lias and that of the Devonian and Silurian rocks lying beneath the coal measures that were a freshwater or swamp formation. The St. Cassian fauna contained a mixture of fossil genera, some common to the St. Cassian beds and the older formations, others characteristic of the St. Cassian beds, and still others common to the St. Cassian beds and the later Jurassic and Cretaceous formations. See Lyell, *Elements,* 6th ed., 1865, pp. 429–36.

60. Marine equivalents of the Muschelkalk and Bunter beds (Middle and Lower Triassic) had not then been discovered.

61. George Berkeley (1685–1754), Bishop of Cloyne, Ireland, in his *An Essay Towards a New Theory of Vision,* Dublin, Pepyat, 1709, 187 pp., and his *A Treatise concerning the Principles of Human Knowledge,* Dublin, Pepyat, 1710, 214 pp., denied the real existence of the external world and said that it existed only as a construct in our minds formed from our sense impressions.

62. Hipparchus, Greek astronomer, born at Nicaea in Bithynia about 190 B.C., made astronomical observations between 161 and 126 B.C.

63. "Everything falls headlong into a worse state."

64. This specimen of Pteraspis found by Mr. Lee was significant to Lyell as the oldest known fossil fish because it was found in a bed of the Lower Ludlow formation (Silurian), whereas the oldest fossil fish previously known had been from the Upper Ludlow. Lyell, *Elements,* 6th ed., 1865, p. 552.

65. Georges Cuvier, *Recherches sur les ossemens fossiles,* 4th ed., 10 vols. and 2 vol. atlas, Paris, Baillière, Levrault, Crochard, Roret, 1834–36, atlas, vol. 1, plate 49.

66. Ibid., atlas, vol. 1, plate 47.

67. See note 57, above.

68. This unconnected reference to the *Philosophical Magazine* (which had earlier been the *Annals of Philosophy*) may have been a memorandum by Lyell to himself. In the July issue of the *Philosophical Magazine,* Lyell published a reply to some criticisms by C. Piazzi Smyth of Lyell's views on the origin of Teneriffe and other volcanoes in the Canary Islands. See Charles Lyell, "Remarks on Professor C. Piazzi Smyth's supposed proofs of the submarine origin of Teneriffe and other Volcanic cones in the Canaries," *Phil. Mag.,* 1859, *18,* 20–22.

69. Heinrich G. Bronn, *Untersuchungen über die Entwickelungs -Gesetze der organischen Welt während der Bildungs-Zeit unserer Erd-Oberfläche. Eine von der Französischen Akademie im Jahre 1857 gekrönte Preisschrift . . . Deutsch herausgegeben,* Stuttgart, Schweizerbart, 1858, 502 pp.

70. Ibid.

71. Sir Philip de Malpas Grey Egerton (1806–81), member of Parliament and amateur palaeontologist, built up a valuable collection of fossil fishes, which he bequeathed to the British Museum of Natural History.

72. Robert Etheridge (1819–1903), geologist.

73. John Carrick-Moore, English geologist of the mid-nineteenth century.

74. This appears to be a reference to Joseph A. H. Bosquet, *Recherches paléontologiques sur le terrain Tertiaire du Limbourg Néerlandais,* Amsterdam, Koninklijke Akademie van Wettenschappen, Verslagen en Medecleelingen, 1859, vol. 7, 30 pp. Joseph Augustin Hubert Bosquet (1813–80) was a palaeontologist of the Netherlands.

75. This seems to be a reference to persons whom Lyell had observed during his travels in America.

76. This marks the end of the continuous extract from Lyell's letter to Thomas H. Huxley, June 17, 1859. The original of this letter is in the Huxley mss. There are frequent differences of wording so that the extract is more a paraphrase than an accurate quotation.

77. Samuel S. Haldemann "Enumeration of the recent Freshwater Mollusca which are common to North America and Europe; with observations on species and their distribution," *J. Nat. Hist. Soc. Boston,* 1843–44, *4,* 468–84.

In his letter to Huxley, Lyell continued:

I stumbled yesterday on a paper in the Boston Journal of Nat.[l] Hist.[y] for January 1844 by S. S. Haldemann in which the transmutation theory is defended in a spirit & with a skill that appears to me to deprive Wallace of much of the originality of his two Essays.

I knew Haldemann in 1845, a good conchologist. Do look at his paper "On recent freshwater mollusca," etc. He is controverting my arguments against Lamarck in my 'Principles of geology'. I quite forgot his paper when I re-edited the Principles & when at the Linn. Soc.[y] I made so much of Wallace.

Samuel Steman Haldemann (1812–80) was an American naturalist and geologist who worked on the geological surveys of New Jersey and Pennsylvania.

78. Lyell to Huxley, June 17, 1859, continued:

If you know of any work of Geoffroy St. Hilaire in which I could get at the first enunciation of the Progressive Development & transmutation theory I sh.[d] be glad to have a reference. I conceive that Lamarck was the first to bring it forward systematically & to "go the whole orang"

believe me
ever truly yours
Cha Lyell

Étienne Geoffroy Saint-Hilaire (1772–1884), French zoologist, had insisted on the unity of type in the animal kingdom and especially during the later part of his life had been a leading advocate of the theory of transmutation of species.

79. Thomas Davidson (1817–85). See note 5, above.

80. S. P. Woodward, *A Manual of the Mollusca; or, a rudi-*

mentary treatise of recent and fossil shells, 3 vols., London, Weale, 1854–56, 486 pp., *2,* 214.

> Of all shell-fish the Brachiopoda enjoy the greatest range both of climate, and depth, and time; they are found in tropical and polar seas; in pools left by the ebbing tide, and at the greatest depths hitherto explored by the dredge. At present only 70 recent species are known; but many more will probably be found in the deep-sea, which these shells mostly inhabit. The number of living species is already greater than has been discovered in any *secondary* stratum, but the vast abundance of fossil *specimens* has made them seem more important than the living types, which are still rare in the cabinets of collectors, though far from being so in the sea.

81. John Crawford, M.D. (1873–68), orientalist, had spent many years in Malaya and adjacent countries. In 1852, he published a *Dictionary of Malay Language* and, in 1856, a *Descriptive Dictionary of the Indian Islands and adjacent countries,* in addition to many scientific papers on the languages and races of the Far East.

82. Michael Faraday, "Observations on the Education of the Judgement. A lecture delivered at the Royal Institution of Great Britain," reprinted in E. L. Youmans, ed., *The Culture Demanded by Modern Life,* New York, Appleton, 1893, pp. 185–224, p. 214.

83. Thomas Story Spedding (1800–70), barrister and country gentleman of Keswick, Cumberland. Spedding had been a friend of Lyell from the time that they were both law students in London during the 1820s.

84. The reference on p. 91 of the Index Book is to Ormsby Macknight Mitchel (1809–62), *The Orbs of Heaven, or, the Planetary and Stellar Worlds,* 5th ed., London, Cooke, 1854, p. 220.

85. Louis Agassiz, *An Essay on Classification,* London, Longman, Brown, Green, Longmans, and Roberts, and Trubner, 1859, 381 pp., p. 10:

> If it can be proved that man has not invented, but only traced, this systematic arrangement in nature; that these relations and proportions, which exist throughout the animal and vegetable world, have an intellectual, an ideal connection, in the mind of the Creator; that this plan of creation . . . has not grown out of the necessary action of physical laws, but was the free conception of the Almighty Intellect, matured in his thought before it was manifest in tangible external forms.

Agassiz thought he could prove that classification had a real existence in nature and Creation for him, therefore, was a translation of the matured thoughts of the Creator into 'tangible external forms'.

86. The second volume of the *Principles* (1832) is largely devoted to a discussion of the particular set of environmental factors, both physical and biological, on which the continued existence of each species is dependent. The inevitable change in the environment that occurred as a result of long-term geological changes, Lyell showed, required the extinction of species. Moreover, the extinction of one species tended to drag into extinction others that were dependent on it. The total long-term effect was, therefore, the successive extinction of species.

On February 20, 1836, John Herschel wrote to Lyell from the Cape of Good Hope to thank him for a copy of the fourth edition of the *Principles* (1835) and referred to the question of the replacement of extinct species by new ones. The question was thus placed before Lyell in 1836, although it was also very probably on his mind in 1831 when he was writing the second volume. Sometime during this period when he was pondering the origin of new species, he must also have considered the question of the origin of man.

87. Adam Sedgwick, *A Discourse on the Studies of the University of Cambridge,* 5th ed., London, Parker, 1850.

88. We have been unable to identify this quotation.

89. Pieter Harting (1812–85), *Die Vorweltlichen Schopfungen verglichen mit der gegen wartigen . . . aus dem Holländischen übersetzt von J. C. A. Martin. Mit einem vorwort von M. J. Schleiden,* Leipzig, 1859, p. 77.

90. Thomas Witham Atkinson (1799–1861), architect and traveler, had traveled extensively in eastern Russia and Siberia and had published accounts of his travels.

91. Pieter Harting, *Vorweltlichen Schopfungen,* 1859, pp. 263–64.

92. Happisburgh was a hamlet on the Norfolk coast between Cromer and Great Yarmouth. Lyell probably referred to Happisburgh, because Happisboro seems nonexistent.

93. Joseph Prestwich, "On the occurrence of flint-implements," 1860, p. 302. Cf. Joseph Prestwich and John Brown, "On a Fossiliferous Drift near Salisbury," *Quart. J. Geol. Soc.,* 1855, *11,* 101–07, p. 102.

94. William Shakespeare, *Henry IV,* part 1, act 3, scene 1, lines 13–17:

> At my nativity
> The front of heaven was full of fiery shapes,
> Of burning cressets; and at my birth
> The frame and huge foundation of the earth
> Shaked like a coward.

95. Ibid.

96. Karl Ernst von Baer (1792–1876), Esthonian naturalist and embryologist, whose most important early discovery was that of the mammalian ovum. Von Baer showed that the embryos of the different classes of vertebrates were very similar in their early states, but became progressively more dissimilar as development proceeded.

97. Thomas H. Huxley, "On certain Zoological Arguments commonly adduced in favour of the hypothesis of the Progressive Development of Animal Life in Time," *Ann. & Mag. Nat. Hist.,* 1855, *16,* 69–72. On p. 71, Huxley cites Karl E. von Baer, *Untersuchungen über die Entwickelungsgeschichte der Fische, nebst einem Andrange über die Schwimmblase,* Leipzig, Vogel, 1835, 52 pp.

98. This quotation and the quotations immediately following appear to be from a lecture or article by Huxley that we have not been able to identify and that does not seem to be included in the published lists of Huxley's writings.

99. What William Harvey placed on the frontispiece of his *De generatione animalium,* London, 1651, were the words "Ex ovo omnia."

100. Josiah C. Nott, "Acclimation; or, the comparative influence of climate, endemic and epidemic diseases, on the races of man," in Josiah C. Nott and George R. Gliddon, *Indigenous Races of the Earth,* Philadelphia, Lippincott, 1857, pp. 353–401. The discussion on p. 361 deals with Étienne Geoffroy Saint-Hilaire's theory of progressive development or evolution and with the origin of all the races of man from a common source.

101. Alfred Maury, "On the distribution and classification of tongues—their relation to the geographical distribution of races; and on the inductions which may be drawn from these relations," in Josiah C. Nott and George R. Gliddon, *Indigenous Races of the Earth,* 1857, pp. 25–86.

102. Manichaeism is the doctrine attributed to *Manes* or *Mani-*

chaeus who founded a religious sect in Persia in the third century A.D. It held that good and evil reigned in the world as equal powers, in conflict one with the other, that Satan was coeternal with God, and that man had originally been created by Satan or the power of evil, but contained a spark of the light of God.

103. Alphonse Louis Pierre Pyramus de Candolle (1806–93), *Géographie botanique raisonèe,* Paris, 1855, 1365 pp.

104. Heinrich Georg Bronn, "II. Tabell über die numerische Vertheilung der fossilen Organismen-Arten in der Reihe der Gebirgs-Bildungen," *Untersuchungen über die Entwicklungs,* 1858, p. 19.

105. Ibid., pp. 204–09.

106. Jacques Boucher de Crèvecoeur de Perthes (1788–1868), *Antiquités Celtiques et Antédiluviens. Mémoire sur l'industrie primitive et les arts à leur origine . . . ,* 3 vols., Paris, Treuttel et Wurtz, 1847–64, vol. 1.

107. This is Lyell's first notebook entry after reading the proofs of Darwin's *Origin of Species.*

108. Trans.: "the insatiable variety of Nature."

109. Alfred Wills, "The Passage of the Fenêtre de Salena, from the Col de Balme to the Val Ferret, by the Glacier du Tour, the Glacier de Trient, and the Glacier de Salena," ed., John Ball, *Peaks, Passes and Glaciers. A series of Excursions by Members of the Alpine Club,* London, Longman, Green, Longman and Roberts, 1859, pp. 1–27.

110. James David Forbes (1809–68), professor of natural philosophy at the University of Edinburgh, had become famous for his studies of glaciers and for his theory that "a glacier is an imperfect fluid or viscous body which is urged down slopes of a certain inclination by the mutual pressure of its parts."

111. John Tyndall (1820–93), English scientist, took his Ph.D. in chemistry and physics at Marburg, Germany, in 1850, and, in 1853, was appointed professor of natural philosophy at the Royal Institution, London. In two papers read before the Royal Society in 1857 and 1858, Tyndall presented a new theory to account for the motion of glaciers, based on his observations of the internal melting and refreezing of ice exposed to sunlight. The expansion that occurred on refreezing would cause a minute movement of the whole glacier mass. The publication of Tyndall's theory provoked a

controversy with Professor James David Forbes and a rash of books and papers discussed in reviews mentioned by Lyell.

112. [James D. Forbes], "Art. I. Glaciers and Glacier Theories," *Nat. Rev.,* 1859, *9,* 1–30, p. 20:

> Professor Forbes is fairly entitled to contend that the phrases 'bruising and reattachment', 'incipient fissures reunited by the simple effects of time and cohession', used by him, are but the vaguer antetypes of the more definite and philosophical terms 'fracture and regelation'.

113. William Hopkins (1793–1866), mathematician and geologist of Cambridge University, had written on the mechanics of earth movements and on the motion of glaciers.

114. Sir David Brewster (1781–1868), Scottish natural philosopher, was at this time at St. Andrews University, but in 1859 was appointed principal of the University of Edinburgh. See [David Brewster], "Art. V. [review of works on glaciers]," *N. Brit. Rev.,* 1859, *31,* 89–124.

115. This letter was in reply to one from Darwin to Lyell, October 25, 1859.

116. Peter Simon Pallas (1741–1811), German naturalist, was invited in 1768 by Catherine the Great to come to St. Petersburg. He spent from 1768 to 1774 traveling through the Ural Mountains, the steppes of eastern Russia, and portions of Siberia. Afterward, he settled in the Crimea to study his collections and write an account of his explorations. In his letter to Lyell of October 25, 1859, Darwin had explained Pallas' view that after a long period of domestication species tend to lose their sterility when crossed with other closely related domesticated species.

117. William Benjamin Carpenter (1813–85), naturalist, was appointed registrar of the University of London in 1856. He was best known for his *Principles of General and Comparative Physiology,* London, Churchill, 1839.

118. Joseph Gärtner (1732–91), German botanist.

119. Lyell seems to have had in mind John Obadiah Westwood (1805–93). Westwood was keeper of the entomological collection and library at Oxford University where in 1861 he was appointed Hope Professor of Zoology.

120. Heinrich G. Bronn, *Untersuchungen über die Entwickelungsgesetze,* 1858, p. 9.

121. Owen had carried out extensive research on the fossil reptiles of the Wealden. See Richard Owen, *Monograph on the Fossil Reptilia of the Wealden and Purbeck Formations,* 4 vols., London, Palaeontographical Society, 1853–64.

122. John Stuart Mill, "Dr. Whewell on Moral Philosophy," *Dissertations and Discussions, Political, Philosophical and Historical; reprinted chiefly from the Edinburgh and Westminster Reviews,* 2 vols., London, Parker, 1859, *2,* 450–509, p. 452.

123. Charles Darwin, *On the Origin of Species by means of natural selection, or, the preservation of favoured races in the struggle for life,* London, Murray, 1859, pp. 344–45:

> We can understand how the spreading of the dominant forms of life, which are those that oftenest vary, will in the long run tend to people the world with allied, but modified descendants . . . On the other hand, all the chief laws of palaeontology plainly proclaim, as it seems to me, that species have been produced by ordinary generation; old forms having been supplanted by new and improved forms of life, produced by the laws of variation still acting round us, and preserved by Natural Selection.

124. This letter is in reply to one from Darwin on October 21, 1859, Darwin-Lyell mss.

125. Bransby Cooper (1792–1853) was the nephew of the famous surgeon Sir Astley Cooper (1768–1841).

Scientific Journal
No. V

1860

Recent
Pliocene
Miocene
Eocene
Cretaceous
Wealden
Oolite
Lias, p. 10
Trias
Permian
Carboniferous
Devonian, p. 14
Silurian
Cambrian
Volcanos & Earthq.[s]

Recent & Post Tertiary [p. 3]

Pliocene[1] [p. 4]

Busk finds Crag Bryozoa different from what w.[d] have been anticipated as regards proport.[n] of recent forms.

December 29, 1858

Banksia fruit is under basalt in Australia. J. Hooker—thinks it older than Cave-animals of Australia.[2]

Miocene [p. 5]

Eocene [p. 6]

Nummulites, says Rup[t] Jones (Jan.[y] 10, 1859), are of two types, N. levigatus & N. variolarius or rather N. planulata.[3] This last is still living—Williamson cites a British with other foreign ones—Red Sea? [4]

Cretaceous [p. 7]

Wealden [p. 8]

Oolite [p. 9]

Oolitic Flora. The absence of ordinary dicotyledons more remarkable than in Coal, for here we cannot suppose it due to peculiarity of Stations. It is also striking that the *Indian* Oolitic flora without them & without *Palms*. This in the tropics, always, assuming the E. Indian (Glossopteris Browniana) Coal to be oolitic.

Ferns & Cycads characterize the E. Indian oolitic Coal flora.

December 31, 1858

Inferior Oolite

An Araucaria (J. Hooker) not distinguishable from Norfolk Isl.[d] Pine-fruit perfect—Somersetshire.

Lindley had given an Araucaria from the Lias but only (the leaves?) not the fruit.[5]

January 9, 1859

Equisetum Columnare

C. Bunbury[6] says that J. Phillips[7] has shown that it is not even at the bottom of the Oolite, tho' Heer[8] says exclusively triassic.

December 18, 1858 [p. 10]

Lias

Mr. Etheridge[9] says that Mr. Higgins[10] of [Turvey?] Dented has the jaw of Microlestes? from bone bed near Bristol in wh. both liassic & triassic fossils, Hybodus, Saurichthys.

December 29, 1858 [p. 11]

Trias

Cheirotherium footprints may not be Labyrinthodon, says Huxley.[11]

December 31, 1858

Stangeria, says C. Bunbury, not the least like Taeniopteris of Richmond. None of the ferns less like a Cycad—Neuropteris being only known as a fern might, as C. Bunbury has said in his paper, be a Cycad for aught we know.

If Equisetum columnare [be] rightly named, it occurs above the Lias in Yorksh.—in Scarborough & in the Keuper in Europe.

Voltzia is a Bunter plant & in the Bunter flora is quite diff.[t] hitherto from the Keuper. Voltzia, therefore, in the Richm.[d] flora is strange.

Pecopteris Whitbyensis as seen by C. Bunbury very imperfect specimen. In identifying it the Richm.[d] specimen with the Whitby, C. B. "followed W. B. Roger's[12] lead". p. 19.

Permian [p. 12]

The Proteosaurus a highly organ.[d] reptile, the beginning of Saurian.

December 1858

[With] Huxley: Callipteris sinuate

Ad. Brong.[niart] Tab. des genres.[13] The fern I gave C. Bunbury from Autun, characteristic of the Roth Codligrade of Germany—in collect.^{n} of Berlin from Saxony.

If the Autun psarrolites are from beds (washed out) above this it is thoroughly in favor of the Permian origin of the psarrolites.

December 23, 1858 [p. 13]

Carboniferous Labyrinthodon

The least develop.^{d} of the genus or family Archegosaurus, the most rudimentary form—Huxley.

Much more frog than saurian. Owen.

More right in this than H. v. Meyer.[14]

Lindley

Article on *Coal Plants* in the Penny Cyclopedia by Lindley in wh. perishable nature of dicotyled.^{s} as compared with *ferns* treated of—[15]

December 20, 1858 [p. 14]

Devonian

Pegasus a fish of the Syngnathus family as strange as Pterichthys.

Ostraceon as singular as Coccosteus, remarks Huxley.

January 5, 1859

Flora

J. W. Dawson's paper Prototaxus—Gaspe[16]

Psilophyton, scalariform vessels, lycopod.^{s} Lepidodendron

Knorrian coal form more obscure—

Absence of Sigillaria (remarks Salter) characteristic of European & American Devonian as contrasted with Coal.

Silurian [p. 15]

Foraminifera, green earth of same genera as recent, R. Jones. Crystallaria? Globigerina? Protalia Polymorphina?

Murchison, 2d Ed. *Silurian.* p. 574. U.[17]

See Plates in Trans. Berlin Acad.[y] R. Jones thinks the above recent generic forms may be identifiable.

April 30 [1859] [p. 16]

Cambrian

Lingula Davisii, Salter[18] says, can now be shown by the hinge to differ generically from the true Lingula, nearer to Obolus, between L. & Obolus (Obolina Davisi?).

In table of brachiopods Davidson sh.[d] have distinguished Lower & Upper Silurian, says Salter.[19]

There are now 175 species in Cambrian or in Barrande's Primordial & only one species identical with Silurian.

Out of 38 Primordial genera only 12 identical with Silurian, a less link than between Permian & Trias & much less than between Cretaceous & Tertiary.

Orthis, a genus common to Cambrian & Lower Silurian.

Out of 24 genera of Trilobites, Barrande has only 3 common to same two formations & only one species Agnostus pisiformis.

Earthquakes & Volcanos [p. 17]

December 23, 1858 [p. 18]

Progressive Development

Cephalopoda of oldest rocks. Huxley says that some of the living Cuttle-fish have no hard parts. Those which I saw with fins for swimming have.

It is not true that in the embryonic develop.[t] of Cephalopods Dibranchiate is first devel.[d] & then Tetrabranchiate

The Odonteuths & Belemnites are as high develop.[ts] as any living cephalopods of the Spirula or pen-&-ink fish divis.[s]. The first start in the Lias is of several cephalopodous families.

Proteosaurus, see p. 12, [of this journal] antea-Ophidians.

Ψ Suppose that Ophidians had been chiefly ancient & Deinosaurs living tribes of reptiles, this w.[d] have been triumphantly used in favour of developm.[t], but if so, the opposite state of things sh.[d] weigh ag.[t] develop.[t] so far as the Reptiles are concerned.

Trias [continued] from p. 11 [p. 19]

Stangeria living at Kew—turnip-like root unlike fern—rhizome—fructifi.[cn] of a Cycas—(or Zamia?). Venation as of a fern showing small value of this character, taken alone, had been described from the leaves sent from the Cape by —— ——[?] as a fern & he had studied ferns well. C. Bunbury.

December 31, 1858

Neufeld? near Basle as well as certain quarries at Beyreuth formerly taken for Lias but now known as Keuper have changed ideas about triassic flora & the Richmond V.[a] Case.

Ophidians

None, says Huxley, in the 80 Triassic reptiles of H. v. Meyer, yet many of aquatic habits in tropical region—(some even salt-water?) see P. 18 for Ψ

Index to Notes on "Creation" [p. 20]

Hooker, Galapagos, one island with most of the Indigenous, Index B.[k] 1–12 p.[20]

Species, MacAndrew, as to no British Isle producing species wh. are indigenous whereas Madeira & P.[o] S.[o] do, *ib.* p. 29 [21]

Wallace, no newer pliocene isl.[d] has in it extinct genera. *ib.* 31 [22]

Wollaston, on species, generic & specific centres. *ib.* 54 [23]

Buist, mean height of land & depth of sea. *ib.* 56 [24]

Whewell, Unity of Plan of organic Creation. *ib.* 62 [25]

Final Causes

Cuvier ascended from science to higher speculat.[s]

Lawrence—influence of shape of brain.[26] Species not to be defined. Man, races of—Man & Orang.

[Pages 21–25 were left blank.]

December 6, 1859 [p. 26]

Creations

May not creation consist of four powers or principles:

1.st The principle of divergence from parent types or the variety-making power.

2.[dly] Natural Selection directing & concentrating the same & working together with hybridization of varieties or clearly allied species

3.[dly] The hereditary principle of transmission of acquired peculiarities.

4.[thly] The progressive tendency to more complex organizations, both physical & spiritual, material & immaterial.

The most transcendent of all these is the first & last; in this is the Pythagorean or Platonic doctrine of the eternal ideas—the great plan or model preexisting in the eternal mind; it is not change or the sportiveness of Nature, but morphological laws [p. 27] which must determine not only the occasional advance but the details of structure. Selection presupposes the [creative?]

design, this fiat to have gone before—transmission by generation merely follows the great act.

The doctrine of Plato and of the nominal lists is most like the conception of a system of development of a preconceived plan, that of Aristotle and the realists is more like the perpetual intervention of special providence, as often as a gap is to be filled up by the creation of a new species.

Brutalizing Theory

That organization which pervaded Man w.[d] have enabl.[d] Harvey to discover the circulation of the blood, Bell, the theory of the nerves—Linnaeus, his views as to animals & plants—Cuvier, his laws of correlation of structure—all including the anatomy of Man as well as now. What is there that sh.[d] really lower the human being, if he be found to be united with this corporeal framework—if he share the same wonderful structure & instincts, the same dependence on the natural laws of the universe. Why sh.[d] the generative link be wanting—the reproductive system.

[p. 28] It is probable that we do not so much overestimate the human as underrate the organic world. If we may have diseases in common so we may have lactation, & may be linked by generation.

Men of Genius

What do we expect to discover of the portentous when this being first entered on the stage, that

> the frame of heaven was full of fiery shapes
> & that the earth trembled,[27]

but no, we find that the new born infant created no greater a sensation than if his mother cat had kittens & he had never been born.

It was in the ordinary way of generation. Whether the step from the highest instinct & intelligence of the most advanced of the lower animals, some extinct species perhaps, to the humblest in intellect of the races of Man, which may have existed, may not also have been made by the same machinery, is a question.

Consequences of Transmutation

Our horizon is enlarged & a wider survey is obtained over the relation of our system to the past organic world. Tracts v. 2, p. 215, (XVII).[28] Some of our old perplexities are solved or lessened, such as rudimentary organs—the death of species—the imperfections of some instincts, but others are intensified. [p. 29] The new difficulties are exceptions in the great whole.

At first the new difficulty perplexes us, but it soon takes its place as one of those exceptions which do not affect the great argument. We find Man admitted to comprehend somewhat more of the working of the terrestrial system—& the relations of its parts—to scan a larger portion of time, to speculate with Plato & Pythagoras on the eternal ideas of the omnipotent presence thro' time & outlasting not only individuals but species, not only species wh. may endure for thousands of centuries, but genera, orders, classes. Do such insights into the infinite tend to lower Man in his estimate of himself—the notion that things material are the phantoms, things immaterial, the realities of this world?

Progressionists

[1stly They] allow that the chain of organic beings began millions of years ago.

2dly. That it went on from simple to more complex in time—from invertebrate to vertebrate, from bird to fish to reptile, r. to bird, bird to marsupial, m. to placental—pachyderm to quadrumanus—gorilla to Man.

3dly. Morphologically & psychologically identical save only in degree.

[p. 30] 4thly. Departure always from most ancient to newer types & these always approximating not only in complexity but in form to existing species.

5thly. Man was the last & crowning operation of this continuous or slightly interrupted chain, the last links of which are missing, though always filling up.

6thly. That the chain is connected together by embryological development ending in Man.

7thly. That the First Cause has followed certain laws in ushering in the successive species, genera & classes into the earth —laws of morphology, also geographical.

8thly. That the formation of permanent varieties or races in common to the human & other species & as marked a character.

9thly. That the same struggle for existence governs Man, subject alike to be cut off in infancy, not longer lived than many subject to destruction by inorganic revolution & earthquakes, epidemics.

[p. 31] 10thly. Yet, tho' so connected, he cannot have obeyed a law of generation. The Unity must stop short when it is suggested that he entered the earth, as one individual of higher endowments, succeeds one of less intellectual or moral capacity.

It is clear that to make the stand here is illogical. They also think that Man is distinct from the brute, essentially belongs to a different system of things, as I suggested in the *Principles* 1st Ed., & who doubt the progressive develop.[t] from the inferior up to Man, the last & crowning operation, they may consistently make a stand.

But the progression in time being granted & the great fact that Man came last, being claimed as the climax of the whole, no objection can normally be made on the ground that the formation of new species by indefinite divergence takes away the gap between Man & Brute.

Dignity of Man [p. 32]

It is somewhat irreverent to fear that any discovery will derogate from our conception of the dignity of the Creator. The real apprehension, however, if the truth be told, the sensitiveness is founded on this, that the dignity of Man is at stake.

It is the genealogy of Man which is rendered less imposing. And if it were lowered, if we were led to estimate less highly the position of Man in the Universe, no doubt it would be a loss—it w.[d] render our hopes less elevated.

Progressionists

If they find the first fish very perfect, they are triumphant, for it required a special creation. It cannot have come by transmut.[n] genealogically from a preceding vertebrate type. But if [p. 33] on the other hand they had met with an Amphioxus in the oldest formations, this w.[d] be claimed in confirming the development theory, the beginning in an embryonic form—a vertebrate type without vertebrae, eyes, head or tail.

If the first reptile, the Archegosaurus, have a notochord which is persistent, it shows how the beginning of that type approached the fish. Had it been a crocodile it w.[d] prove like the first fish that the type of reptiles came perfect from the hand of its Creator—Sedgwick, p. 225, *Discourse*[29]—perfect fish first.

December 8, 1859 [p. 34]

Alarm

We are accustomed at every new step in astronomy, ethnology & geology to arrive at some conclusion which staggers us. It must always be so, because as the history of the world is one of progress intellectually, it is always emerging from a state of comparative darkness & superstition in which some errors are intermingled with the highest truths of religion—when any of them are modified we cannot but feel alarm & fear. To imagine that our conception of the Deity will be lowered, that in his works we shall ever discover anything derogatory to the highest conception of his attributes might almost provoke a saint. But we may bring truths to light which may shake our confidence in some one of our articles of faith in which [p. 35] our estimate of the dignity of man is founded. We may feel inclined to say "if this be truth", etc.[30]

If such species was created separately the idea is magnificent, beyond the utmost stretch of the human imagination, but not so overpowering as the idea of a machinery by wh. all shall be evolved out of one primeval egg, the sustaining power of the deity giving a continued action to the same—neither one nor the other can be comprehended. We can only admire.

But would the genealogical law justly derogate from our elevated notion of the relation of Man to the universe, to living & inorganic things, around us? Here is the rub. Let us look it steadily in the face. Now, before it is proved. When it is only possible, perhaps probable, that it will [p. 36] be established.

How many millions die every century before childbirth, during & soon after.

Hundreds of millions in infancy. How many tens of thousands in each century of adults are in a state of imbecility which amounts to a want of responsibility. The philosopher & the theologian cannot or ought not to shut his eyes to such an astounding fact. Neither, after he has given it due attention & granted the enigma to exist, can it alter his conviction of the high intellectual powers & the moral responsibilities of the vast majority of mankind.

If most of the human race their faculties are in a very low state of civilization bordering on barbarism & some are savages —this does not check our [p. 37] aspirations or make us disbelieve a future state. Why if we sh.[d] find that for tens of thousands of years savage nations preceded the civilized sh.[d] we wonder. Or if, what may never be proved, instinct by a leap by which a passage was at once made from irrational to rational, from mere instinct to the humblest reason, which c.[d] know right from wrong, if instead of such a leap a transitional state occurred between the two, what riddle c.[d] this present beyond that of the contemporary difficulties which disturb us with doubts, but which we justly overlook as exceptional, if advanced to refute the claims of man to a high position in the universe.

Creative Addition, Sedgwick, 215 [p. 38] [31]

Were these additions made at successive geological periods by one continuous genealogical succession, or by a power analogous to that by which life was first introduced into the planet? Was the creative force that modified the organic world the same as that which modifies species daily & causes new varieties or was it manifested by some wholly distinct law, for a law is apparent in both cases.

How can we understand Creation without development & consequently apart from considerations of time?

If an acorn be called into existence it may contain doubtless [p. 39] the germ in it not only of an oak, but of those several varieties of oak which are so marked that botanists are not agreed as to whether or no they sh.[d] rank as species.

But if this acorn perishes before it has germinated then varieties or species cannot be said to have ever been created on the globe.

If the first human race which flourished on this planet was of very humble capacity, intellectually estimated & incapable of producing men of the highest capacity except by changes brought about by the lapse of ages & when some new race was formed the creation of Man in his full maturity of power must date long after the [p. 40] original stock appeared.

And why should we exclude the idea of time from such development when we know as a fact that parents of ordinary ability may have offspring possessing high genius or moral faculties. The mechanism of generative succession does not preclude development of that most exalted of all orders with which moral & spiritual endowments are bound up.

The Tertiary species would never have had any existence had not the planet endured so long. If they came in by generative succession, still we must suppose the [p. 41] continuance of the first originating power, this permanent action of what that power may be.

We cannot in reference to an eternal first Cause, to whom time has no intelligible relation, to whom the past, present & future are the same, distinguish between Creation instantaneous or developed in time; but if we grant that not only varieties but species are made to succeed each other in the way of ordinary generation by successive deviation from a parent type, then must we imagine the creative force to pervade time & to be a part of this principle [p. 42] of divergence. And this is particularly evident when the divergence implies not only variation, but addition, whether of physical or instinctive, or intellectual power—in no case more markedly than when a son of very extraordinary mental powers is born of parents of moderate faculties.

It may often have been a matter of wonder to reflect how such a power as that of Man first entered the planet—the greatest event which ever occurred in its history. But if we beheld some great lawgiver or teacher exerting a powerful influence over his generation, & regarded by them & their successors as almost a divinity, & curiosity was [p. 43] awakened to trace him to his origin, & ascertain in what moment he had entered on the scene of this life, it w.[d] be found that no portents accompanied his nativity. Some might be invented & believed by a credulous age but if the truth were known it w.[d] not be found that the Earth trembled or that "the front of heaven was full of fiery shapes" but that Nature was as insensible to the birth as if any one of the lower animals had at the same moment pupped or kittened, & the great prophet, teacher, or philosopher had never been born.

If then we should hereafter be able to prove by the evidence in favour of progressive development from mere instinct to a mixture of [p. 44] instinct & intelligence & from this again to a higher intelligence in the order of time. 2.[dly] by showing, that species have been produced in succession by generative acts, both in the case of animals & plants, & thirdly that it is in the highest degree improbable that Man should be exceptional on this head seeing that he is made subject to the same laws, in so many cases where it might have been imagined that he would have been favoured. If, I say, we sh.[d] find that Man is a part of the same series or organic developments, we cannot suppose that he would depart from the Unity of the Plan.

The races of Man approach as near to species or those of plants [p. 45] or of animals.

Thought & Stature

If it could be shown that in the animal world instinct & ability were much more invariable in the individuals of the same stock or family than stature & physical attributes, then it might be said that the derivation of successive increments of intelligence in the way of generation were improbable.

But in the case of the human race it is just the reverse. That a man of gigantic intellect sh.[d] be born of parents of moderate capacity is more likely than that one of gigantic stature sh.[d] be born of a dwarf.

We cannot expect that the genealogical principle will prevail in regard to forms & not to instincts. If the latter, & even human intelligence, [p. 46] varies from one generation to another then no amount of difference between the intelligence of species can be advanced as an argument against the common bond of descent, because we see increments of mind indicating greater 'leaps' than Nature makes in regard to corporeal attributes.

The great step has been made when once we admit what is called historical development. The successive coming in of higher & higher grades of organization & intelligence, Man being the last link of the chain, a chain unavoidably broken & interrupted & which may forever remain so, but which is always becoming somewhat less defective.

When we range the Vertebrata in a series tracing them from a fish to the Gyrencephala & place the Quadrumana at the head of the last [p. 47] & Man highest in the whole series & latest in time, when we say that a creative power following a plan gave origin to these in succession, departing farther & farther from the first types in successive geological periods & admit even that the least advanced of the human races approaches somewhat nearer, say even a very little, to the highest of the quadrumana in intelligence, how can we stop short & deny the probability that the mode of originating the new forms & degrees of intelligence is one which gives rise before our eyes to new rational characters, new individual powers, & turns of mind, new trains of thought & variations from a parent type of a far higher order than the difference in cunning of two species of the dog or other tribe of mammalia. [p. 48] After admitting all this, is it not to strain at a gnat & swallow a camel if the progressionist is shocked at the theory of transmutation?

Dignity of Man, p. 32 [32]

It would imply a want of faith in the wisdom of the Omnipotent if we feared that by investigation of his works we sh.[d] arrive, by exerting the reason He has given us, at conclusions less & less elevated of his attributes. And if the truth were confessed & the apprehension lurking in the depth of the soul were drawn out, if the secret misgivings were divulged, it is not any fear of such consequences that alarms the philosophers or the theologist, but fear lest the dignity of Man in the relation to the Universe sh.[d] be lowered by establishing a nearer link of union between him & the inferior animals—then his conscious feeling of superiority, his hopes & aspirations [p. 49] naturally lead him to indulge—aspirations without which he becomes in his own eyes a creature of the moment, incapable of earnest thought & of sacrificing for the good of others, doomed to be as ephemeral & insignificant in himself.

December 11, 1859

Progressive Development & Transmutation Tend to Same End

They who advocate most strenuously the gradual ascent from the simplest to the most complex forms are usually the most adverse to the origin of new species by descent. While the latter [do not], Mr. Darwin at least does not, nor Prof. Huxley, embrace the development in time, or at least mistrust the data on which it is built as being as yet far too imperfect & hypothetical.

Yet every step made towards confirming the one theory tends to strengthen the other & to render it more probable.

[p. 50] If the doctrine of descent be embraced, Man becomes what the progressionists desire, the crowning operation of a scheme which it has taken millions of years or perhaps of ages to develop gradually.

The successive links of the chain are assumed to be interrupted by the one hypothesis in Nature, by the other only in Nature's records. But in both theories, there is the same chronological advance; in both the Creative power is assumed to work according to a progressive plan; in both we have the fish at one end of the Vertebrate chain & Man at the other & Man preceded by quadrumana having, as in the case of the Miocene Dryopithecus, the same number of bones, 242, all of them mostly homologous with Man including even the same number of vertebrae in the rudimentary tail as Man.[33]

[p. 51] Can we doubt that there is some deeper meaning in this strict affinity in structure beyond what is called the symmetry of Nature—the genealogical hypothesis affords at present the only explanation.

December 11, 1859

Descent

It is not the Creative Power respecting which we are making an hypothesis, nor about its nature, but whether that power works by law & if so is that law that of hereditary transmission or descent. Is there any blood relationship between the species of two successive epochs—or only of a part of them as De Candolle[34] supposes in the case of plants?

Grant that two races, the European & the Australian, may be of unequal intellectual capacity & that the least advanced may be the [p. 52] oldest. Has there been a blood-relationship between them or have the two races had distinct original stocks?

This question may be discussed without approaching the essence of the race producing power. If the originating influence pervades successive generations in a species in which there are many varieties, then its action cannot be limited to what may be called the birth of the species, suppose a national character, national features, inflexion of the voice, to take ages to form so may it be with a species.

December 9, 1859

Connexion of Progression & Transmutation

I have long had a suspicion, growing gradually to a conviction that [p. 53] if ever the development in time could be established with any approach to that completeness which its most strenuous advocates claim for it, the transmutation hypothesis would also prove true. If I had any reluctance to embrace the one, I felt it equally to the other. In fact what can it matter if we place the animal classes, orders, genera & species, in a series accord.[g] to their successive grades & see the ranks filling up, the more we investigate & come to the opinion that the chain was progressive in time—leading up from simple to complex, from instinct to intelligence, & thence to reason, whether the machinery employed in development resemble that by which higher intellectual & moral natures are at certain rare intervals brought into the world, or some hitherto unknown & unguessed at method [p. 54] be adopted? What difference can it make if already the method of generation with the suckling of the young, a period of gestation longer than some mammalia, shorter than others—exactly coinciding with some—& all the other essentials of growth, liability to disease, to lose a majority before reaching the adult state, the common lot of the other vertebrates exposed in like manner to perish by convulsions of the inorganic world. Why will this analogy of the progressive development be once adopted rather with eagerness & popular favor, sh.[d] the stand be made just when there was least reason to expect anything exceptional.

They, however, who do not embrace the doctrines of transmutation for any species, may consistently resist it in the case of Man. But if they admit it for varieties & are so led as to accord it to the majority of [p. 55] species they will then find it impossible to withhold it from the rest & this development in time will lend itself to the same.

Passage from Irrational

After all, if the whole scheme was disclosed to us & we had no break in the record, it might perhaps appear that this passage from the irrational to the rational was by a leap. There is nothing unphilosophical in the conjecture. The enigma psychologically considered (in a point of view) may be, or might prove to be, less than one with which we are every day familiar. Millions of riddles of profound darkness surround us. We need not go to the Bushmen or Australians, but to those healthy adults whose capacities are undeveloped in civilized & advanced races. The history of [the] globe may contain no [lack of] such embarrassing problems if we desire to dwell on them. Habitually [p. 56] they who have faith in the goodness of the Deity regard these difficulties without dismay as exceptional, for such they are, but Geology cannot be expected to add to them—except in so far as every wider survey of the whole scheme offers more to puzzle as well as to admire & understand, with the satisfaction, however, of feeling that such is the vastness of the scheme that exception sh.^{d} less disturb us.

If this conviction that we are responsible & have a destiny reaching beyond the present is not diminished by such undeniable facts before our eyes, how much less if we only suspect others in the remote past.

December 21, 1859

Progression in Plants

The complexity of coniferae of the Coal or Devonian & of Sigillaria & Lepidodendron is not lessened by the absence of any accompanying angiosperms.

Retrogression

May not reptilia & [p. 57] cephalopoda progress & retrogress more than once in geological time?

Letter to Dr. Hooker, December 19, 1859 [35]

My dear Hooker,

I have just finished the reading of your splendid Essay on the origin of species as illustrated by your wide botanical experience & think it goes very far to raise the variety-making hypothesis to the rank of a theory as accounting for the manner in which new species enter the world.[36] Certainly De Candolle's book[37] was like the old doctrine of those who only [p. 58] called in spontaneous generation for explaining those cases where they were unable to trace the origin to an egg or seed. Nevertheless the extent to which he granted nearly half of what he really believed to be true species, to have been derivative in the way of varieties, was calculated to lead philoso.[l] & logical minds to go the whole length of transmutation.

I thought your way of putting it very clear & the style luminous, the acknowledgment of R. Brown[38] p. v. handsomely done.

The number of grand generalizations is really stupendous when [p. 59] one considers the vast number of species to which they relate, such as the excess of unstable over stable forms; the limitation of genera & orders owing to extinction & of species by destruction of varieties—the non-reversion to wild stocks which struck one as very new & important—p. IX, the centripetal tendency of hybridization—species being realities even under the new view.

The equality of distribution, [p.] xiii, of the Acot[yledons]—mono,—Dicot[yledon][s] is very wonderful if the Dicotyle.[s] angiosperms are geologically so modern. [p. 60] The gymnogens ought like the marsupial quadrupeds to have kept some one country to themselves from the Oolitic period in order to make the case of plants parallel to that of animals. Tho' it must be owned that we do not yet know whether in Tertiary times there may not have been a rich placental fauna in the Australian area.

The two first notes of p. vii are very interesting & show what grand speculations & results the 'creation by variation' is capable of suggesting & one day of establishing. [p. 61] The 3.[d] note in same page is useful for it seemed to me that both Darwin & Wallace spoke too much of varieties under domestication or horticulture as if they were artificial products instead of being what Nature might & w.[d] yield under certain natural conditions.

What you say of Naudin makes me wish to get his work or papers tho' the note at p. xi seemed to me rather vague & unpromising about a *"function dans l'ensemble des choses."* [39]

I am much puzzled, p. xv, at the sinking islands having so few [p. 62] species & peculiar generic types. A newly risen island ought to have fewest.[40] But the atolls mark the spots where mountainous islands & continents once were. [Are they] the last remaining rocks of sunken & ancient centres of creation? I suppose that the stations supplied by the atoll are unlike any which existed in the old & new submerged nucleus? & therefore all the previously established plants perished.

But the rising land retains whatever new creations spring up in it as the stations are [p. 63] always multiplying in number & diversity of character.

The facts & views in p. xvii are wonderfully suggestive & grand as is all about the glacial migrations.

The geological chapter is only too short. I should like you to have seen at your leisure Heer's collection at Zürich, with him to explain it. Not but that he requires checking for his ultra-progressionist doctrines, but then he has made out more by fruits corroborating leaves than appears as yet in his publications.

[p. 64] The remark at p. xx about the Lias insects is very significant for if they imply the existence of plants other than gymnosperms & ferns, then in all the other formations older than the Weald, we may expect the same undiscovered species & orders of plants to lie concealed.

The Characeae you made appear first in Cretaceous [p.] xx. This you may defend out of my books until I went with E. Forbes in separating the [p. 65] Purbeck from the Cretaceous part of the old 'Wealden'.

But now that the Purbeck are Upper Oolite & that there are 'gyrogonites' in Middle Purbeck (see *Manual* 5th ed., p. 299) a correction is required.[41]

P. xxii you say that all the principal families of paleozoic plants still exist. I fear the progressionists like Agassiz will complain (& Bronn) that you have not fairly brought out the absence in the early beds of the Dicotyledonous angiosperms.

But what you say [p. 66] of the Coniferae & very highly developed Lycopods, etc., in the Coal, implying so vast an antecedent lapse of time to allow of progression from the simplest to the most complex, appears to me quite unanswerable & never before so well & strongly put.

[p.] xxiii How Bronn & Agassiz will fume about the Coniferae & other arguments in this page. The last footnote especially, for tho' if they knew their own interest, they would allow of time, antesilurian time without stint, all the progressionists are too [p. 67] eager to have proofs of a beginning to be liberally inclined in that direction.

[p.] xxiv Retrogression [is] also imperfectly reasoned on in animals. It is admitted in reptiles on comparing the living with the extinct of Oolitic period. But why not grant that in the animal & vegetable kingdom there may have been several periods in succession, & wide apart, of progression & retrogression. Yet this would greatly put out the progressionists for the carboniferous reptiles might then be less developed than the Triassic [p. 68] simply because they had degenerated from Silurian ones.

Dawson of Montreal has just found 4 species of reptile in an erect tree in Nova Scotia, 3 of them of higher grade than any one previously found in the Coal, & above 50 specimens of pupa—no other land shells being known between this & the Eocene! [42]

C.L.

[January 1860] [p. 69]

H. C. Watson on Darwin's Origin of Species,
Received, January 1860 [43]

1. Admitted the probability of a gradual variation from parent to progeny, from ancestors to descendants, to an unlimited extent, if traced through a vast series of successions.

2. Also, that among the countless numbers of organic beings produced, a process of natural selection will in the long run determine which kinds shall die out and which shall continue to produce descendants, themselves in turn more or less changing still.

3. Also, that a constant divergence from the lineal types is thus effected because variation is [p. 70] necessarily divergence, & because no extremely ancient type appears to have been exactly continued to the present epoch.

4. Such divergence nepotally onwards, necessarily implies an equal convergence ancestrally backwards; which must therefore

be admitted, although unproved except for slight variations and brief periods.

5. Constant convergence backwards must lead at length to one single prototype, which itself must have been originated in some mode different from the mode in which all its successors have been produced; because that prototype could have had no [p. 71] organic parent.

6. Constant convergence onwards from that prototype must have led successively to types. 1–10–100–1000–10,000–100,000–1,000,000. We may assume the million to represent the number of specific or quasi-specific types now existing on the earth.

7. Are we now at or near the end of the divergent variations? It would seem not. No grounds are shown for supposing that variation from parent to progeny has ceased, or that it is slower or less complete than heretofore. And the theory of the 'Origin' proceeds on an assumption that present facts [p. 72] evidence a continuance of variation now as in eras past.

8. Therefore this figure of 1,000,000 may, and will, for aught we can see to the contrary, go on indefinitely increasing until the Earth has its present million increased to millions 2–3–4–5–6–7–8–9–10– etc., etc.

9. Eventually, in this course, there will be as many millions of species as of beings or pairs of beings; every one or pair still tending to vary. If it be argued that 'natural selection' will prevent this vast increase of specific types, at any one time in existence, the answer is, why did it not arrest them [p. 73] at 10,000 or 100,000? Why now, rather than then? Or why at 2 millions rather than at 4 million? etc., etc.

10. Thus, whether we go ancestrally backwards, to arrive at one prototype, not produced like all the rest that have succeeded it; or go nepotally diverging onwards to a termination in countless millions of species,—in either course we arrive at something so totally different from what is now observed, so utterly incapable of proof, so inconceivable as a reality, that it must seem in itself an improbability amounting to an absurdity almost.

[p. 74] Ought we not to seek carefully among present facts or events for some counter-action in nature, adequate to save us from being forced upon this almost *reductio ad absurdum* in either direction, either backwards or forwards? I think that the counter-action can be seen and specified.

12. Would it not be better to reason up to no beginning, & to no end, than thus to make out a seeming absurdity by arriving at results totally different from, if not absolutely inconsistent with, the present course of natural events as seen [p. 75] in our own small span of time?

13. It would appear that the difficulty could be met & neutralized by the hypothesis or inference that individuals *converge* into orders, genera, species, *as well as diverge* into species, genera, orders, through nepotal descent.

14. No doubt that hypothesis (rejected in the 'Origin') is far more difficult to support by evidence. But various facts in nature seem explicable only on this assumption. And it tends to remove other difficulties which might be urged against the theory of the 'Origin' and which nepotal divergence by itself seems unequal to explain,—for example the intricately [p. 76] complex manner in which characters are crossed and combined in species, the effect of which would likely be to throw the same species into different groups, not subordinate to each other, if classification were attempted on the hypothesis that any character in common between two species must have been derived from some common ancestor.

Answer by C. Darwin to Mr. H. C. Watson's Objection[44] [p. 77]

I remember putting the case to myself almost in your words, —Why does not every individual become converted into distinct species? The answer, which I framed, and the discussion on this curious point, which I have somewhere in M.S. is nearly to the following effect: Although the number of species supported over equal areas under apparently nearly similar conditions is now very different in different quarters of the world; and although we [p. 78] are far from knowing that the most prolific area is fully stocked with species, as perhaps may be inferred if not so from some European plants having become naturalized even at Cape of Good Hope, yet Geology shows us, at least within the whole immense Tertiary period, that the number of species of shells & probably of mammals has not increased. How is this? As far as mere physical conditions are concerned (i.e. heat, moisture, height, etc., etc.) it seems probable that a sufficient number of species [p. 79] would soon become sufficiently adapted to the most diversified conditions, but the relation of organic being to organic being, I fully admit is far more important, and as the principle of divergence goes on augmenting the number of species in any area, or in the world, the relation between being and being will become more and more complex, and there seems at first no limit to the amount of possible diversification of structure and therefore no limit to the number of species. But I think the following causes will tend to limit the number (1) the *amount* [p. 80] of life (not number of specific forms) supportable on any area cannot be illimitable [*sic*], depending so largely as it does on physical conditions; and therefore where very many species are supported, each or nearly each will be few in individuals; and any species with scanty numbers will be liable to extermination from accidental fluctuations in seasons and number of enemies. The process of *exter*mination would in such case be rapid, whereas the process of production of new species would always be slow. [p. 81] Just imagine the extreme case of as many species as individuals in England, and the first severe winter or very dry summer would exterminate forever thousands on thousands of species. (2) I suspect that when any species becomes very rare, close interbreeding will tend to exterminate it; at least authors have thought that this comes into play in accounting for the deterioration of Aurochs in Lithuania and of Red Deer in Scotland, Bears in Norway, etc. (3) As far as animals are concerned some species owe their origin to being adapted to prey [p. 82] on some one other being, but if this other one being be rare, it will not have any advantage to an animal to have been produced in close relation to it, whilst so poor in number, and therefore it will not have been produced. (4) When any species becomes few in number, the process of modification will be slower for the chance of favourable variations arising will be lessened. Therefore if we suppose all or most of the species in any area to be very numerous and consequently each poor in [p. 83] individuals, the proofs of modification and of giving birth to new forms will probably be retarded. (5) And this I am inclined to think is the most important element: a dominant species which has already beaten many competitors will tend to beat more; the relation of organism to organism being more important than relation to physical conditions; and Alph. de Candolle has shown that those species which spread widely tend generally to spread *very* widely, and consequently they will tend to supplant several species in several areas [p. 84] and thus check the inordinate increase of specific forms throughout the world. Hooker has shown that in S.E. corner of Australia, where there are many invaders, (i.e. dominant forms) from different quarters of the world, the endemic Australian species have apparently been greatly reduced.

How much weight to attribute to these several causes I do not pretend to assign; but conjointly I think they must [p. 85] limit the constant tendency to augmentation of specific forms. What do you think? I should like to hear.

With respect to 'convergence' I daresay, it has occurred, but I should think on a very limited scale, (owing to strong principle of inheritance retaining resemblance to parents) and only in case of closely related forms. In case of varieties and subvarieties of cultivated plants, I dare say (I have vague remembrance that I have heard such a case) that the offspring of [p. 86] one variety might easily and so closely resemble the offspring of a distinct variety as to be classed as a subvariety of it, instead of its own parent. When thinking of analogical or adaptive characters, as of external resemblance of whale and fish, it occurred to me that the same cause acting on two closely related forms (i.e. those which closely resembled each other from inheritance from a common parent) might confound [p. 87] them together so closely that they would be (*falsely* in my opinion) classed in the same group. But I should require pretty strong evidence before I should be willing to admit that this had often occurred.

C. Darwin

Repugnance to Quadrumanous Genealogy

To a naturalist not deeply imbued with ancestral prejudices, derivation from certain rude savages, if mankind began with a low grade of Man, might be as displeasing as to come from a Chimpanzee. To see not the external but the wonderful internal structure, it is not the moral but the anatomical configuration in which the analogy lies. Why should we shrink from connection with such [p. 88] structure? If as progressionists we admit that in the series of limbs the differentiation is so trifling as to relate to whether there be 13 or 12 pairs of ribs, one extra carpal bone in each hand, joints in which one of the higher quadrumana—may differ from another, why sh.[d] we not be prepared to go a step farther & to imagine that the links or filiation is not some undiscovered one, but a revealed instead of an unrevealed 'deep meaning', a law which it is permitted us to comprehend rather than a mystery wholly hidden from us, a mode of operation only so far beyond the reach of our faculties to comprehend, as it relates to the originating cause to which the beginning of the material & spiritual universe may be due, but intelligible so far as the evolution [p. 89] of continuous events on this planet is not a sealed book.

Will the distance be lessened if all the intermediate links of the chain be supplied, w.[h] may one day conduct us genealogically from the quadrumana to Man? Or if this record remain forever imperfect, will that destroy the close connexion between what remains & Man? The great fact, that the brain of the Bushman, as Tiedemann[45] shows, leads toward the brain of the Simiadae, implies a connexion between want of intelligence & structural assimilation, so does the prognathous tendency of the Australian & negro race.

Each race of Man has its place, like the inferior animals.

Adjustment [p. 90]

During the period of change or adjustment when the circumst.[s] are altering fast yet not so fast as to cause extirpation, many varieties flourish & form & some are gradually put out of existence. Then comes a stationary period—very few invasions & immigrations of new species—no isthmus worn thru or promontory united to main land or intermediate island found to help migration.

A settled state of things, adjustment takes place—species get into a set [condition] & remain more permanent for part of a geol.[1] period than they did before for a few thousand years as equilibrium maintained for great numbers of species—tho' some must be always varying.

Man's Origin from Organic Inferio^r Being [p. 91]

If the time sh.^d come when it must be acknowledged that some one of their wonderful structure, in wh. Harvey might as readily have descri.^d—the circul.^n of the blood, or Bell the true theory of the nerves, as in the human subject, some one of the lost species by wh. a passage was effected, if it c.^d be found that the Deity instead of fashioning man out of inorganic matter, not of clay or mud, had caused him to evolve out of a being, physically speaking, almost as highly organized as himself, the general public w.^d after a few generations marvel at the shudders [to] which such an idea gave rise.

Instinct of Future State

Those aspirations which spring not from the inductive inferences of the Reason, but from some provision in Man's moral Nature, is [a] human instinct higher & more certain than either revelation or external sources of information.

Greg, p. 300 [46] [p. 92]

It is not from enquiries into the physical world, present or past, that we gain an insight into the spiritual; we may arrive at conclusions unwelcome to our speculations.

Extract from Lecture by Agassiz, Boston, January 1860[47]

What has the whale in the arctic regions to do with the lion or the tiger in the tropical Indies? There is no possible connection between them; and yet they are built respectively according to one & the same idea. There is behind them & anterior to their existence, a thought. There is a design [p. 93] according to which they were built, which must have been conceived before they were called into existence; otherwise these things could not be related in this general manner. Whenever we study the general relations of animals, we study more than the affinities of beasts. We study the manner in which it has pleased the Creator to express his thoughts in living realities; and that is the value of that study for intellectual Man; for while he traced these thoughts as revealed in nature, he must be conscious that he feels, [p. 94] and attempts as far as it is possible for the limited mind of man to analyze the thoughts of the Creator, to approach if possible into the counsels that preceded the calling into existence of this world with its inhabitants, and there lies really the moral value of the study of nature; for it makes us acquainted with the Creator in a manner in which we cannot learn [of] Him otherwise. As the Author of Nature, we must study Him in the revelation of nature in that which is living before our eyes.

[p. 95] Bell's discovery (in 1809) of the distinctness of the nerves of motion and sensation—Sat.[y] Rev.[w] [48]

Letter from Rev. Leonard Jenyns
to C. Darwin, January 4, 1860[49]

My dear Darwin—

I have read your interesting book[50] with all carefulness as you enjoined,—have gleaned a great deal from it, & consider it one of the most valuable contributions to Nat. Hist. Literature of the present day. Perhaps you may like to know what I think of your particular theory, tho' you will doubtless have many opinions offered far more deserving to be [p. 96] weighed than mine. As you are aware, I am no stickler for the multitude of so called species created by so many naturalists of late years, & I always thought the time was not distant when, after the brain-splitting process had been carried to its utmost length—some at least would see the necessity of retracing their steps, & again uniting a large number of the forms they had so carefully separated.

But I frankly confess I did not look for any such large assemblages of species to be brought together in this way as the descendants from one & the same stock, [p. 97] similar to that you have attempted in your volume. By this you will see that I embrace your theory in part, but hardly to the full extent to which you carry it. Still I allow you have made out a very strong case, and I will not pretend to say what future researchers in the same direction may not ultimately establish. I can quite fall in with the view that those fossil animals which so closely resemble their living representatives at the present day, are in fact the progenitors of these last;—such indeed has been my opinion for many years, tho' a contrary one, I [p. 98] know had been adopted by many of our first Geologists & Naturalists. I can also well believe that whole families have had a common parentage at some remote period of the past, & that the same may have been the case, reasoning analogically (tho' this is not always safe in Nat. Hist. speculations),—with groups of even higher denomination. But I cannot think that *all* the difficulties which stand in the way of so extensive a generalization have been entirely got over. It seems to me that if "all organic beings that have ever lived on this earth, had descended from some one primordial form",[51] [p. 99] as you seem to think possible, we should find (either among the fossil or living species) the same connecting links, & 'fine gradations' between the highest groups, that we *do* find among the lower, & which renders classification & definition so difficult. Thus birds as a class may be shown to be all more or less closely connected, with only here & there a saltus which we cannot hedge over: the same perhaps of all vertebrate animals—if we had all the lost & unknown living forms at hand to fill up the gaps;—but when we come to another [p. 100] quite distinct type of organization—the Annulosa, for instance, how little of intermediate between this & the vertebrate, or between either of these & the molluscous, etc. Still more what an hiatus between vegetables & animals, tho' just at their respective origins, or where the vital powers seem least active, they are scarcely separable by any distinguishing characters. But why should this never be the case afterwards—& not merely at the beginning of things—if they originally sprung from the same primordial germ?

[p. 101] One great difficulty to my mind in the way of your theory is the fact of the existence of Man. I was beginning to think you had entirely passed over this question, till almost in the last page I find you saying that "light will be thrown on the origin of man & his history".[52] By this I suppose is meant that he is to be considered a modified & no doubt *greatly* improved orang! I doubt if this will find acceptance with the generality of readers. I am not one of those in the habit of mixing up questions of Science & Scripture, but I can hardly [p. 102] see what sense or meaning is to be attached to *Gen.*: 2.7, & yet more to ver. 21, 22,[53] of the same chapter, giving an account of the creation of woman,—if the human species at least has not been created independently of other animals, but merely came into the world by ordinary descent from previously existing races—whatever those races may be supposed to have been. Neither can I easily bring myself to the idea that man's reasoning faculties & above all his *moral sense* c.[d] ever have been obtained from irrational progenitors, by mere [p. 103] natural selection—acting however gradually & for whatever length of time that may be required. This seems to me doing away altogether with the Divine Image that forms the insurmountable distinction between man & brutes.

February 19, 1860

Language

Never can start into being in an adult state any more than species.
May last very long or a short period.
One may outlive all its contemporaries.
May be thought to come from heaven whether in the beginning or at the confusion of tongues.
But there was a time when no modern language now spoken existed.
And when all that existed then are now extinct.
There are *two* ways of a language dying out, by transmutation into another so that the descendant c.[d] not understand their ancestor if they came to life again;
or, secondly, by extermination of the last speakers of it, or their natural death.

> But may there be also two ways of [p. 104] languages coming in as well as of going out?

If a pair are isolated they w.[d] still have a few words & tho' these may be lost the connecting links w.[d] never have been quite lost. But in truth the creative or inventive power would always continue & be developed as intellect & science made progress—& more new words may be added when a language is already rich in terms, than was spoken for thousands of years by the first rude tribes when the nation was founded or drew its inheritance.

The innate power suggests new words & systematizes them by grammatical laws & those of syntax. This power is the same as that which was manifested in a less marked degree by the first rude savages. It is ever evolving fresh wonders. When we have got rid of the Garden of Eden as a myth & the tower of Babel & show by what steps new languages grow or divide into dialects or are partially mixed by crossing or [p. 105] deviate into such distinct forms as that their affinities are scarce recognizable we are still as far as ever from knowing how this marvellous power is produced—this wonderful weapon of thought. The miracle performed in our own time is greater than we could have ever witnessed at any remote epoch to which our science carries us back.

Dugald Stewart—on language as an instrument of thought.[54]

Progressive Development

This theory instead of being mere naturalism, instead of banishing & distancing the supreme intelligence from the work of creation, instead of denying the intellectual causation of all the phenomena of change, instead of separating the mental & natural phenomena, requires a constant increasing intensity of miraculous power, a succession as it were of avatars, & at each visit a larger portion of the divine intelligence visiting the earth [p. 106] in a material form.

A man of genius, as Asa Gray[55] has said, may construct a theory of the universe, more scientific at once & more theistic than any one which has yet prevailed. More will be explained, but the inexplicable will more & more augment in kind & degree, that which we can never hope to comprehend will expand in a still faster ratio.

Bronn, *Wahl der leben's weise*[56]

Pressure of population on means of existence drives individ.^{s} to choose new ways of living; this shifting for themselves by new ways, which varieties adopt, leads to their perpetuation as they prevail, if favoured—by new natural conditions of food and climate.

Monstrosities

See Harvey's case of begonia in *Gardener's Chronicle*.[57]

Also of Aspicarpa, Malpighiaceae, Darwin's book, also Campanula rotundifolia; Alph. De Candolle *Botanique geographique*[58]

Progression [p. 107]

Aspicarpa Darwin [*Origin* p.] 417

Perfect & degraded flowers in the same plant may imply that the common notions of rank are unfounded or, that here is a transitional case. Both kinds of flowers (see Harvey Gardener's Chronicle)[59] ripen their fruit & are therefore equally effective.

Herbert Spencer to C. Darwin, February 22, 1860 [60]

H. S.'s own doctrine of 'evolution' put on so satisfactory a basis.

"You have wrought a considerable modific.^{n} in the views I held. While having the same general conception of the relation of species, genera, orders, etc., as gradually arising by differentiation & divergence like the branches of a tree & while regarding these cumulative modifications as wholly due to the influence of surrounding circums.^{s} I was under the erroneous impression that the sole cause was adaptation to [p. 108] changing conditions of existence brought about by habit, using the phrase condition of existence in its widest sense as including climate, food, & contact with other organisms (for general statement of this view see Essay pp. 41, 45.)[61] But you have convinced me that throughout a great proportion of cases, direct adaptation does not explain the facts, but that they are explained only by adaptation through Natural Selection.

Many (etc.) must have been struck with the fact that among all races of organisms the tendency was for the best individ.[s] only to survive & that so the goodness of the race was preserved.

[p. 109] I have in Essay on Population etc. remarked this as a cause of improvement among mankind.[62]

But I & every one overlooked the selection of 'spontaneous' variations without which I think you have clearly shown that many of the phenomena are insoluble.

You have shown that the doctrine furnishes explanations to phenomena otherwise inexplicable. However the argument may as yet fall short of direct demonstration, yet the indirect demonst.[n] is to me conclusive. I take it to be incredible that so many different kinds of evidence sh.[d] coincide in supporting a doctrine that was untrue.

Pictet[63] to C. Darwin, Geneva,
February 19, 1860 [p. 110]

Your book makes science young, clear, elevated, but no facts to prove that principle that slight modification multiplied by any factor of time, no matter how long, could reach the character of families, etc., or c.[d] produce profound modific.[s] in organization.

Progression

If true we can no more predict the future from the past & present than a person who had traced the develop.[t] of an infant to an adolescent could tell what to expect in the adult or whether it w.[d] afterwards go on developing or stop suddenly or gradually decay.

Who shall say that the system of things which Geology interprets & cannot trace back to the beginning was the first ever witnessed on the planet. To speculate on former [p. 111] systems may be a waste of thought & useless to assume that there never was an antecedent one, may be unphilosophical. That the series of organic changes of which we have examined many stages was the first, even if we say the beginning of it, is an arbitrary dogma. It is like taking for granted that the remotest nebula, hitherto reached by the telescope, marks the utmost limits of the space filled with worlds.

Tentative Creation

To suppose the throwing out of varieties to be creation feeling its way, a delegated power often failing, but in the main succeeding, is the idea which reconciles to many minds the action of Secondary Laws. But in truth this is as inexplicable as direct acts of creation, only it is more consistent with analogy of all other things. Disturbing Causes allowed to interfere & apparently to thwart the object in view. The [p. 112] existence of evil, of imperfection, the interference of some adverse influence in the physical like the moral obliquity of the human soul.

Extract of Letter from A. C. Ramsay to C. Darwin, February 21, 1860 [64]

I had for years been using to myself the terms mutability of species, and transmutation of species, for want of better words to express the feeling, (amounting to a conviction) that in Time species passed insensibly into each other instead of being produced by separate acts of [p. 113] creation. But the idea implied by your principle of Natural Selection taken in connection with the struggle for existence is very different indeed from the vague gropings towards the light in which myself and others with various degrees of hesitation indulged. My faith, such as it was, was founded on the following considerations. The disputes among Naturalists as to what constitutes a species, the wranglings as to whether such & such were distinct species or varieties, [p. 114] the acknowledgment of permanent varieties, the existence of specific & still more of generic centres, of representative species in space, & in time in a geological sense, & the links (especially the rudimentary organs) that bind genera together, especially in the higher Mammalia, when living & fossil forms are considered together. The succession of small miracles required to produce certain species in a formation just a very little different from those in the preceding formation went sadly against my [p. 115] mental stomach & I never could reconcile myself to the idea of a Creator making a number of small experiments as it were, of feeling his way in the work of Creation. And when asked where are the perfect gradations, that show the passage of fossil species, I have always maintained that the record as shown in the succession of formations was too imperfect for us to find this & that it is not till an aberrant branch, multiplied to a vast extent, that we were likely to find fossilized any of its [p. 116] representatives at all.

March 3, 1860

Coscinopora globosa, a number of, from a tumulus near Bedford, shown me by Tennant,[65] bored through exactly as in the Amiens or S.[t] Acheul specimens.

Oscillations of Level

Submerged forests all round the Bristol Channel—Huxley—[66]

See Sedgwick & Murchison on raised beach near Bideford.[67]

Prof. Phillips thought the trunks would not stand if exposed to sea but at Amroth Huxley saw (near Tenbigh) stumps 2 f.[t] high, 1000 years old, bored by pholades.[68]

Oscillations

Ramsay says the submergence was in Wales 2300 feet, see in his book & a paper in Geolog.[l] Journal.[69]

Antiquity of Man [p. 117]

Throws great light on extermination both of animals, & in Denmark, of trees.

Hoxne[70]

The flint hatchets from H. shown us by Lady C. Kerryson[71] are like [those of] Amiens but quite sharp pointed, not broken, 3 of them as if never used—found 10 foot down (at least the last found) from the surface. A tooth of horse, position not known—Rib of large animal (qy. what?) 5 f.[t] below the flint hatchet.

Elevation of Mountain

Stür[72] informed me that the marine miocene of Vienna (Upper Mio.) [Notebook] 215 p. 93, ascends as we approach Sömmering Alp. So Lower Miocene highest in the Alps as in the —— —Eocene also—Miocene of Asia Minor highest in neighb.[d] of Loess, highest near Alps? See N. B. 215, p. 130, Schotter[73]

Elephants in Scotland

Drift, northern, implies submergence & therefore no land animals found in it, if any found they are marine shells. In N. America or U.S. mastodon *over* drift.

March 10, 1860 [p. 118]

With Darwin, Down: Progress[74]

Savages may make constant progress for 1000.[s] of years as in Stone Period in Denmark acquiring & transmitting to their posterity increased powers of observation, of eye sight, of physical strength, of endurance, of observation shown by remembering localities, by tracing the footprints of animals & man.

Natural selection is at work, not in giving advantage to those who have the finest brain, or that which we call genius, but in qualities best fitted for savage life. This son of a man of genius, consid.[d], says Whateley,[75] so stupid that they sent him to Australia in despair, turned out a first rate colonist.

One fool may cause others to be like him according to atavism in families where talent prevails.

Brain is inherited like other advantages.

The Fuegians showed a number [p. 119] of traits like Europeans when they came to live among them, more than Darwin could have anticipated.

A slow but steady progress may be made by a hunter tribe for 10,000 years & yet they may remain ignorant of the use of metals. Many arts must have been known to the shell-mound race in Denmark to enable them to catch fish & yet they were always ignorant of metal.

Defective Instincts

When we substitute law for direct interference & speak of an instinct being at fault, we evade the great difficulty which consists in defects (for such they are according to our best judgment) occurring in the works of an all perfect & omnipotent being. That an insect, a fly, sh.[d] be attracted by the flower called a [Venus] fly trap, by its resembling the smell of putrid meat, & so be lured to its death is the misdirection of an insect [instinct] given for the good of the creature—If we consid.[d] every act as one of special [p. 120] intervention, we find ourselves in the presence of the great & perplexing difficulty, the existence of evil & if the naturalist cannot solve this problem, so neither can the theologian.

The great question of free will & fate & predestination lies at the bottom of all this & they, who hope to explain the enigma which the creation of species & instincts involves, may hope to solve all the metaphysical riddles which in the present state of our knowledge are mysteries by no means diminishing in magnitude as we advance.

Perfection of Organization

Von Baer's standard is "the amount of differentiation of the different parts" (in the adult state adds Darwin) "& their specialization for different functions", or as [Henri] Milne Edwards w.[d] express it, "the completeness of the division of physiological labour".[76]

But in fish some rank the shark, others the teleostean fish as most perfect, [p. 121] the former as being most like reptiles, the latter as being most strictly fish-like & most unlike other vertebrate orders.

Plants, some think the highest to be those wh. have sepals, petals, stamens & pistils fully devel.[d] in each flower, highest other plants when the several organs are much modif.[d] & somewhat reduced in number—latter, says Darwin, best opinion.

Retrogression in situations where certain organs superfluous. *Ib.*[77]

Why so many simple beings still? Lamarck introd.[d] spontaneous generations, but Darwin's natural selection implies no innate tendency to perfection & development is necessary.

Rhizopods, or infusoria, foraminifera, have remained for several geolog.[l] periods unchanged. [p.] 3.

Brain, if warm blood more highly organized, why then do not cetacea compete with fish. p. 4.

Higher organization w.[d] often be of disservice—more delicate organs more easily put out of order.

March 10 [1860] [p. 122]

Site of Oceans

Same to some extent in Miocene Period.

If Straits of Celebes being deep causes partition of two faunas it proves how much oceans of greater depth must have been persistent. Where sea shallow, mammalia of islands same as of continents, where deep even if channel is narrow they are different, or none.

Galapagos = S.[t] Helena, Madeira.

Sedgwick's Objections[78]

Is there any proof that the large cephalopods of the Palaeozoic rocks were gradually driven out of the old Fauna by the incursion of more highly organized & more powerful animals of a kindred organizat.[n] & function.

Is there any proof on known geological facts that the strongest & most highly organiz.[d] Palaeozoic fishes disappeared from the old fauna thro' any like causes?

Are the reptiles that have been discovered among the palaeozoic rocks of a low organic type? Is there any proof that the highly organized & powerful reptiles of the Secondary period were driven out of the fauna by any process of develop.[t] & natural selection?

[p. 123] Is there any proof that the large extinct mammals of S. America & New Holland became by any conceivable process of Natural Selection the progenitors of the existing Edentata & Marsupialia of those countries.

Is there a shadow of proof from the ethnographical & physical history of Man that any one of his oldest varieties was derived from a quadrumanous progenitor?

Is there any *physical indication* of an enormous interval of geological time between the appearance of the Permian fauna & that of the Muschelkalk or between the fauna of the Muschelkalk & that of the Lias?

Is there any proof derived from Natural Hist.[y] that Time has in itself, & independently of varying conditions, a tendency to change an existing fauna?

March 10, 1860 [p. 124]

Asa Gray on C. Darwin[79]

A strong point in favour of D.'s theory is that species are not (yet) now placed in countries best adapted to them. This makes strongly in favour of local origination & hardly less so for C.D.'s particular theory.

Agassiz will have it that not only no tertiary species, or forms of which they are the progenitors, exist now, but even that no organism of the age of the mastodon is not extant—that so far as terrestrial organisms go, there was a time between the last tertiary & the drift period, & also between that & the present when no living thing existed. So he will consistently have it that the men who made the flint knives in the Somme valley [p. 125] belong to none of the present 30 species of Men!

Jukes[80] [to Darwin] (February 27, 1860)

I believe in the perfect indefiniteness & frequently the vast length of the interval between the formation of any one bed & that which rests upon it.

I believe any amount of beds may be intercalated almost anywhere not only between any two formations, but any where in any format.ⁿ & that the ordinary ideas of Geologists as to contemporaneity will have to be greatly expanded.

I am therefore prepared to treat with very little respect the argument ag.ᵗ you derived from missing links.

When you have anything approaching to a continuous series of beds, you have the links; moreover, when you get species ranging through a great series of beds [p. 126] they do vary. Calymene Blumenbachii, for instance, is found in both Upper & Lower Silurian, but Salter says that those found in Lower Silurian vary so much from the Wenlock ones that he has doubted their being the same species & was inclined to give them another name. So Baillie[81] & I have arrived at the conviction that Spirifer Verneuilli, S. disjunctus & S. strictus are mere varieties of a species, that found in the Lower [Silurian] beds having been [p. 127] christened Verneuilli. So Terebratula hastata has about 30 synonyms & it is very likely that some of the varieties are of greatly diff.ᵗ ages.

Your theory accounts for just so much of progressive improvement as we see has taken place, i.e. an improvement & a multiplication & a complexity in the whole, but not necessarily absolute in every step or every instance, & relative but not always absolute.

Gwyn Jeffreys[82]

who differs from E. Forbes
[Godwyn] Austen as to size of mollusca—
See MacAndrew's answer to G.[odwyn] Austen in ——[83]

Darwin thinks that Barrande's distinct provinces of Primordial [genera] may be partly owing to the confounding

March 13, 1860 [p. 128]

With Darwin: Extension of Same

America	Norway
A	A
a	a
B	B
b	b
C	C
c	c

Thus B may have had same species as B in Norway, but if you compare b with B Norway you get distinct species—b. N. being a modification of B. America.

Hooker's Galapagos paper good. Littoral mollusca must have means of migration unknown to us especially the viviparous ones.

The Scilly islands may have derived their shells during subsidence, every part being shore by turn between main land & Scilly isles.

Dwarfing

—takes place in plants & animals as you go north but Darwin knows of no cases going [p. 129] South.

But MacAndrew mentions cases of mollusca where northern species dwindle in going south.

Antagonism of Gwyn Jeffreys & of Watson to E. Forbes' opinions.[84]

Latham[85]

Man conquers man, & occupant displaces occupant on the earth's surface. By this means forms & varieties which once existed become extinct. The more this extinction takes place, the greater is the obliteration of those transitional & intermediate forms which connect extreme types, [p. 130] and the greater this obliteration, the stronger the lines of demarcation between geographically contiguous families. Hence a variational modification of a group of individuals simulates a difference of species, forms which were once wide apart being brought into justaposition by means of the annihilation of the intervening transitions.

Down, March 13, 1860

Valleys in Chalk

The subangular & angular accumulation of flint in Darent dry valley, see N.B. 220, p. 80,[86] & Greenstreet Green & all those near Down in the chalk, some of them posterior to sandpipes as in pit A near Down, no pipes as at B, because says J. Lubbock[87] valley is posterior [see Fig. 10], [p. 131] imply,

Fig. 10

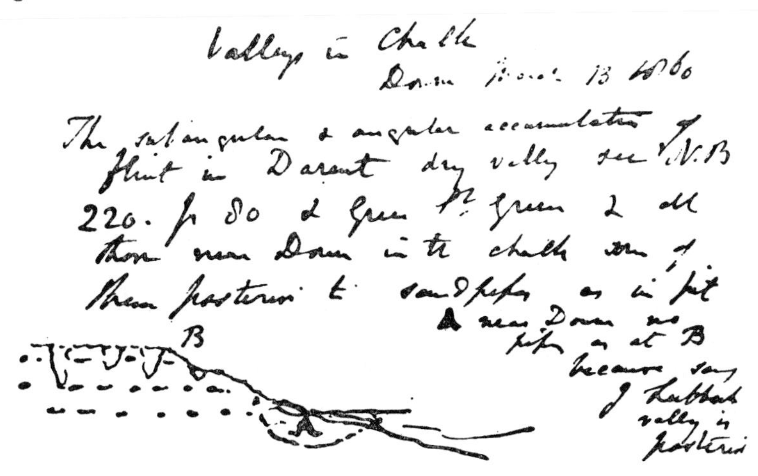

accord.[g] to Trimmer,[88] subaerial excavation. So must the Amiens flint? The escarpments of N. Downs must be posterior to the stream of flints which ascends from Greenstreet Green? If so how can the escarpment be of marine origin?, unless the sea cut away [see Fig. 11] the escarpment after the mammoth gravel A was formed. Submergence after 2.d elephantine period & last point of elevation of the Weald?

Fig. 11

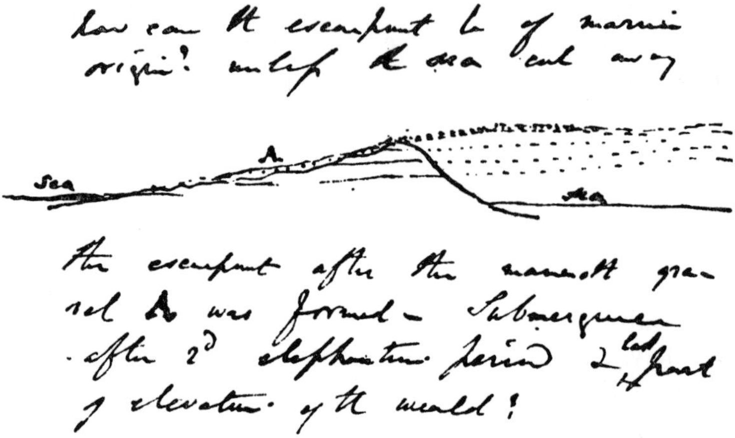

Greeks

The Greeks were the only civilized people of antiquity who unlike the Egyptians, Chinese, Hindoos, had no regular organized priesthood. Only certain families had oracles & temples, but not the State, so that there was more freedom for new schools of philosophy, less Danger than in Magna Graecia & Etruria of setting up new opinions & doctrines, moral or political. [p. 132] This education of Greece continued in a republican or constitutional form for 500 years, see Grote.[89] During this period there was time for craniological development & cerebral improvement as well as for increases of beauty of form so much adored by the Greeks. Both then, the improved intellect & the beauty, remain after 2000 years & tho' the moral dignity & favour of the Turk may be superior, years of oppression have not prevented the Greek race from retaining their superiority in other respects.

It may be said that the popula.n of Greece was small compared to modern nations, even if we take in all their colonies—

also that the northern nations who overran the Roman Empire displayed after 14 centuries of darkness such an aptitude to adopt & profit by the inheritance of the classic writers of antiquity as seem to show a fitness for improvement which their previous education in the north had not prepared them for. [p. 133] The northern [peoples] had a degrading superstit.[n], [made] human sacrifices, were drunkards, wild hunter tribes.

The Greeks were disciplined by a keen competition of rival races, much selection owing to that, emulation, freedom of thought, less persecution for new ideas & opinions than elsewhere, a less rich endowment of certain doctrines deemed to be infallible. Nothing like the Spanish Inquisition selecting every original thinker & burning him. Mr. Grote thinks that in no modern community w.[d] a teacher go on like Socrates for 50 years inculcating truths distasteful to the higher power. It might be impossible now even at Boston in New England.

An occasional prosecution might occur at Athens, but not systematic or regularly organized, nothing comparable to the Inquisition.

March 16, 1860 [p. 134]

Ancient Land

If there were 3 miles of sinking in the Carbonif.[s] era, how could Darwin infer that all was low land—at that epoch?

The whole time the coal was found at level of sea or near it. Whence came the sand & mud & in P.[a] the pebbles of the lower coal. [?]

How could there be so much sinking at successive periods & yet land left for trees of oolite, of Wealden, etc. N.B. 151, pp. 27–30.[90]

Brain of Negro

Owen said that in Jul. Caesar's time, brain of ancient Briton may have been small.

Letter from Boucher de Perthes, Abbeville, March 16, 1860 [91]

I have just sent you a little box of fluviatile & land shells from Menchecourt. They come from two different beds. You will [p. 135] find them separated in the box by a layer of cotton. At the bottom of the box is a detailed ticket of the lowest bed. These shells which are almost all fluviatile are from the 'sable aisse' & at a depth of 8, 9, or 10 meters, which is I believe over the level of the Somme. At a certain point we have found a great number of them & perfectly uninjured & of large size. They belong to species now living in our marshes, only they appear to be stronger. In my opinion they have been born & died in the place where they are found & they have died in consequence of the incursion of a current, which brought in the 'sable aisse' or blue [sand] under [p. 136] which & in which they are found.

I have said in my 1.st vol. of *Antiquities* that agglomerates of hard sand, which must have been the result of animal decomposition, seemed to prove that the elephants, whose bones were found, must have arrived there with their flesh on.[92] In fact these bones are neither worn nor rolled, one might believe that these elephants were swallowed up in this marsh, even before the torrent or any other sudden or successive cause brought down the deposit of sand in which they are found above the cut flints & the rude flints. This in my opinion would support your system, that the beds of Menchecourt are not [p. 137] the effect of a convulsion of the elements or of a catyclysm, but of a long space of time. The primitive soil must then have been the bottom of a lake or a marsh resting on the bed of chalk, a bed of tuff would be formed at the bottom of the water, as we now see formed at the Somme, the shells which I send you could have lived in this water. The coarse silex with the hatchets & a part of the bones could be brought in by a current of water or a torrent, living animals would, as now happens, have fallen into this marsh or this hole & perished. The sea which at high tides, at that time entered the Somme, would have brought in the marine shells which are occasionally met with below or above the white sand. [p.

138] After a lapse of time the bed of 'sable aisse or blue' could have covered the marsh & later the bed of 'sable gras or jaune', then the clay, then the rolled silex or 'terrain de transport.' Lastly the vegetable soil would by degrees have covered the whole. Doubtless ages or perhaps thousands of ages may have been necessary to bring about this result. This is, I believe, the way in which you look upon the thing & these shells, which you are going to examine, if they —— in situ, as I believe, will materially confirm what you have theoretically thought. I have also found some marine [p. 139] shells, but few & they do not come from the same place as the shells I send you—but they belong to the same beds. The existence of these marine shells at Menchecourt cannot, however, be doubted. Mess.s Davis,[93] Daillant,[94] Butenor,[95] Prestwich,[96] etc. have found them. I am having them sought for & I will give them to you when you come to Abbeville.

Deluge

If the Mediterranean, blown by a westerly gale, overtopped the barriers & flooded the Dead Sea, hills 1300 f.t high might be submerged especially if a channel were cut, & then if the wind changed, the channel might be blocked up by a sand bar & evaporation & wind might dry up the country, & the hills reappear.

Ireland, No Fossil Elephants [p. 14]

Are there any old alluvia in Ireland with E. antiquus & Rh[inoceros.]. leptorhinus? Or any caves with them & Ursus spelaeus & Hyaenas.

Was Ireland never united to England, save after drift period & that in the time of Megaceros, but when mammoth rare or extinct?

Why no elephants in Scotland in fossil state?

Is there any proof of a preglacial connexion between Ireland & the continent except E. Forbes's supposed Lusitanian flora in West of Ireland?

Did the sea of the northern drift, when it polished & furrowed the hills & dales, sweep clean away all the old elephantine alluvia? and if so, why did not the caves still [p. 141] preserve memorials of a more ancient fauna?

It may be said that if Ireland was ever land in preglacial time, it must have had some land animals in it.

How far north in Scotland have fossil elephants been found? In Behring's Straits & in Siberia they go as far north as lat. 60° to 70°. Why not in Scotland?

Did the submergence of the glacial period obliterate them? Was a fossil mammoth found in digging Caledonia Canal?

No proofs in Perthsh[ire], Kincardine & Forfarsh.[ire].

Are the most modern signs of land in Scotland the Mull plant-beds of D.[uke] of Argyl?[97]

Does the form of Moraines & distribut.ⁿ of glacial boulders [p. 142] & the parallel roads of Glen-Roy demonstrate a gradual submergence of preexisting land afterwards re-elevated.

Are there no cave fossils to prove antiquity of land north of [the] Tweed? letter to Geikie, March 20, 1860.[98]

In Behring's straits or in cliffs near & in Siberia in lat.ˢ 60° & 70° multitudes of Elephants so that it is not the being so far north that has made Elephants scarce in Ireland, Wales? & Scotland. And if it be said that in the same region tertiary strata are wanting, still we find, in superficial gravel in England, far away from tertiary formations & in caves, many extinct quadrupeds.

Searles Wood[99]

—objected to Darwin's theory that he had never found a transitional series between two tertiary species. I asked him, if he had done [p. 143] so, whether he would not have concluded that the species, previously regarded by him as distinct, ought to have been united as one. He confessed that he should have done so.

The proof therefore required by the objectors to Darwin is in the nature of things most difficult to find & when found will not produce the slightest effect in convincing the opponent. Of this we have evidence in the fact that we have living plants which, if the intermediate links were gone & the extreme varieties left, w.[d] rank as species but wh. do not, because the connecting varieties are not extinct.

Hippopotamus—Folkestone

The N. Downs were such as they are now when Hipp.[s] bed was deposited, therefore difficult to believe that the Greenstreet Green Mammoths belonged to a valley newer [p. 144] than the escarpment of it. Davis, See N.B. 175, p. 3 [100]

Natural Selection Deified

If we take the three attributes of the deity of the Hindoo Triad, the Creator, Brahma, the preserver or sustainer, Vishnu, & the destroyer, Siva, Natural Selection will be a combination of the two last but without the first, or the creative power, we cannot conceive the others having any function.

The destroy[ing] force is selection, the sustaining [force] preserves things, whether in the individual or successive generations the same, but in order that life sh.[d] exist where there was none before, that instinct sh.[d] be superadded to new sensation & mind in the course of time, t[hus] instinct w.[d] acquire a power of creation constantly at [p. 145] work & this is not selection, but creation, the variety-making not the destroying, or continuing by inheritance, power. Nothing new w.[d] appear if there were not the creative force.

March 28, 1860

With Lieutenant Cowell [101]

Dead Sea 1311 f.[t] below Mediterr.[n]
1300 " deep
Lowest part of barrier between it & Mediterranean 350 f.[t] of limestone.
Eight small streams enter it beside the Jordan.

The Jordan 8 f.[t] deep, 100 wide, runs 3 1/2 miles an hour.
Yet evaporation [is] such that the lake remains of same height only varying a few inches.
Dead Sea 60 miles long, 7 or 8 (?) wide.
Birds swim on it & fly over it.
Difficult to get your head under such dense water.

April 17, 1860 [p. 146]

Origin of Species

We may assume the existence of all great classes & orders & if with.[t] pretending to explain their origin it can be shown that thus modific.[n] of these by gradual changes w.[d] have prod.[d] other allied species according to a law of Creation & that this law, however mysterious & incomprehensible to us in its general nature, is nevertheless one like the variety-making power, a vast step will have been gained.

If Mr. Darwin has made it much more probable that there is this genealogical & maternal link, this blood relationship between the species of successive geological faunas, the step is immense altho' the nature of the Causation is as impenetrable a mystery as that of ordinary generation.

When we substitute for whole divisions of the animal kingdom the [p. 147] law of ordinary generation, the origin *ab ovo* for what has been termed spontaneous generation, we have made a vast step, humanly speaking, in zoological science, altho' the fact that after a cycle of changes the individual dies & then is replaced by a germ-cell, which tho' simple in structure, so far as our microscope can detect, contains within itself the embryo of a being like its parents having some of their individual peculiarities transmitted to it. All this may be more marvelous than the old hypothesis: farther removed, as hopeless[ly] distant from a solution as the idea of springing up from inorganic elements, nevertheless, no one can deny that we have advanced far [p. 148] beyond our predecessors by having in certain wide departments, in insects for example, proved that every one, with.[t] except.[n] arises from an egg, & so of intestinal worms.

In order to evade the real question it is objected that we cannot derive all from one egg. If an hypothesis which attributes all the living species as the descendants of the Plio. & then of the Mio. & then of the Eocene, then again from the Cretaceous & then from the oolite, the percentage of generic forms being always a third or 35 per cent, as Pictet has said, then we may fairly for a time reason out the question without perplexing ourselves ab.[t] the beginning of organic life.

It is as if when we reason of the origin of the structure of volcanoes of successive periods, whether by [p. 149] upheaval or by the outpouring of inclined & originally sloping beds of lava & ashes, I were met with the question, what was the state of [the] earth's nucleus in the beginning. Was it solid or fluid, a highly interesting question to which possibly science may one day discover an answer, but it belongs to cosmogony rather than to geology.

Our present business is of a humbler grade. We must be content with trying to explain the manner in which certain changes have been brought about in the organic & inorganic world & may be satisfied if we can find out something of the material laws by wh. they are gradually effected. The initial formation of the planet, or as of its first living inhabitants, are grander subjects.

April 20, 1860 [p. 150]

Creation, Dawson's Archaea, p. 52 [102]

So far from the work of Creation being removed from us, we are far more brought near the presence of the Spiritual Creator than if we had lived at any prior period. Once admit progress to be the law & Man an integral part of the same series of developments & the manifest.[n] of power is greater as time runs on.

April 25 [1860]

With Darwin

Mammalia & Birds have not been studied like plants over wide enough areas so as to show the relative variability of genera small & great.

The Grouse, Tetrao scoticus, now thought by many ornithologists to be only a variety of a Norwegian Species!

Insects, large & small genera in part gained from Wollaston's own book by Darwin, & yet W. thinks the small genera vary as much. [p. 151] Waterhouse says that the smaller species of mammalia are like arrested forms of larger animals, dwarfed forms of bigger ones of same type. This w.[d] lead to degeneration theory.

Volition

As in the human dispensation certain physical & moral laws are ordained & every result must be foreseen & foreknown & what seems chance is governed by system, yet we have to admit free will tho' we cannot reconcile it with the admission of law & predestination, so we may have to admit that we can never understand a reserve of volition in the supreme creative or variety making power by which [p. 152] every individual is made to vary from every other individual & by which not only degeneracy but progressive develop.[t] is ensured or effected in spite of controlling laws such as those of generation.

Man, Exception, Carpenter, *Medico-chirurgical* [*Review*]—p. 404, April 1860 [103]

"The psychical capacity of Man separates him from the nearest mammalian type so as to indicate his origin from a distinct creation." [104]

If this be true we cannot argue from rudimentary organs, from affinities & homologies, from Unity of Type, from embryology in favour of the link of common descent. Grant the miracle & why not grant all minor ones? If in the type where [p. 153] there are the same 242 bones in the Orang & all homologically connected, you cannot concede a connexion of a material kind, certainly no other sh.[d] be granted.

The difference of brain is so great between the European & the Bushman as to show in this, as well as in mental capacity, a wide range. The difference in the brain of a Bushman & a gorilla is, says Prof. Huxley, less than between an Orang & a Lemur.

The degree of intelligence of Man may be so superior to that of any of the inferior animals that we may assign him psychologically to a diff.[t] class or perhaps Kingdom of Nature, but this cannot entitle us to shut our eyes to the physical connexion. The same mammalian type, so many cut off [p. 154] in infancy & some instinct, & the intelligence apart from instincts, analogous to the reason of Brutes, no exemption from diseases, same circulation of the blood & nervous system, embryo accord.[g] to Von Baer identical (refer to V. B. instead of Agassiz as Darwin has done). The same memory, anger, pugnacity, jealousy, dreams, suffering of pain, liability to monstrosity, races, accommodation to climates.

If we give up the type as an independent creation not linked with any preceding one by a material bond, all the best arguments in Mr. Darwin's & Wallace's & Lamarck's work fall to the ground.

April 24, 1860

Monstrosities

Suppose we tried to obtain evidence of a 6-fingered race of men from fossil remains or of extinct races of Man analogous to Negro, or Australian, what labour would it not require, yet the [p. 155] missing intermediate links are asked for as if we ought to have found them.

Preface

I have felt for many years that if the doctrine of development in time c.[d] be established, as contended for by Prof.

Sedgwick among others, by Lamarck who sketched it out, advocated by those, down to Bronn[105], who were particularly opposed to the doctrine of transmutation, that it at once w.[d] remove my chief point of resistance & repugnance, the identi.[ty] of Man with the inferior animals or the transit.[n] from the latter to the human race.

I stated in all Ed.[s] of the *Principles* that Man seemed to be the beginning of a distinct series & if so, ought not to be taken as the last link of a chain of gradual & successive developments. They, who advocated the doctrine that the advancement led from the fish to Man, but that there were such gaps as that the concession of such an ascending series did not involve the approximation of the human & inferior mammalian type, have always appeared to [p. 156] me vastly to underrate the amount of our ignorance. I have always had to protest ag.[t] their finality doctrine. If the smaller intelligence of the ornithorhinchus gave place to the marsupial & this to the eleph[an].[t] & chimpanzee & Man's coming last was the highest point of the same series of elevation in rank & specialization in function, the ordinary & steady progress of removing or lessening one hiatus after another must end in assimilating the two series of instincts with a little reason, & reason with a little instinct.

Hence I have always felt disposed to make my stand ag.[t] too hasty an assumption of the law of progression, seeing the proof also of degradation as in the case of the reptiles, & of the continual discovery of higher & higher types to remoter periods of the past & the appearance of very highly organized invertebrate types such as [p. 157] cephalopoda in very old rocks, our ignorance of land animals in older strata, the great imperfection of our knowledge especially of air breathers.

But the fact of Man having come last I never doubted or attempted to gainsay—& I consider those who labour so hard to establish the point, coupling it with embryological theory of development, as not consistent with themselves. They appeal to Man as having been the last created, as a proof of progress from invertebrate to vertebrate & take advantage of the undoubted defectiveness of our palaeontological informat.[n] to deny the probability of the links wh. w.[d] unite Man & the inferior animals.

The time doubtless of the last step even if it embrace a great part of the Newer Pliocene is very short, but we have no right to consider that we have yet more than opened a page of the last volume of the anthropomorphs. [p. 158] If all the arguments of embryology & rudimentary organs, & the succession of proximating allied forms in time & those derived from classification & homology & geographic distribution—if all these have any weight with us in deciding in favour of Darwin's view, in preference to any other hypothesis yet advanced, then it is quite impossible to exclude Man from the same series of affinities because, in structure, embryological develop.^{t}—formation of strongly marked races, structural affinity, cranial, dental, & others he is most closely allied. Resist this evidence & we may safely reject so much of the transmutat.^{n} theory as to rely on some law wholly different from that of blood relationship to prevail & to explain all the difficulties we [p. 159] encounter. We may then be sure that new types arise, if from some secondary law, from one which necessitates no derivation from preexisting organic forms. For no favour has been shown to Man. No ground for imagining that a departure has been made in order to prevent his moral & intellectual nature from being bruised in the conflict with the preordained & previously prevailing laws of the inorganic & organic world, inexorable laws not only of mortality (his longevity not being greater than that of the Elephant) but of the loss of the majority before attaining the adult state.

Extinction

In Preface show that I gave the first systematic proof of the struggle for existence ensuing during the changes of the organic & inorganic world, the successive extinction of animals & the only mode of extinction, which I did [p. 160] not admit, is that by transmutation, which being granted by no means precludes or supersedes the other, which is a proved *vera causa* while the transmutat.^{n} is still only the most probable hypothesis as yet advanced. "All plants at war," cited by me from De Candolle.[106]

I established the cause, which resembles the extinction of a language by death of the last survivor who speaks it, but not the loss of a language by the dying out & modification of some characters, etymology & pronunciation changing, & by the inventive power of Man introducing new terms, which corresponds to the ever-present creative energy of the supreme cause & which seems to augment in its manifestation in intensity & rank, just as the latent inventions of terms & forms of speech in a language rise in abstract meaning & dignity & when [p. 161] united to the production of the simple & complex beginning of rude tongues, may often resemble the combination of some embryonic & rudimentary organs & functions of the lower type with the more exalted spiritual & physiological developments of the higher.

Let us be satisfied with the simple, humble, material, matter-of-fact question of ascent by generation, as ethnologists may of the filiation of languages & leave speculations as to the creative or formative force & its Nature. Just as we may put aside whether the power to invent an abstract term is an emanation from heaven or 'inspiration', or a force evolved out of the multiplication of instincts, as Herbert Spencer might assume. Ethnologists may or may not agree with such metaphysical speculation & yet be more fully employed in working out the problem of the growth & variation of languages & their extinction [p. 162] & the relation of the dead to the living.

Agassiz has thrown out as a conjecture, rather than a theory, that certain distinct races of Man may have originated in nations of individuals, man being a social animal. Now if this were true we might speculate on an advanced race suddenly springing up, higher in mental power than the European & inventing at once a language more complex & richer than any of the preexisting languages. One fact in confirmation of this course of Nature w.[d] overthrow all the reasonings in wh. Marx,[107] Müller,[108] Bunsen,[109] Humboldt[110] & others have indulged on the past history & growth of languages. But such a new tongue would probably want those signs of crude & embryonic developm.[t] which are detected in the construction of old tongues. Its rules w.[d] be more regular, nothing like those terms compared by Darwin to rudimentary organs, where the derivation & pronunciation quite differed. Nevertheless if the race improved & [p. 163] advanced, it might enrich & alter its language & then future ethnologists w.^{d} have that starting point of a new type which some suppose Man & each of the types of the great classes or subclasses to have furnished.

We must not be too ambitious, but be content to make one great step at a time, to waive all expectation of explaining the creation or origination of Instinct & Reason, or even of our morphological types, & to be satisfied with a comparatively slight step, the connexion or independence of the successive sets of species which inhabit the planet, with an antecedent nearly allied set or assemblage by the bond of descent & blood-relationship.

Are or must we abandon the test or the idea of blood-relationship as peculiarly marking a species or does it extend far wider & even indefinitely so that it may embrace genera, orders & even classes. [p. 164] There is yet another limit which it w.^{d} be well to set to our ambition & that is all speculation as to the beginning of life whether it was by one or many types. This forces us to form conjectures on a period respecting which we have no facts whatever. It is like the question of the condensation of a primitive nebula, the primaeval fluidity & temperature of the globe, & other hypotheses which may be of legitimate scientific interest, but which a geologist, who has more than enough to explain as deduction from positive fact, may well dispense with.

We sh.^{d} dwell less & lay inferior stress even on the oldest palaeozoic faunas which are still rich in species & avoid any speculation on the remote & less perfectly known glimmerings of life in ante-Cambrian or pre-Silurian strata.

Having eliminated therefore questions of the origin of life & the number of primordial [p. 165] types & 2.^{ly} the question of the nature of the Creative Force or Power—we may most profitably try the question of the derivation of successive species from preexisting ones & the question of progress. This latter is most in favour of new types because it introduces an increasing intensity of power & thereby prepares the mind for that want of uniform action of the governing force, which may be compared to a supernatural intervention. But we do not believe in the interference, direct & irregular, of the supreme power in the ordinary course of Nature in human times, because of the progress which is making in human develop.t & the possible modification of the human organization, cranial & other, in consequence of such developm.t of intellectual power & in that same sense we may continue to believe in the uniform action of the same laws even if compelled to grant progression from the simple to the complex.

Distinctness of Man [p. 166]

Dr. Carpenter admits as possible that all birds may have come from one copy & yet Man separately created.

Medico-Chir. p. 404, Apr. 1860 [111]

Man therefore more distinct than a mere Order of Mammalia!

He also says that the existence of rudimentary organs is one of the best arguments in favour of Darwin. But the os coccygi is a rudiment? Any rudiment in Man is teleologically the least explicable, & when we most want a final cause. For in proportion as we deem the psychical character peculiar we might expect independence of structure, & if not, then the separate creation of all species of other classes follows as an *a fortiori* probability.

Progress [p. 167]

What solace can it be to one who is told that the individuals now living are merely a transient phase of the organic world leading on to a future perfection—a link in a progressive scheme, that if our ancestry were without souls we shall at least have a posterity so superior that they may look on our humble & low position, very much as we do on the Fuegian or native Australian, with pity.

But it must be admitted that to all who believe in the immortality of their own souls, the former existence of inferior or the future prevalence of higher races can neither enhance nor extenuate their own prospects. The image of progress, of generation following generation, may be a visible manifestation to us of the aptitude of the soul & spiritual Nature of Man for expansion [p. 168] & elevation. Suppose there had been nothing but degeneracy from an original superior type or absolute uniformity & incapability of alteration as in certain instincts of wild animals, so far as we can observe them, which seen in several 1000 yrs., during which Man has made great progress, to be stationary. Would not that have been a more discouraging fact to the aspiration of Man than that wh. an indefinite progressive theory & improvability of races opens to us.

The analogy of the present & future world would best satisfy our longings if both were characterized by indefinite progress & there is nothing in this new revelation to repel this idea & much to sustain it.

[p. 169] Some will feel that if by lineage we are inseparably linked with species without souls, this enfeebles our prospects.

But is there not the same passage from infants to adults, if not from European to Bushman, or from the idiot to the human individual, just a degree higher removed? to say nothing of madness.

What comfort is it to believe that after 50 or 500 centuries there will be a race so superior as to look back historically on the present with the same pity which we may feel for the Fuegians. The answer is that to have descendants of high & noble lineage cannot degrade us, more especially if they are of the same species, or if there being no such reality as species, they are of the same corporeal & spiritual stock. Every one would feel lowered if all the men of genius had never existed. If so, the future multiplication of higher [p. 170] minds can only cast lustre on & raise our estimate of our own dignity.

Variation

The most striking fact is that plants may continue unchanged for many generations & that if at length they begin to vary they go on.

Suppose the interval before & after the Glacial Period which may mean 200,000 years & that, in a stratum superimposed, the same shells are found & a species is so much modified in the interval in consequence of change of place, migration, cold, remigration, that we hesitate to ascribe it to the same species, it may not have varied for an antecedent millions of years & then in a very short interval, geologically speaking, a mere fraction of one period, it may have changed so as to be scarcely, if with propriety, identifiable.

Progress [p. 171]

I have for the last 30 years consid.[d] the subject of progressive develop.[t] from the lowest to the highest vertebrate, the theory of the gradual introduction by a creative force evidently acting by rules & accommodating successive sets of creatures to the changes of the organic world as a theory that led naturally to that of transmutation. It has always been declared by those who advocate the hypothesis, that Man was the crowning operation of the series of creative acts. Some like Hugh Miller have delighted to trace the increase of the proportional quantity of brain from the marsupial to the placental & thence to the dog & Man, others to show that the anthropomorphic species most like Man come last, that the post-pliocene Gorilla was of higher grade than the Miocene Gibbon & so forth.

Now all these attempts, to a geologist who has lived to see the [p. 172] frequent filling up of gaps, such as that between the Paleozoic & Triassic groups now made out see ch. —— p. —— —& then discovery of new members of the Tertiary & Cretaceous system such as the *Craie pisolitic* & the Thanet Sands p. —— —& the fact that in all these cases the organic remains of the intercalated series are of an intermediate character between the under & over-lying formations, are attempting to establish a more & more graduated series of creations between the humblest instincts of the fish to the dawn of intelligence & reason in the Elephant, dog & Chimpanzee. When we find that the brain in diff.[t] races of Man is modified in the direction of a nearer resemblance [p. 173] of the least advanced members of the higher races & of the lowest races in the direction of resembling more the Apes, we cannot but perceive that the progressive system tends more & more daily to prove first that a secondary law of creation has been & is in force, 2.[ly] that the force tho' occasionally admitting of retrogression is in the main progressive, 3.[dly] that it embraces man in its system of advancement in time & that there is a connexion between approach to the human structure & a higher intelligence, for the reason of Brutes differs only in degree & not in kind.

When all the doctrines of the progressionist are granted, the great step towards what is called the material view of the creation has been made, & the transmutationist enters the field after the moral objections have been removed, the growth of instinct into intelligence & finally reason has been [p. 174] conceded. To halt then, when the numerous & independent classes of phenomena, which transmut.^{n} can explain w.^{d} be thereby wholly unaccounted for, is to lose the gain while the whole abandonment of an old notion is accomplished.

When we contemplate all the consequences of granting that the geological series of extinct animals shows a gradual scale of developm.^{t} ending in Man, & behold the unconcern with which the orthodox lay down this conclusion, often far overstating the evidence on which it depends, insisting on it with dogmatic positiveness, & witness the repugnance with which the same authors hear of the possibility of the law of creation embracing a lineal descent from the anterior geolo.^{l} races, including the most highly organized & those now living, including Man, we may certainly feel surprised at the want of foresight & infer that they [p. 175] cannot see far before them or contemplate the logical deduction from their premises.

Superior & Inferior Races of Man

It may be contended that the lowest races may occasionally give origin to a man of great intelligence showing that under favourable circumstances a branch of such a stock might originate a higher race & in this way a still higher. All may belong to one species, may descend from a single pair, but to imagine that this can be done in one or two generations under any circumstances is not consistent with experience.

It follows that intellectual endowments (or aptitudes) are to a certain [extent] transmissible by generation, may accumulate & multiply & if all the existing races are capable of improvement, they may have been preceded [p. 176] by others less advanced 50,000 years ago, less perfect in their psychological develop.[t] tho' their vision, smell, & strength may have been such that they c.[d] maintain themselves ag.[t] the wild animals.

May 2, 1860

Perfectibility: Letter of C.L. to Darwin

It is small comfort or consolation to me, who feels that Lamarck or Darwin have lessened the dignity of their ancestry, making them out to be with.[t] souls, to be told, "Never mind, you will be succeeded in unbroken lineal descent by angels who, like the Superior Beings spoken of by Pope,

'Will show a Newton as we show an ape.' " [112]

The case of the theologians may be desperate, but is there not some way by which the philosopher aspiring to a higher destiny may come to look upon the progressive system [p. 177] as more consonant than any other with his hopes.

It may perhaps show a tendency in all spiritual natures to advance, & even at a geometrical ratio, & thus if individuality be not a dream, if it be something real which cannot be lost & annihilated by death, & on that the aspirations of all men must depend, then a system of decay & degradation—or even a stationary aspect of the universe w.[d] be less promising & pleasing than one of endless variety & improvement. Taking the inner light & intuition as guiding Man to a belief in the immortality of the soul, the picture of progressive advancement in a long series of generations & species with only occasional retrograde movements, would not be unconsolatory. It involves no doubt the link of common descent with inferior & soulless beings, but this exists around us & is no new enigma. To those alone who have been in the [p. 178] habit of closing their eyes to the phenomena of every day life & of human nature, can this riddle be for the first time revealed by Geology. Imbecility, idiotcy, madness—all the grades between the imbecile & the man of some, but limited capacity—inferior races—the fact of hundreds of millions in each century dying before they have passed in infancy the line which clearly can be claimed as parting responsible from irresponsible beings, whether they have passed from this life as responsible beings or not, undeveloped—the case from the mere embryo to the new born.

Let every geologist freely grant that these are astounding facts. Who has not anxiously reflected on them, if he has thought at all—who ought to be surprised if in lifting up the veil & seeing more of Nature's past history, the same enigma reappears?—It is what you sh.[d] have expected, more of the same kind of perplexing doubts.

"In doubt to deem himself a God or beast." [113]

[p. 179] It is worthy of remark that, when the poet (writing *Essay on Man* a century ago?) enumerated the various sciences which that wondrous creature Man had entered upon, go measure, weigh air & state the tides, he omits the insight gained into the past history of the earth & its inhabitants, but he represents Man, nevertheless, as being as much as now.

In doubt to deem himself a God or Beast.[114]

It did not require that his remote genealogy sh.[d] be pried into in order to raise the question.

No mention is made of attempts to unfold Nature's laws in the remote past—but the realities of the present period of the world was enough, without any speculation on past genealogies to make him sketch man as a being. "In doubt," etc.

INDEX [p. 180]

4. Busk on Crag Bryozoa
   Banksia under basalt older than Cave-Animals. Hooker
6. Nummulites, planulate type still living. R. Jones
9. Oolitic flora—absence of Dicotyledons more remarkable than in Coal, for here variety of stations.
" Ferns & Cycads in Indian oolitic flora in tropics.
" Oolite Araucaria is true. J. Hooker.

", 11. Equisetum columnare.
10. Microlestes of trias.
11. Stangeria of Richmond V.[a] Coal.
" Voltzia in Richmond V.[a].
" Pecopteris Whitbyensis
12. Proteosaurus of Permian, highly organized.
" Autun fern like Permian one
13. Archegosaurus rudimentary form.
" Lindley, on Coal plants, Penny Cyclopedia, Ferns.
14. Devonian fish not stranger than living, Huxley.
" " flora, Dawson on Prototaxus.
15. Foraminifera, Silurian antiquity of some living generic forms. R. Jones.
16. Lingula Davisii, new genus. Cambrian.
" Cambrian has now 175 species & fewer generic links with Silurian than between Cretaceous & Tertiary.

[p. 181]

18. Cephalopoda, Huxley, bearing on progressive development.
" Geological age of ophidians & Deinosaurians against progressive developm.[t].
19. Stangeria at Kew, fern-like venation.
" Ophidian now in H. v. Meyer, 80 triassic ones.
20. Index to Notes on 'Creation'.
26. Creation, speculation on laws of.
27. Plato's ideas. Reality.
" Man not lowered if descended from the lower animals, Harvey's discovery.
28. Man of genius—earth trembled.
" Transmutation reconciles us to death of species.
29. This new difficulty soon takes its place among others allied to it.
" Progressionists, points in which they verge towards transmutation.
31. " illogical for them to object to transmut.[n] on usual grounds.
32., 48. Dignity of Man, not of Creator at stake.

33., 34., 35. {Progressionists say 1.st created fish very perfect yet of 1.st reptile rudimentary. They also exult.
36. Millions die in infancy—difficulty not new.
37. Jump from irrational to rational.
38. Creation cannot be understood without development.
39. If an acorn remains undevelop.[d] an oak is never created. Time an essential element in Geological creation.

[p. 182]

42. Creation most manifested when offspring intellectually superior to parent.
43. The "earth did not tremble".
44. Man may be part of same series of organic developm.[ts]
45. Intellect more variable than stature.
46. 'Historical development' the great admission.
47. Deviations of the highest order are the moral & 'turn of thought' yet these by descent.
49. Development & transmutation closely allied.
50. In both progressive plan, as Man last.
51. Some deep meaning in homology of Man & Ape.
" The Nature of the Creative Power is something separate from the question of the bond of hereditary descent (& from Natural Selection).
52. Races may be produced apart from 'descent'.
" My growing conviction that progression w.[d] involve transmutation.
53. If an ascending series, why not by some machinery as by which intellect passes in augmented power?
54. Man subject to same laws as animals.
55. Rational & irrational, leap from.
56. Complexity of Coniferae in Coal positive, tho' dicotyledons absent.
" Reptiles, if they or other class retrograde once, why not oftener?

[p. 183]

57. Hooker, J., Letter of C.L. on his Essay on transmutat.[n]
69. Watson, H. C., on Darwin's *Origin of Species*.

77. Darwin's answer to same—extinction.
87. Quadrumanous genealogy—anatomical resemblance.
89. Brain of Bushman.
90. Set of species—stationary & varying periods of the same.
91. Future state—inner light.
93. Agassiz, on species being thoughts of the Creator.
95. Jenyns' Letter to Darwin on "Origin etc."
103. Languages die out in two ways, like species—other analogies.
105. Progress implies continual intensifica.[n] of miracle-working power.
106. Begonia monstrosity.
107. Spencer, Herbert, letter to Darwin on *Origin.*
" " 'Spontaneous variations'.
110. Pictet to C. Darwin on "Origin of Species".
111. Beginning not proved.
" Tentative Creation.
112. Ramsay, A. C., letter to Darwin on "Origin of Sp."
116. Coscinopora in tumulus, Tennant.
" Submerged forests, Bristol channel.

[p. 184]

117. Hoxne flints.
" Mountains rise more than sea coast, Stür on Miocene of Sömmering Alp.
" Elephants, fossil rare in Scotland.
118. Savages, progress of, with Darwin.
119. Defects in instincts.
120. Progress means specializing of organs.
121. Many simple forms as of old.
122. Oceans changing their sites.
122. Sedgwick, his printed objections to Darwin's theory.
124. Asa Gray on Agassiz on Darwin.
125. Jukes' letter to Darwin, assenting.
127. Darwin on Barrande's palaeozoic provinces being chronolog.[l] divergences.
129. Latham, on Man exterminating races of Man.

130. Valleys in chalk at Down recently excavated, Lubbock with me.
131. Greeks advanced because no organized Priesthood.
134. Dana on non-existence of continents in Coal Period—answers to.
" Britons in Julius Caesar's time small brain.
134. Boucher de Perthes on Abbeville gravel, origin of.

[p. 185]

139. Dead Sea & cause of deluges. (145)
140. Elephants fossil few in Ireland, why?, & Scotland.
142. Searles Wood on Darwin on "Origin".
143. Hippopot.[s] Folkestone & escarpment.
144. Natural Selection & Hindoo Trinity.
145. Dead Sea & deluge. Cowell.
146. Bond of Descent of Darwin great step even if origin not explain.[d]
150. Small & large genera varying, Darwin as to objections.
151. Volition apart from law in acts of Creation as Human Free Will.
152., 166. Carpenter makes Man's soul a ground for separating him from law of transmutat.[n]
154. Man cannot be excepted, if so, nearly all Darwin arguments fail.
155. Preface, notes for, on Develop.[t] & Transmut.[n], progression.
166. Carpenter, on Man being exception, & rudimentary organs.
167. Immortality of Soul & progress.
170. Variation not always at same rate.
171. Progress accord.[g] to law, H. Miller.
175. Inferior Races of Man.
177. To look on Newton as we view an ape.

Notes

1. George Busk, *A Monograph of the fossil Polyzoa of the Crag,* London, Palaeontographical Society, 1859, *13,* 136 pp.

2. "Many of the tertiary fossil plants of Australia would seem to be very closely allied to existing ones; these include the *Casuarina* cones of Flinders island, the *Banksia* and *Araucaria* wood of Tasmania, the *Banksia* Cones of Victoria (which seem identical with those of *B. ericifolia,* though buried under many feet of trap)." Joseph Dalton Hooker, "Introductory Essay" in *The Botany of the Antarctic Voyage of the H.M. Discovery Ships Erebus and Terror,* vol. 3, *Flora Tasmaniae,* London, Reeve, 1860, 359 pp., p. ci. The source of Hooker's information was Joseph Beet Jukes, *A sketch of the physical structure of Australia, so far as it is at present known,* London, 1850, 95 pp.

3. T. R. Jones, *A Monograph of the Tertiary Entomostraca of England,* London, Palaeontographical Society, 1856, *10,* 68 pp. Thomas Rupert Jones (1819–1911) was assistant secretary to the Geological Society of London and a leading authority on micropalaentology.

4. W. C. Williamson, *On the recent Foraminifera of Great Britain,* London, Ray Society, 1858, *20,* 107 pp., pp. 36–39. William Crawford Williamson (1816–95), English naturalist, studied medicine and in 1851 was appointed professor of natural history at Owens College, Manchester.

5. John Lindley and William Hutton, *The Fossil Flora of Great Britain; or figures and descriptions of the Vegetable Remains found in a fossil state in this country,* 3 vols., London, Ridgway, 1831–37.

6. Sir Charles James Fox Bunbury (1809–86), palaeobotanist, was Lyell's brother-in-law. This was almost certainly a personal communication, probably in conversation.

7. John Phillips (1800–74), the nephew of William Smith, "the father of English geology," was appointed professor of geology at Oxford University in 1853.

8. Oswald Robert Heer (1809–83), palaeobotanist, was professor of natural history at the University of Zurich.

9. Robert Etheridge (1819–1903), English geologist.

10. Charles Longuet Higgins (1806–85), "benefactor" of Turvey, was educated at Trinity College, Cambridge, and thereafter studied law and later medicine. He was not admitted to the bar, but practiced medicine at Turvey until his father's death, when he used his inheritance to renovate all the buildings in the town. He also lectured on natural history and other subjects.

11. Thomas Henry Huxley (1825–95), zoologist and educator.

12. William B. Rogers, "On the relations of the New Red Sandstone of the Connecticut Valley and the coal-bearing rocks of Eastern Virginia and North Carolina," *Amer. J. Sci.,* 1855, *19,* 123–25.

13. Adolphe T. Brongniart, *Tableau des genres des Végétaux fossiles considérés sous le point de vue de leur classification botanique et de leur distribution géologique,* Paris, 1849, 127 pp. Adolphe T. Brongniart (1801–76), the son of Alexandre Brongniart, was a professor at the Museum of Natural History, Paris.

14. In 1844, the first reptile to be found as early as the Carboniferous period was discovered by Christian Friedrich Hermann von Meyer in the Coal of Bavaria. In 1847, three more fossil reptiles, each of a different species, were found in the coal field at Saarbruck and were described by Georg August Goldfuss as members of a new genus, Archegosaurus ("Über das älteste der mit Bestimmtheit erkannten Reptilien, einen Krokodilier und einige neue fossile Fische aus der Steinkohlen-Formation," Leonard and Bronn, eds., *N. Jahrb. mineral, geogn. geol., und petrefakt.,* 1847, *15,* 400–04). Von Meyer discussed the natural affinities of Archegosaurus and suggested that they were related more closely to Labrynthodon, a genus of Amphibia than to the reptiles (C. F. Hermann von Meyer, "Über den Archegosaurus der Steinkohlenformation," *Palaeontographica,* 1851, *1,* 209–15). Cf. Thomas H. Huxley, "On Dasyceps Bucklandi (Labyrinthodon Bucklandi, Lloyd)," *Mem. Geol. Surv. Great Britain,* 1859, pp. 52–56; Thomas H. Huxley, "On a fragment of a lower jaw of a large Labyrinthodont from Cubbington," *Mem. Geol. Surv. Great Britain,* 1859, pp. 56–57.

15. "Coal Plants," *The Penny Cyclopedia of the Society for the Diffusion of Useful Knowledge,* 27 vols., London, Knight, 1837, *7,* 290–96, p. 294:

Professor Lindley has recently proved that plants are capable of enduring suspension in water in very different degrees, some resisting a long suspension almost without change, others rapidly decomposing and disappearing. One hundred and seventy-seven plants were thrown into a vessel containing fresh water; among them were species belonging to the natural orders of which the Flora of the coal-measures consists, and also to the common orders, which, from their general dispersion over the globe at the present day, it might have been expected should be found there. In two years one hundred and twenty-one species had entirely disappeared; and of the fifty-six which still remained, the most perfect specimens were those of coniferous plants, palms, Lycopodiaceae and the like; thus showing in the clearest manner that the meagre character of the Coal Flora may be owing to the different capabilities of different plants of resisting destruction in water. The same experiment accounts for the want of fructification in fossil ferns; for it showed that one of the consequences of long immersion in water is a destruction of the fructification of those plants.

16. John W. Dawson, "On fossil plants from the Devonian rocks of Canada," *Quart. J. Geol. Soc.*, 1859, *15*, 477–88.

17. Roderick I. Murchison, *Siluria. The history of the oldest known rocks containing organic remains, with a brief sketch of the distribution of gold over the earth,* 2d ed., London, Murray, 1854.

18. John William Salter (1820–69), geologist, in 1846 had joined the Geological Survey of Great Britain as an assistant and in 1854 succeeded Edward Forbes as palaeontologist to the survey. He gave Lyell assistance from time to time with palaeontological questions.

19. Thomas Davidson (1817–85), palaeontologist, who made extensive studies on British fossil Brachiopods.

20. Lyell's Index Book, p. 12, refers to Joseph D. Hooker, "An Enumeration of the Plants of the Galapagos Archipelago," *Trans. Linn. Soc. London,* 1847, *20,* 163–262, and includes the following notes:

Most of the indigenous spec.[s] of plants restricted
to one islet. p. 239
Paucity of Monocotyledons
Agents of transport, birds, etc. p. 253
Representatives of species " 259
Want of atmospheric storms to
transport species top " "

21. Index Book, p. 29, refers to Robert MacAndrew, "On the geographical distribution of Testaceous Mollusca in the North-east Atlantic and neighbouring seas," *Liverpool Lit. Phil. Soc. Proc.* 1853–54, *8*, 8–57, pp. 44–50, and includes the following notes:

p. 44 All nearly common to Madeira & Canaries of marine mollusca—i.e. 85 per cent. qy. if Madeira [waters] more dredged, percentage w.[d] be greater?

44 Acephala less locomotive but more range than gastropoda.

47 Curve to the West from Cornwall to W. of Iceland & Hebrides of many shells not in the channels of N. sea

48 Few marine species confined to limited areas (qy. including littoral sp.?)

49 Saxicava arctica most cosmopolite of shells.

50 No British isle produces species peculiar to itself, but each isle of Madeira, Canaries & Azores—even the Desertas.

27 Mogador shells half of them British.

22. Index Book, p. 31, refers to Alfred R. Wallace, "On the law which has regulated the Introduction of New Species," *Ann. Mag. Nat. Hist.*, 1855, ser. 2, *16*, 184–96, p. 188. Here Wallace wrote:

> On the other hand no example is known of an island which can be proved geologically to be of very recent origin (late in the Tertiary for instance), and yet possesses generic or family groups, or even many species peculiar to itself.

Lyell's complete notes on Wallace's paper are given in Scientific Journal I, note 1, pp. 65–66.

23. Index Book, p. 54, refers to T. Vernon Wollaston, *On the Variation of Species with especial reference to the Insecta; followed by an inquiry into the nature of genera,* London, Van Voorst, 1856, 206 pp., p. 184, where Wollaston wrote:

> I have before announced my conviction that *generic areas* have a real existence in Nature's scheme; and that, consequently, where species which are so intimately allied that they can with difficulty be distinguished, prevail, there is presumptive reason to suspect . . . that the areas which they now colonize were once connected by an intervening land,—or, in other words, that the migrations of the latter were brought about, through ordinary diffusive powers, from specific centres within a moderate distance of each other.

24. Index Book, p. 56, refers to remarks by George Buist at the ordinary monthly meeting of the Bombay Geographical Society on

Thursday, August 17, 1854. Buist said in part; "Proceedings of the Bombay Geographical Society," *Trans. Bombay Geogr. Soc.,* 1856, *12,* i–xciii, p. xix.

> The mean level of the whole solid land above that of the sea is 1,000 feet,—that is, were our mountain masses smoothed down, and our valleys and sea brought up to one general tableland, its surface would be 1,000 feet above that of the ocean . . . The mean depth of the ocean, again,—that is, of its basin, were this scooped out and smoothed till it resembled a tank or cistern is about 22,000 feet or four miles.

George Buist (1805–60), Anglo-Indian journalist and amateur scientist, was an unpaid inspector of the astronomical, magnetic, and meteorological observatories of Bombay and a journalist on the *Bombay Times.*

25. Index Book, p. 62, refers to William Whewell, *History of the inductive sciences,* 2d ed., 3 vols., London, Parker, 1847, *3,* 457.

26. Index Book, p. 63, refers to William Lawrence, *Lectures on physiology, zoology and the natural history of man, delivered at the Royal College of Surgeons,* London, Callow, 1819, 579 pp. This book was published in a long series of later editions, and it is not clear to which edition Lyell may have been referring.

27. Shakespeare, *Henry IV,* part 1, act 3, scene 1, line 13.

28. It seems to have been Lyell's custom to bind together offprints and reprints of scientific and scholarly articles that he received from scientific colleagues and friends and to refer to this series of bound volumes as tracts.

29. Adam Sedgwick, *Discourse on the Studies of the University of Cambridge,* 5th ed., London, Parker, 1850, p. ccxxv:

> Leave the Class of Fishes and pass on to the scale of Reptiles, dead and living. Is this scale chronological? Is the last collective Reptile Fauna the highest, shewing the last and best development from Nature's womb. Not so.

30. The quotation is from a letter by Charlotte Brontë to an unidentified correspondent, dated February 11, 1851, discussing Henry George Atkinson and Harriet Martineau, *Letters on the Laws of Man's Nature and Development,* Boston, Mondum, 1851, 396 pp. See Scientific Journal No. III, note 35, pp. 210–11.

31. Sedgwick, *Discourse,* 1850, p. ccxv:

> In every successive *Fauna* of Geology we find the same kind of

animal subordination we meet with now in the living world; and the very earliest Genera and Orders were not organically inferior to the Genera and Orders of this day which we derive from corresponding grades in the scale of Nature. Nay, sometimes the primeval Genera and Orders are organically superior to their corresponding types in the living world.

32. This refers back to the discussion on p. 32 of this journal; see our page 330.

33. Lartet had found the fossil remains of Dryopithecus in beds of the Faluns of the Loire Valley, of Miocene age, near Saint Gaudens. See Éduard Lartet, "Note sur un grand singe fossile qui se rattache au groupe des singes supérieurs," *Comptes. Rendus. Acad. Sci.,* 1856, *43,* 219–23.

34. Alphonse de Candolle, *Géographie botanique raisonée,* 2 vols. (continuous pagination), Paris and Geneva, Masson and Kessman, 1855, 1365 pp., p. 1067 et seq.:

> Nous devons reconnaître la probabilité que certaines espèces sont très anciennes, qu'elles ont traversé plusieurs révolutions géologiques.
>
> D'un autre côté, nous ignorons si toutes les espèces actuelles ont paru en même temps. Peut-être elles se sont succédé, soit qu'elles aient dérivé d'anciennes espèces à des époques successives, soit qu'elles aient été créées successivement par une cause surnaturelle.

35. The original of this letter is among the Hooker mss. The letter, as given in the notebook, is in Lady Lyell's handwriting and is copied exactly from the original, except for a brief closing paragraph. See note 42 below.

36. Joseph D. Hooker, "Introductory Essay" to *Flora Tasmaniae,* 1860. The essay is dated Kew, November 4, 1859.

37. De Candolle, *Géographie botanique.*

38. Hooker, "Introductory Essay" to *Flora Tasmaniae,* p. ii. Hooker here pays tribute to the value of Robert Brown's study of the flora of Australia published in the Appendix to Matthew Flinder's *A voyage to Terra Australia; undertaken for the purpose of completing the discovery of that vast country and prosecuted in the years 1801, 1802 and 1803,* 2 vols., London, Nicol, 1814.

Hooker wrote:

At the time of its publication, not half the plants now described were discovered, vast areas were yet unexplored, and far too little was known of the vegetation of the neighbouring islands to admit of the Australian Flora being studied in its relation to that of other countries. Nevertheless we are indebted to Brown's powers of generalization for a plan of the entire Flora, constructed out of fragmentary collections from its different districts, which requires but little correction from our increased knowledge, though necessarily very considerable amplification. Although he could not show the extent and exact nature of its affinities, he could predict many of them, and by his detection of the representatives of plants of other countries under the masks of structural peculiarity which disguise them in Australia, he long ago gave us the key to the solution of some of those great problems of distribution and variation, which were then hardly propounded, but which are now prominent branches of inquiry with every philosophical naturalist.

39. Charles Naudin, "Observations concernant quelques plantes hybrides qui ont étés cultivées au Muséum," *Ann. Sci. Nat.,* 1858, ser. 4, *9,* 257–78.

40. Hooker, "Introductory Essay," to *Flora Tasmaniae,* 1860, p. xv:

I may however indicate as a general result, that I find the sinking islands, those (so determined by Darwin's able investigations) characterized as atolls, or as having barrier reefs, to contain comparatively fewer species and fewer peculiar generic types than those which are rising.

41. Charles Lyell, *A Manual of Elementary Geology,* 5th ed., London, Murray, 1855, p. 296:

Thick siliceous beds of chert occur in the Middle Purbeck filled with mollusca and cyprides of the genera already enumerated, in a beautiful state of preservation, often converted into chalcedony. Among these Prof. Forbes met with gyrogonites (the spore vessels of Charae), plants never until 1851 discovered in rocks older than Eocene.

Lyell made a slight slip in his page reference.

42. Lyell concluded his letter to Hooker with the following paragraph, omitted in the journal:

I read your Essay when staying with the Van der Weyers &

he ordered it & will I am sure appreciate [it] in a way that few of our literary men can.

believe me
ever truly yrs
Cha Lyell

[P.S.] I should be glad of references to Naudin.

Sylvain van de Weyer was a Belgian scholar who frequently visited England. The Lyells were visiting at his house in Belgium.

43. Watson evidently outlined these points to Charles Darwin in a letter which Darwin transmitted to Lyell. He summarizes his position in the supplement to his *Cybele britannica.* Hewett C. Watson, *Cybele britannica; or, British Plants and their geographical relations,* 4 vols., London 1847–59, and Hewett C. Watson, *Part First of a Supplement to the Cybele Britannica,* London, privately printed, 1860, 119 pp., pp. 30–33.

Hewett Cottrell Watson (1804–81), botanist, had sailed in 1842 with the *Styx* to the Azores, where he collected the plants of these islands. His principal interest was the geographical distribution of plants.

44. We have not been able to locate the original of this letter.

45. Friedrich Tiedemann, *Icones cerebri Simiarum et quorundam Mammalium rariorum,* Heidelberg, 1821, 55 pp.

46. William Rathbone Greg, *The creed of Christendom; its foundations and superstructure,* London, Chapman, 1851, 307 pp.

47. We have not been able to find where this lecture by Louis Agassiz was reported.

48. Anonymous, "Life of Sir Charles Bell," review of Amédée Pichot, *Sir Charles Bell, Histoire de sa Vie et de ses Travaux,* Paris, 1858, *Sat. Rev.,* 1860, *9,* 189–90, p. 189.

49. This letter is not in the collections of Charles Darwin's correspondence in the Cambridge University library.

50. Charles Darwin, *On the Origin of Species,* London, Murray, 1859, 490 pp.

51. Ibid., p. 484: "Probably all the organic beings which have ever lived on this earth descended from some one primordial form."

52. Ibid., p. 488.

53. "And the Lord God formed man of the dust of the ground

and breathed into his nostrils the breath of life; and man became a living soul" (Genesis II:7). "And the Lord God caused a deep sleep to fall upon Adam, and he slept: and he took one of his ribs, and closed up the flesh instead thereof. And the rib which the Lord God had taken from man, made he a woman, and brought her unto Adam" (Genesis II:21–22).

54. Dugald Stewart, "Inferences with respect to the Use of Language as an instrument of Thought, and the Errors in Reasoning to which it occasionally gives rise," *Elements of the Philosophy of the Human Mind,* 5th ed., 2 vols., London, Cadell and Davies, 1814, *1,* 197–203.

55. Asa Gray (1810–88) was appointed professor of botany at Harvard University in 1842 and from then until his death was the leading botanist in the United States. Gray reviewed Darwin's *Origin of Species* in the *American Journal of Science.* See Asa Gray, "Review of Darwin's Theory on the Origin of Species by Means of Natural Selection," *Amer. J. Sci.,* 1860, ser. 2, *29,* 153–84.

56. Heinrich Georg Bronn (1800–62), palaeontologist of Freiburg, was later a professor at the University of Heidelberg.

The phrase "Wahl der lebens weise" (trans.: "Choice of mode of life") was chosen by Bronn, who prepared the first German translation of Darwin's *Origin of Species,* as the German equivalent for Darwin's term "natural selection." Darwin objected to the German phrase, pointing out that it held Lamarckian implications. He wrote to Bronn that he had referred the question to several scientific men, one of whom was evidently Lyell. See Charles Darwin to Heinrich G. Bronn, February 14, 1860, in Charles Darwin, *Life and Letters,* Francis Darwin, ed., 3 vols., London, Murray, 1887, *2,* 278–79.

57. In a review of Darwin's *Origin of Species* in the *Gardener's Chronicle,* 1860, *20,* 145–46, William H. Harvey describes a case of abnormal flowers in *Begonia frigida* and showed that the flower "sports" were so different from a normal Begonia that they might have belonged to a separate natural family of plants. Joseph D. Hooker replied to Harvey's criticism in a subsequent issue (*Gard. Chron.,* 1860, *20,* 170–71), pointing out that although the mature flower "sports" were very different from the normal Begonia flowers, they could be traced back through their development to stages common with those of the normal Begonia. Cf. Darwin, *Life and Letters,* 1887, *2,* 274–76.

William Henry Harvey (1811–66) was an English botanist who in

1835 went to Capetown where for several years he studied the botany of South Africa. After his return to England in 1844, he became an acknowledged authority on algae and was appointed curator of the herbarium at Trinity College, Dublin, and, in 1856, professor of botany there.

58. Alphonse de Candolle points out that the polypetalous Campanula rotundifolia, which had been reported from the mountains of the canton of Neuchâtel, had not been rediscovered. Alphonse de Candolle, *Géographie botanique raisonée,* 1855, p. 1143.

59. See note 57, above.

60. Herbert Spencer to Charles Darwin, February 22, 1860. This letter is not in the Darwin correspondence in the Cambridge University Library.

61. Herbert Spencer, "Progress: its law and cause," in Herbert Spencer, *Essays; Scientific, political and speculative,* ser. 1 & 2, London, 1858–63. In the edition we have used (New York, Appleton, 1892), this essay is in vol. 1, pp. 8–62. The discussion Spencer refers to is on pp. 49–53.

62. Herbert Spencer, "A Theory of Population deduced from the General Law of Animal Fertility," *Westminster Rev.* (Amer. ed.), 1852, *57,* 250–68, esp. pp. 266–68.

63. François Jules Pictet to Charles Darwin, February 19, 1860. This letter is not in the Darwin correspondence in the Cambridge University Library.

64. Andrew Crombie Ramsay to Charles Darwin, February 21, 1860. This letter is not in the Darwin correspondence in the Cambridge University Library.

65. This is probably a reference to James Tennant (1808–81), mineralogist, who was professor of mineralogy at King's College, London.

66. This reference was based on personal communication from Huxley. See Charles Lyell to Thomas H. Huxley, March 4, 1860, Huxley mss.:

> I think you told me that you observed some two or three years ago on the coast of Bristol Channel or Devonsh.? signs of alternate upheaval and depression proved by the shells of raised beach.
>
> I should like to have your case whether published or not to cite.

67. Adam Sedgwick and Roderick I. Murchison, "Description of

a raised beach in Barnstaple Bay on the north-west coast of Devonshire," *Trans. Geol. Soc. London,* 1843, ser. 2, *5,* 279–86.

68. See note 66 above.

69. We were not able to identify the reference Lyell had in mind.

70. In 1804, John Frere had reported to the Society of Antiquaries of London the discovery of flint tools in a pit dug to obtain brick clay at Hoxne near Diss in Suffolk. After Joseph Prestwich had returned from confirming Boucher de Perthes's discovery of flint tools in undisturbed gravel beds at Abbeville, in France, his attention was drawn to Frere's earlier find in England. In October 1860, Lyell visited Hoxne to examine the beds in which the flints were found.

71. Lady Kerrison was the wife of Sir Edward Kerrison, Baronet (1821–?) of Hoxne Hall near Bungay, Suffolk, who had preserved many of the flint implements from the pit at Hoxne.

72. Dionys Rudolf Josef Stür (1827–93), Austrian geologist.

73. In September 1856, Lyell had visited Austria where he had accompanied Stür and other Austrian geologists on field excursions.

74. Lyell spent the weekend of March 9–13, 1860 at Charles Darwin's home, Down House, Down, Kent.

75. Lyell is probably referring to Richard Whateley (1787–1863), Archbishop of Dublin.

76. Henri Milne-Edwards (1800–85), *Leçons sur la physiologie et l'anatomie comparée de l'Homme et des Animaux faites à la Faculté des sciences de Paris,* 13 vols. Paris, Masson, 1857–81.

77. Ibid.

78. Adam Sedgwick, *Discourse,* 1850, pp. lxi–lxviii.

79. Asa Gray, "Review of Darwin's Theory . . . ," 1860.

80. Joseph Beete Jukes (1811–69), geologist, who went in 1839 to Newfoundland to make a geological survey of the island. From 1842 to 1846, he served as naturalist aboard *H.M.S. Fly,* sailing the Torres Strait and seas about the north of Australia. After 1846, he worked on the Geological Survey of Great Britain and, in 1850, was appointed director of the Irish branch of the Survey.

81. William Hellier Bailey (1819–88) was palaeontologist on the staff of the Irish branch of the Geological Survey and Jukes's assistant.

82. John Gwyn Jeffries (1809–85), conchologist, first studied law and for several years practiced as a solicitor at Swansea. In 1856, he was called to the bar and moved to London, where he practiced as a barrister and carried on his scientific interests.

83. We have been unable to determine the paper Lyell had in mind.

84. Apparently, both John G. Jeffries and Hewett C. Watson objected to the views Edward Forbes put forth in his paper on polarity. See Edward Forbes, "On the Manifestation of Polarity in the Distribution of Organized Beings in Time," *Roy. Inst. Proc.*, 1851–54, *1*, 428–33.

85. Robert G. Latham, *Opuscula, Essays chiefly philological and ethnographical,* London, Williams and Norgate, 1860, 418 pp., p. 140. The quotation is from an essay "On the subjectivity of certain classes in ethnology," first published in the *Philosophical Magazine,* May 1853.

Robert Gordon Latham, M.D. (1812–88), ethnologist, was educated at Eton and King's College, Cambridge. After a few years in medicine, he devoted himself entirely to philology and ethnology, subjects on which he wrote extensively.

86. The entry in Notebook 220, p. 80, reads:

> Falconer
>
> Barwell Cave near Weston-super-Mare & other caves near Mendip Hills, Bristol—Elephas primigenius rare, most of the Eleph. in caves [are] E. antiquus.
>
> The valley near Darent shows subangular flint gravel with often level surface. How different must have been the form of the valley & quantity of water in the old time of the flint gravel.

87. Sir John Lubbock, later Lord Avebury (1834–1913), was a banker and member of Parliament who was also an enthusiastic amateur geologist and naturalist. At this time, he lived with his parents at High Elms, Down, Kent, and was a neighbor and friend of Charles Darwin.

88. Joshua Trimmer (1795–1857), geologist.

89. George Grote (1794–1871), political economist and historian of Greece, began life in his father's bank, took part in the founding of the London University, and was elected to Parliament. In 1846, he published the first two volumes of his *History of Greece,* a twelve-volume work completed in 1856.

90. Lyell's entries on pp. 27–30 of Notebook 151 are miscellaneous, but the ones to which he seems to be referring deal with the elevation and submergence of Wales at different geological periods. On p. 30 is the following note:

> Wales was land after the magnesian or dolomitic conglom. period but the Mendips were cut away before the New Red period or by the early new red sandst. sea.
> Wales was perhaps land during tertiary period & then submerged during boulder period nearly all of it.

91. The original of this letter is in the Lyell mss.

92. Jacques Boucher de Crèvecôeur de Perthes, *Antiquités Celtiques et Antédiluviennes,* 3 vols., Paris, Treuttel and Wurtz, 1847–64, *1,* 224–25. Boucher de Perthes quotes as the observation of M. Baillon that when fossil bones are found arranged as if they had been deposited while still articulated together, the sand around them formed a very hard mass.

93. Joseph Barnard Davis (1801–81), craniologist, went in the summer of 1820 as surgeon on a whaling ship to the Arctic. He later practiced as a surgeon at Shelton Hanley in Staffordshire and made a large collection of human skulls. In 1856, he began with Dr. John Thurman (1810–73) to publish *Crania Britannica: Delineations and descriptions of the skulls of the aboriginal and early inhabitants of the British islands; with notices of their other remains,* 2 vols. London, privately printed, 1865.

94. We have been unable to identify Daillant.

95. We have been unable to identify Butenor.

96. Joseph Prestwich (1812–96), geologist.

97. George Douglas Campbell, eighth Duke of Argyll (1823–1900), "On tertiary Leaf-beds in the Isle of Mull," *Quart. J. Geol. Soc.,* 1851, *7,* 89–103.

98. Archibald Geikie (1835–1924). Because this letter from Lyell to Geikie of March 20, 1860 is not among the Geikie papers at the University of Edinburgh Library, it must be presumed lost.

99. Searles Valentine Wood (1798–1880), geologist, who studied for many years and wrote definitive monographs on the fossil Mollusca of the Crag.

100. The entry on p. 3 of Notebook 175 describes, with the aid of a diagram, the strata forming the escarpment at Folkestone and

shows a "bone bed" as a superficial deposit overlying the outcrop of the Lower Greensand on a low plateau, about 100 feet above sea level and below the mass of the Chalk Down, about 700 feet high. The great Chalk escarpment is obviously older than the thin superficial "bone bed."

101. We have been unable to identify Lieutenant Cowell.

102. John William Dawson (1820–99), *Archaia; or studies of the cosmogony and natural history of the Hebrew Scriptures,* London, Sampson Low, 1860, 400 pp.

103. William B. Carpenter, "Review V," of Darwin, Wallace, Baden-Powell, and Hooker, *Brit. & For. Med. Chir. Rev.,* 1860, *25,* 367–404, p. 404:

> So too there seems to us so much in the psychical capacity of Man, however degraded, to separate him from the nearest of the Mammalian class, that we can far more easily believe him to have originated by a distinct creation, than suppose him to have had a common ancestry with the Chimpanzee, and to have been separated from it by a series of progressive modifications.

104. Ibid., p. 404.

105. Heinrich Georg Bronn (1800–62), palaeontologist of Heidelberg.

106. Charles Lyell, *Principles,* 11th ed., 1872, *2,* 439.

107. This is probably a reference to Carl Friedrich Heinrich Marx (1796–1877), professor of medicine and medical historian at Göttingen, who edited the anthropological works of Johann Freidrich Blumenbach.

108. Friedrich Max Müller (1823–1900), "Comparative Mythology," in *Oxford Essays contributed by Members of the University,* 4 vols., London, Parker, 1855–58.

109. Christian Carl Josias Bunsen (1791–1860), *Christianity and Mankind, their beginnings and prospects,* 7 vols., London, Longman, Brown, Green and Longmans, 1854.

110. Carl Wilhelm von Humboldt (1767–1835), *Über die Verschiedenheit des menschlichen Sprachbaues und ihren Einfluss auf die geistige Entwickelung des Menschengeschlechts,* Berlin, 1836.

111. William B. Carpenter, "Review V," 1860, p. 404:

> Supposing for the sake of argument, that we concede to Mr. Darwin that all Birds have descended from one common stock,

—and we cannot see that there is any essential improbability in such an idea . . . —yet it by no means thence follows that Birds and Reptiles, or Birds and Mammals, should have had a common ancestry.

112. Alexander Pope (1688–1744), *An Essay on Man in four epistles to a friend,* epistle 2, line 8.

113. Ibid.

114. Ibid.

Scientific Journal
No. VI

May 3, 1860

With Prestwich: Amiens Flints

[See Fig. 12.] The sand with angular flints, corresponding to the Diluvium Rouge [M] of Hebert,[2] is over all & also under the Peat. At Menchecourt seen rapidly to descend the valley. This No. 4. attributed by Prestwich to a transient flood which

Fig. 12

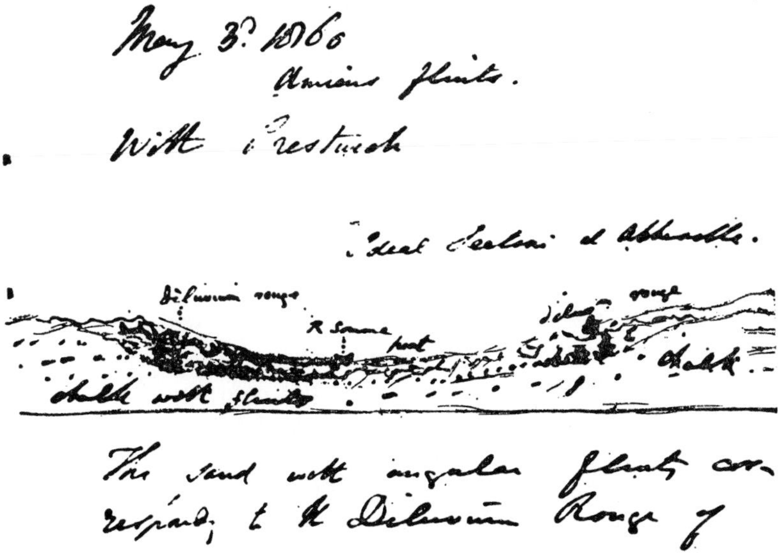

succeeded to the brickearth or Loess Period. Whether the water of this inundation was salt or fresh he has no means of judging, but absence of shells [p. 2] not enough to assure him that it may not be marine.

As to Macclesfield drift with marine shells, he suggests that it may be an older marine gravel of glacial period & the other alluvial strata spread over the country, which Binney[3] found unfossilif.^{s} may be of more modern & supra-marine origin & contain the wreck of the older marine drift remains.

This may agree well with Symonds'[4] notion about Cropthorne where the estuary beds examined by Strickland[5] seem more modern than a gravel of the drift period. [See Fig. 13.]

Fig. 13

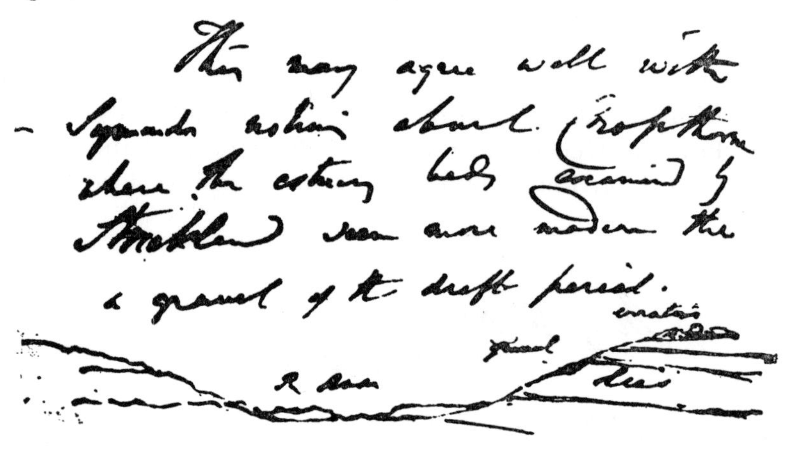

[p. 3] Rhinoceros bone found at Cropthorne with marine shells in it.

Amiens Flints

Prestwich remarks that those worked flints which have most blows are of the most regular forms, whereas in the case of flints broken by Natural causes it is just the contrary. They are of the most irregular form where they have been oftenest chipped & broken.

The Cyrena was found in the bed (from which the marine shells come?) with a flint knife, the human workmanship of which Prestwich no longer doubts. It was taken out at Menchecourt in his presence.

Species [p. 4]

The question of blood relationship or not sh.^{d} be kept distinct from two others, 1.st the original types when life was introduced for the first time into the planet, 2.dly the nature of the variety-making or creative or formative power. With both these mysterious questions I shall at present have nothing to do. Whether there was one or a hundred initial types, or at how remote a period they may have been formed, before the earliest signs of organic remains yet discovered—or whether the creative power be governed by a fixed law like that of gravitation or vegetable growth or by some power more akin to the free-will of Man which, tho' under the control of laws natural & psychological, we believe & cannot help believing to be free.

We may leave these transcendental questions aside & confine ourselves to a more humble one, whether the creative power in which I conceive design & supreme intelligence is manifested, is or is not governed by the ordinary laws [p. 5] of continuous generative succession by which new sets of individuals & often new permanent varieties of old stocks successively appear on the earth.

Whether is this hypothesis more probable than that of the creation of new species independently, of any connexion with pre-existing living organic forms on the earth, or if connected, having a relationship different from that which links a remote & distinct variety of a species from its prototype.

I have given my reasons fully 30 years ago in the *Principles of Geology* for not embracing this law of continuity by descent of the living from the antecedent Pleistocene beings, & of them from the older Pliocene, Miocene & Eocene Tertiary faunas, as has been long before imagined & propounded by Lamarck. My arguments may be referred to as showing how difficult it is to [p. 6] support the law of creation by continuous descent. How entirely we are without positive proof of the indefinite variability of species so that one may pass into another.

I pointed out however in the 1.st & 9.th Ed.s of my *Principles* that there is a constant extinction of organic beings going on independently of Man's exterminating power (Ch.s ——),[6] that there is among plants as well as animals a continual struggle for life pp. —— & that so long as fluctuation in the animal & vegetable & inorganic world are going on (& they can never cease, the changes being incessant, both geographical, climatal & organic); there must be a gradual dying out of species & unless the world is to be depopulated or to become less varied in its living [p. 7] productions there must be a coming in of new forms or species.

Ought we to incline, after reviewing all the past or geological phenomena & those of the present geographical & structural relations of living beings, to the hypothesis of independent creation, or to that of the endless modifiability of organic forms already in existence? I never doubted that the reproductive power is in full operation, as active now as ever, & I endeavoured to show that, if intermittent, it might not in a few thousand years, still less in a few centuries, be capable of demonstrative proof in the present state of science. A New species, I contended, might be created annually & yet there is no reason to assume that naturalists sh.[d] have been able to establish the reality of so great [p. 8] an event (Ch. —— Ed. —— & Ed.)[7]

In contemplating the solution of this great problem I have always regarded the question of the origin of Man, considered in his relation to the animal world, as one which could not be set aside, for if such a member of the organic world (however exceptional in a psychological point of view, yet so normal in his physiological & structural) could not be assumed to be derivative by descent from any anterior living being, it would be inconsistent after embracing & admitting such a recent miracle, the creation of such a species, by a supernatural interference or by some act of creation governed by laws wholly distinct from those by which the most marked & permanent races of plants & animals are produceable from an original stock [p. 9] it would be gratuitous, after assuming the creation of such a complex being with a structure like that of the mammalia & with some instincts identical with theirs, to grant that less highly organized creatures may have started into being without being developed out of an older series.

The laws of development being from simple to complex, they who believe that Man has been progressive, & that the organic world has been evolved out of simpler elements, must be prepared to be staggered at the idea that the newest & grandest facts? in the universe, of which we have to take cognizance, give the lie to & contradict this inference.

If Man was formed directly out of earth, air & water, or out of clay, we may *a fortiori* conceive beings of simpler structure to [be] of like origin. [p. 10] I adopted the hypothesis of limited variability 30 years ago & have adhered to it, not that we have experience enough to establish such a dogma. Mr. Darwin has written a work which will constitute an era in geology & natural history to show that the rival hypothesis of unlimited variability is the more probable of the two, & that the descendants of common parents may become in the course of ages so unlike each other as to be entitled to rank as a distinct species, from each other or from some of their progenitors.

No one who has not dwelt much on the geological succession of organic beings and the changes which Time has brought about can enter securely on this question.

Does the hypothesis of indefinite modifiability explain a variety of [p. 11] independent phenomena, if we grant it, while that of limited variability is sterile?

Alternate generation has never yet been known to give rise to any new race with characters differing from the progenitor & transmissible by generation. It is not true therefore that we can as yet derive from the phenomena an idea of the origin of a new species not liable to revert to its primitive form.

Darwin explains non-scarcity of simple primordial-like beings to persistence of generic types & possibly degradation.

Progress

First. There is the familiar fact of the evolution from the embryo to the infant & from the infant to the perfect or adult state.

2.dly The progress of man from a savage to a civilized state, cranial development without being at the expense of physical force (see Godron).[8]

3.dly Superior races forming & becoming dominant at the expense of inferior. [p. 12]

4.thly The fact that Man came after other animals which are greatly inferior in their combined physical & psychological forces.

5.thly The very generally received opinion of geologists that the more highly organized beings, such as the placental mammalia, succeeded in their appearance in the earth to the marsupial & that these were less ancient than reptiles, these than fishes, & fishes newer than vertebrates.

6.thly That in the vegetable kingdom there is an approach in like manner to a higher from a simpler & lower organization traceable in time.

7.thly Such a progressive scheme is also supposed to be borne out by the embryological development of the human being & of the higher mammalia. [p. 13]

8.thly In Man, however far removed from the most nearly allied of the lower animals, there is a variation in his different races not only analogous to that of the races & species of the anthropomorphous mammals, but the brain & limbs approach them when the intelligence is less.

Transmutation, from p. 5

Let us not suppose that no step is gained because the mystery of creation of a spiritual & intellectual being is as great as ever.

We may suppose, that in a country which has reached a high state of advancement in Philosophy, it was related that in a neighbouring state a man of great genius [had lived], who as a statesman or poet or prophet exerted a vast influence & caused a revolution in the world, that his birth had been attended by prodigies in external nature & that the ordinary laws of generation had been violated or suspended and that his mode of entering the world had been miraculous.

[p. 14] A commission was sent to enquire whether the unusual phenomenon had been attended by some deviation from the common laws of birth. They might return & report that after full investigation they had satisfied themselves that the tales, of 'the earth having trembled' & how the face of heaven was full of fiery shapes at the time of the nativity of the hero or poet, were unfounded, that no sensation in nature or human society marked his being ushered into the world, that they c.d not learn that his parents had been distinguished for any uncommon faculties, but were ordinary persons & that while some of their children were no more remarkable than themselves others were even deficient.

Now such a report might well surprise & disappoint the curious for there would be nothing unreasonable in imagining that the sudden apparition of such unwonted intellectual power had been marked by some deviation from the ordinary course of [p. 15] every day life, that something supernatural had attended the development of so much spiritual power, that as dwarfs are never born of giants, nor giants of dwarfs, so a leap by which such unusual intellectual stature is suddenly attained might well have prepared the philosophers to look for an extraordinary genesis for inspiration, some token from heaven, some secondary law of Creation, deviating from that according to which the normal succession of individuals, like their parents, is perpetuated, so that they should be relieved from the apparent anomaly of a higher nature being evolved out of an inferior one.

But no one having in him this true philosophic spirit of enquiry w.[d] fail to rejoice that the myths of superstition had been dissipated & that the true mode of the advent of the hero had been ascertained. The origin of the superior [p. 16] mental or moral nature would still be open to conjecture, but the ordinary laws governing the succession of events in the physical universe would be rescued from uncertainty & irregularity. As Naturalists, a great step or great removal of error would have been gained—altho' the psychological enigma might seem greater than ever. If it be found that the cranial development of the hero be remarkable this again w.[d] be a mere variety, nothing more supranatural than the distinction between Man & Man.

So the hypothesis of transmutation w.[d] explain nothing, aid us perhaps little or nothing & yet lead to vast results in other investigations.

The alleged miracle may be believed in good faith.

Imbecility

Is it so unsupportable to have in common with an ancestor of the [p. 17] same species what we cannot but share with contemporary beings of our own human race? The stern law which cuts off a third of those born in infancy may also have prevailed when an inferior race preceded us. But the passage from irrational to rational may have been abrupt.

Deification of Matter

They who deify Matter & Force, or Natural Selection, or any other agency to which they may assign the powers of Nature, by which new forms of instinct & new grades of Mind & of Conscience, are successively introduced into the terrestrial system, may surely allow him who imagines a creative intelligence having a will like Man, a personal Deity designing the evolution of new forms, instinct & intelligences, & as Man amidst the laws unchangeable, which govern the Material & Spiritual World, [p. 18] cannot help believing himself a free agent, so he may conceive the volition of a supreme being as consistent with a system of secondary laws, tho' they may be as incomprehensible to our understanding in our present state.

The tie by descent may prove to be the hidden bond which connects genera & orders & classes together of the Animal Kingdom & if we succeed in establishing the first as highly probable & eventually as a true theory, which no one can reasonably doubt, the creation at successive periods of new beings will still be believed in & ascribable to some power wholly exterior to all this mechanism, some causation different from that of the variety-making power, if we include monstrosities in this. What is this Variety-making power? That is the question. We cannot answer it by saying that it is the power which causes the deviations [p. 19] from an original type when these are in the direction of advancement to a higher grade of being, it must be some power beyond that of secondary mechanism, unless delegated by a supreme artificer.

Languages

As any language has been in some way connected with a preceding one so may every species. Languages may improve & may degenerate, may change with comparative rapidity, or be persistent for indefinite periods—& once extinct can never reappear. All this may be true of species.

But, though languages may all be affiliated, it cannot be said that each new one is simply a modification of its predecessor. It may contain words expressing thoughts & generalizations of a more complex nature, sentiments more exalted, religious ideas more pure than any preceding race of Man required or could comprehend. The same inventive power which struck out the first roots & grammatical constructions [p. 20] continue to act & may be manifested in a higher & intensified way. And so may the Creative Power, which in the earlier ages may have given rise to simple germ cells, to sensation, & then to instinct, & which may, afterwards continuing the same, exert the higher function & cause Mind, reason & higher intellectual powers—not by modification of what preexisted—but by creative additions. There may be a law of continuity as in the languages of Mankind & yet the species—creating, like the word-inventing power, may continue & be destined, as Time rolls on, to display higher deeds both subject to laws & yet exercised by Volition. If it be not irreverent to compare things human with things divine, we may conceive, etc.

It may happen in the course of ages that as some words & forms of speech & of grammar are always growing obsolete & new ones invented, the languages coming from one original stock may have nothing in common to the parent tongue [p. 21] from which they branched off & yet if they could be followed up historically the bond would be discoverable. They may be absolutely unlike in spite of the Unity of type & only be related as the inventions of a similar being, having like thoughts & placed in somewhat similar situations, must have much in common—a similar constitution of Mind—Unity & Continuity, therefore does not of necessity imply a discoverable relationship in every true existing language & it may be so perhaps in regard to some living species, which at times indefinitely remote, sprang from a common progenitor.

Free Agency of Man

We are more certain of it than of the existence of Matter. In a certain sense this power to originate is supernatural—not subject to any of the secondary causes.

[p. 22] He who for the first time contemplated this train of thought, who traced the affinity of Man to the embryo & watched its decay, who recognized the transition from the highest genius & heroism thro' imperceptible gradations to imbecility, who reflected on madness & inherited deformities & was not saddened & unhappy must be almost less than human. There is something great, says Pascal,[9] even in Man's unhappiness—& if we lift up the veil we observe fresh links between the spiritual & intellectual & the brute nature, & find in it much which humbles us & creates anxiety albeit no greater than we knew before in kind, we cannot be surprised. Not that the passage from inferior animal to the rational may ever have been as gradual as what we can now trace in living & surrounding Nature for the passage may have been abrupt as between very different degrees of intelligence in parent & child. But if there be the link of descent it will not prevent the free-agent & responsible being from being sure of his consciousness & spiritual being, nor in [p. 23] any degree lessen to [*sic*] aspirations after immortality, his sense of superiority to the world of irrational beings.

If the philosopher regards these discoveries with indifference, it is rather because they add so little to the enigmas which had already struck him, not that he has no cherished feeling for the dignity of his race. It is because the realities of surrounding Nature are too vivid & demonstrable & palpable to make the more uncertain conclusion respecting the past & a geological link a source of uneasiness. He has gone thro' this stage of unhappiness & if he has come to the conclusion that he is something more than a transient & ephemeral existence, like a bubble in the great ocean of life reflecting for an instant the glories of the universe & then bursting & being resolved into the spiritual ocean, if he has thought so there is nothing in the progressive development system which can disturb his mind.

[p. 24] The existence of evil, of the power to do wrong, are beyond our power to reconcile with the omnipotence of a benevolent creator & yet no Man w.[d] wish not to believe in his free agency & his capability of doing wrong—he w.[d] not wish to be less than Man. So if he desires to exclude the supernatural from his creed, to believe in nothing but Matter & Force, & not in Mind outside of Nature, governed by secondary laws. Why cannot he admit a First Cause sustaining & creating, yet within the limits of certain laws of generation so far as regards the Unity of Plan, without which Man c.^{d} not study Nature & understand it & interpret its records.

The interference of a free agent is a new phase, but the creation out of nothing or without an antetype connected in the way of generation w.^{d} be a breach of the Unity of the Plan. Once prove [p. 25] that all living organic beings, plants & animals, are to come into the world as the offspring of a certain closely allied antetypes & then it is intelligible that Man sh.^{d} not be made an exception, that the law of continuity sh.^{d} be unbroken.

A Man c.^{d} not range above the animals if he had not the power to do right & wrong—yet these animals have some sense of right—of property—shame.

May 12, 1860

Davidson's Table, Brachiopods[10]

Cambrian & Lower Silurian have together 15 genera & only one in common or 6½ per cent.

Lower & Upper Silurian have 22 genera, 12 in common or 54 per cent.

Upper Silurian & Devonian [have] 31 genera of which 16 in common or 52 per cent.

[p. 26] Cretaceous & Tertiary 12 genera, all but one in common, or 91 per cent.

Permian & Trias 18 genera, 8 in common or 44 per cent.

Very small break between the Cretaceous & Tertiary, not near so great as between Upper & Lower Silurian.

Creation

We cannot evade the difficulty if we introduce the creative power acting, whether by a secondary law or by primary & immediate interference of imperfection in the details, monstrosities, abortion, premature death, distortion, insanity, it is simply the one great enigma of the existence of Evil.

But granting this, the question of whether all being foreseen & planned in all its details, the introduct.^{n} of new varieties & by them of species is carried on by the bond of descent [p. 27] or by some other law of discontinuity, or if of continuity this is managed by some law distinct from that of blood relationship.

If we could settle the question as to the possibility of the European & the Negro having descended from one common stock, we might still despair of obtaining proof of the question of any two distinct species. If naturalists are not yet agreed as to the common parentage of races, how can we expect them to be in regard to species?

We shall always have to admit that some power akin to, or identical with, that by which the original germ of life was introduced has gone on acting, & if so, it may be as supernatural as in the beginning, a power above nature, tho' acting by her laws & through them. It may be the work of a free [p. 28] agent like Man but with infinitely higher powers.

All these admissions may be made, perhaps must be made, the miracle of the creative, the sustaining & the destructive powers may be granted or assumed as beyond our comprehension & still there remains another minor question wholly within the future compass of human science, which at present can only be an hypothesis, but wh. may one day be a theory & this is whether part of the system of Nature is that all the races of species, Man included, are derived from a single original stock in the way of generation & secondly whether still further, all species have been [p. 29] connected, the human races inclusive, by the same bond.

Apart from such admissions or laws there may be a creative power, into the mystery of which we shall never penetrate, an originating force to which life, sensation, instinct, intelligence, moral consciousness, reason & all that is spiritual in the soul of Man may be due. The material laws by which such a power may choose to be governed may to a great extent be studied & verified by us & one of these laws may be that new species are part of continuous series of variations in the ordinary way of descent & not by some law [p. 30] wholly distinct, as if a new being were formed at once out of inorganic elements & without any antetype closely allied, of which it may be a modification.

Let us suppose some mighty warrior, statesman & prophet to have sprung up in a distant part of the world & to have raised a whole nation from insignificance & that his birth was declared to have been attended by portents in Nature & to have been wholly distinct from that of ordinary mortals of divine lineage & that the superiority of his genius was acknowledged to be such by many as to lend countenance to the popular belief. The philosophers of a more civilized nation send competent enquirers to ascertain the truth of the tales, unprejudiced men, not averse to believe that extraordinary phenomena may be connected with supernatural [p. 31] events & miracles. They investigate the facts & reports, examining witnesses contemporary of the birth, who report that the tales are fabulous, the earth did not tremble, etc. parents ordinary. Like giant born of dwarf—whence comes the new power—more natural if the tales had been true—more perplexing to find nothing unusual.

It would have been less perplexing if the new phenomenon had, implying some causation of a higher order, some greater power exerted, been attended by a manifestation in the material & organic world of a new order, if there had been some departure from the usual mode of generation, or advent of a new individual. But a great step would have been gained by ascertaining that the miracle, which it may still be called, was not allowed to disturb the ordinary [p. 32] physiological laws of our system & must be matter for the psychologists to speculate upon. There may be, no doubt, some cerebral phenomena in the individual & he might perhaps transmit them to his descendants or to some of them.

Species, p. 99, *Manual* [11]

Nature made them & then broke the die. I have left this passage for whatever theory we adopt of species; it will equally hold true that they cannot reappear when they have once died out.

Enigma

When we can explain the origin of Evil & reconcile it with the goodness of an omnipotent God with which we may well believe it to be reconcilable—when we can reconcile the free agency of Man with the prescience of the Deity & the unvarying laws of Nature, we shall be able to understand the mystery of Creation. But without solving such problems, or even entirely comprehending what a Race or what a species may be, we [p. 33] may gradually arrive at determining in what outward manner & by what machinery the God of Nature causes the succession of forms to people the Earth. If as I have always believed & maintained, this power of creating, originating & destroying is in as full force now as ever; we cannot continue to be attentive observers of the system without discovering some clue to it. We shall find not only the limits of divergence from type, but the abrupt appearance of some new type, the formation of new & permanent races, whether by what is called independent Creation or by ramification from some one pre-existing stock. We shall discover new links & transitional forms, or their absence will strengthen our conviction that they have never existed. [p. 34] The question also of progress in development will be tested.

May 17, 1860

Life

Definition impossible at present, as much so as of spirit, the positive philosophers simply describe certain effects—the others transcendental.

The spiritual [is] like the vital in this respect.

By means of life certain bodies preserve certain specific properties & their individuality, in spite of the loss & successive renewal of the matter composing them & go thro' certain phases belonging to their species.

Cloquet says Life is the faculty of resisting up to a certain point the general laws of Nature.[12] St. Hilaire 101, vol. 2 —— *Hist. Nat. Gen.*, 1859.[13]

Duvernoy,[14] resisting the general laws of *dead matter.*

[p. 35] Flourens says that no one has succeeded in defining

life & we may say of it as LaFontaine s.[d] of *L'impression,* one can only leave it in the bosom of the Deity.[15]

May 18 [1860]

Relative Age of Northern Drift & Welsh Cave Animals: Letter to Prestwich

Wokey Hole & Mendip marble of Mountain Limestone without glacial polishing.

On latitude of Wealden—qy. unmoved when Moel Tryfan & N. Wales went down & up 2400 ft., or at least never submerged.

In Ireland glacial markings on beds in lat. of S. Wales, about?

If S. Wales caverns oscillated like N. Wales, they w.[d] show more signs of submergence under the sea?

What is the relation of Welsh N. drift & mammoth period.

" " " " of same drift with El.[ephas] antiquus & Rhinoc. leptorh. & Hippopot. major period?

[p. 36] Does not the valley of the Somme contain relics of two mammalian periods?

Has Milne[16] got the true Hippopot. tusk from St. Acheul?

Same puzzle in N. America, Mastodon giganteus after the N. drift. Megatherium not proved to be so modern.

Has Woodward [17] determined the Cyrena fluminalis?

May 19, 1860

Microlestes

Triassic mammalia [of] Charles Moore[18] determined by Dr. Falconer.[19]

1. Microlestes, crown without fang, posterior molar upper? jaw like Plieninger's fig. Lyell, *Suppt. to 5th Ed.* p. 18.[20] but with broader

crown & a wider chasm between the marginal lines of tubercles.

[p. 37] 2.dly Another somewhat similar with one fang.

3.dly 4 conical middle incisors of microlestes?

4.^{thly} Shell of crown of larger tooth with a fang probably different genus—mammalian & from same locality.

5.^{thly} Very minute almost microscopic, very perfect 2-fanged tooth or premolar, well pointed cusp as in Amphitherium. Qy. what genus?

6.^{thly} A cervical vertebra very perfect—unequivocally mammalian. The body very small in relation to the great diameter of the spinal canal. I. Geoff. St. Hilaire v. Owen[21]

Bossuet[22] argued that the nearer the apes approach Man the better for the anti-materialists—for it shows all the more that it is the spiritual which distinguishes Man, not the physical.

May 22, 1860 [p. 38]

Creative Power

Some imagine that if the design of the first Cause, manifested in a series of creative acts & exercised with volition, not governed by Fate or by laws controverting the Deity himself, be once admitted, the question of creation by variation is set at rest.

But why should this follow? May not a power strictly allied to, or identical with, that which gave origin to the first types of organic beings in the earth still intervene & yet the laws of variation & of lineal descent be the modes in which this power is displayed.

When first species were made & fitted, some for terrestrial, others for aquatic habits, the laws of vitality must have been accommodated to the previously established laws [p. 39] governing the inorganic creation.

It is perhaps unprofitable to speculate on the beginning as

we do not know whether life was not contemporary with the formation of the planet itself. I know of no geological proof of an 'Azoic Period'. But as it is very common to argue that the same miracle which gave rise to a first creation may continue to be repeated, it may be replied that this originating, designing & creating power may continue to act now as in the beginning in accordance with laws which are fixed by the Will of the Supreme Cause. In the beginning of the introduction of Life upon the earth there was a subjugation to laws. It is true that the vital powers have been defined as opposed to Natural laws or resisting them. [p. 40] Cuvier, I. G. St. Hilaire, vol. 2. p. 73,[23] says that "living bodies seem for a time to resist the laws which govern inanimate matter, in *corps brutes* & even to act on all that surrounds them in a manner entirely contrary to these laws."

But still the first species or genus, whether aquatic or terrestrial or both, whether of few or many types, must have been so constructed as to be in harmony as much as now with all the surrounding condit.[s] of the inorganic world—& consequently subject to their laws. The volition of the Deity may have been limited to the creation of a new cause, certain vital laws which might be regarded by a being foreign to the earth as new & [p. 41] natural at the time, controlling but not subverting Nature.

We may imagine a like reserve of supernatural power capable of introducing sentient beings after plants, the intelligence of the higher animals often mere instinct & the reason of Man in addition to the instincts of brutes & yet all, tho' so many new Causes supervening in harmony with preexisting laws & not subversive of them—each being to the mere laws of vitality what they were to the laws of mere brute matter—controlling & governing, not subverting, supernatural as additions to previously appointed natural laws, but acting thro' & by those laws & in accordance with them.

[p. 42] If we grant to those who believe that the Creative Power, manifested by the progress of the world of organization, sensation & instinct, intelligence & reason is not a mere piece of brute mechanism but something akin to human free will & volition; not a mere engine but in the highest sense the Deity which performs the first lesser miracles of saying let there be life, the God who can say, let there be reason, the moral & intellectual faculties of Man exist, is a higher power than the first or a higher exertion of the one great power.

To grant the lesser miracle & to deny the greater [p. 43] seems unphilosophical.

Why should we not for a generation at least humbly content ourselves with endeavouring patiently to decide whether among the numerous laws to which it has pleased the supreme Cause to subject the manifestations of his creative power on earth, he may not have willed that the coming in of new beings, whether merely living, or endowed also with instinct—whether possessing in a slight degree intelligence approaching to reason, as exercising fully the genius of the most gifted of mankind, should be subject to one common law of descent—that these [p. 44] showed no sudden formation of the higher structures out of earth, air & water, but that they sh.[d] all be modifications of preexisting types.

We may say that we know nothing of the number of aboriginally created types—they belong to a period beyond our investigation but we can enquire which is the most probable hypothesis, whether the calling into existence of higher & higher types has or has not been subject to the ordinary laws of generation, or has been governed by some independent mode of action.

Varieties by degradation may be referred to a secondary law, but progress seems to imply, however gradual, a higher power tho' it may be subject to law. It may not be [p. 45] wholly controlled by preexisting law; the free will of Man & his interference is a New Cause; to impute this to be some kind of Secondary action to which we impute an abortive birth or a variety by degradation seems a contradiction.

There must have been the power which said 'Let there be reason'—we may call this power what we please, the variety-making power, transmutation, Progressive Development, Natural Selection; by whatever name we designate it—it becomes what has been called the First Cause, the Deity. If we ascribe the coming in of the eye & of the brain of Man, the faculties of sight & of reasoning to Natural [p. 46] Selection, we simply deify Natural Selection—& we may as well account for the addition of Life itself to Selection as to refer to it the faculties of the Human Race.

Natural Selection becoming identified with the same power which was manifested in a less high & supernatural way when the first germ of life was created.

If it be said that all this proves that we are beyond our depths in trying to remove the veil of mystery, which envelops creation —granted, but it may be that the Creative Power, which has developed successfully new forms & intelligences & spiritual essences, introduces them by transmutation of previously existing [p. 47] organic types, this is the humbler question to which we may confine outselves. Some are of opinion that the First Cause has not so operated.

Let us suppose that in some country remote from European civilization, a great prophet, statesman, warrior & philosophic law giver, had risen up in a region of a rude nation, had worked a mighty change & improvement in their laws, power & social condition & that it was reported that his birth had been attended by strange portents & had been distinct from the usual way of human origin—that he had been sent from heaven by mysterious agency, in the maturity of his strength & manhood & that splendid prodigies had confirmed his divine origin & mission. To examine into the truth of these stories before it sh.[d] be too late might be the task assigned to some trusted enquirers. They visit the scene of the origin of the prophet & report that they have ascertained from eye witnesses that the stories of miraculous origin are mere fables, that the tales that "the earth trembled at his nativity" or that "the face of heaven was full of fiery shapes," they [p. 48] could prove that the ordinary course of Nature was in no way violated, moreover that the parents, tho' they had ample opportunity of showing it, had they possessed this, were of ordinary capacity, had never evinced any superior endowments, no more than some others of their children & that some of them were even below the average capacity of their race. Now to some whose theory of causation required that every unusual phenomenon was probably attended by some marvelous departure from the every day course of events, this arrangement might be a great disappointment but every report on genius w.[d] require that the truth had been ascertained—this mystery might seem greater than ever, that is, without any departure from the usual & everyday laws of descent, a difference in mental stature exceeding that of the birth of a giant from a dwarf, or a dwarf from a giant, had been witnessed—but it w.[d] be a great point to have discovered so much of the working [p. 49] of the author of the Universe as is disclosed in such a fact—which the enthusiasm or fanaticism of devoted followers might have disguised in a cloud of incense.

That something so intellectually & spiritually superior sh.[d] emanate from commonplace parentage might seem the greater miracle of the two, in fact their theory of causation might be baffled, but to know the truth must reluctantly be admitted to be a gain.

The advent of a higher power, the progress of the race, would not be the less true because not brought about according to their preconceived theories.

What if a series of such advances be secured from something infinitely more humble as an original? Where may this progress end? What if cranial development in the prophet be inherited or inheritable—to a certain extent, where may this end?

It is not enough, if even we believe in progress, to say that if in the beginning [p. 50] we are compelled to grant the miracle of the creation of life, we may be prepared for subsequent miracles, we ought rather to reverse the order if there be a power capable in comparatively modern times of saying, 'Let the moral & intellectual faculties of Man exist', & that power be obeyed, how much more reason is there to believe that at preceding periods the same omnipotent First Cause may have said, 'Let there be the intelligence of the highest of the inferior animals', or at still earlier epochs, 'Let there be instinct —sensation, life'—& even this miracle of, 'Let there be life', may be higher than that, by which brute matter or the laws of the inorganic world, was instituted. [p. 51] The question of progress is a much higher one than that of simple creation, because it implies a series of miracles—the supernatural continually intensified, the origination of New Causes acting in conformity with pre-existing laws.

There may have been a time when the law of generative

descent was first introduced with vitality into the planet before any created thing higher than a plant existed. Such is the speculative opinion of many. It may then have been determined that when sensation, instinct, intelligence & Reason appeared in succession, they sh.[d] submit to this law, this secondary cause, but the addition by the Creative Power of sight, memory, jealousy, the highest brute intelligence, passion, to plant life may be always thro' generation & the variety-making power & not by any new secondary mechanism in the material world.

However natural may be the disappointment [p. 52] of enthusiastic followers & admirers of the prophet, when a less exalted genealogy was reported, yet in the eyes of the reflecting, no real degradation w.[d] be incurred. Whatever may be spiritual & divine in the prophet is not really the less so, tho' it might not accord with the theory of causation in which certain enthusiasts had indulged. The cool looker-on w.[d] not estimate the genius & gifts of the man less highly because of human growth.

If we refer the introduction of Man, his free agency & volition, his moral & intellectual faculties, his reason, to a secondary Cause, why refer the origin of vitality to a higher cause? Why may not both be referable to the Volition of the Deity?

Whatever be the name we assign to the power which created the planet & the laws of inorganic matter & afterwards gave origin to plants & animals, that same power may naturally be evoked as the cause of much higher manifestation of creative power.

May 22, 1860 [p. 53]

Law of Continuity by Descent

We may enquire whether there has been creation by a continuous law & whether the laws imply progressive advance from simple to complex—

Whether time enters into the conception of Creative acts or whether they are instantaneous.

There may be design & a plan & yet one of the many laws by

which unity of plan & continuity may be worked out may be the ordinary laws of generation.

The miracle-working power may be manifested with increasing intensity from the time when the command was given, 'Let there be Life,' to the day on which God said, 'Let there be Reason.'

We may enquire into the link of descent & hereditary transmission & the mode of variation & of growth without pretending to solve [p. 54] the problem of creation.

Just as we may endeavour to investigate the rise & growth of languages & their successive extinction without pretending to define or comprehend what may be the Nature of that inventive power by which Man gives origin to new terms & words—

The metaphysical question in regard to the faculty of imagination is one thing, the laws of grammar & so forth—the humbler enquiry whether every tongue & dialect has been affiliated with some preceding one is another—

Whether one language was the primordial one, or many were formed & spoken simultaneously by several human original stocks may be a question difficult to [p. 55] answer—for which no exact date may be procurable, but this sh.[d] not deter us from enquiring whether, since the time of history & tradition one language has grown out of another & new terms have been successively introduced by the same imaginative power which gave rise to the original terms of the first tongue or tongues—

The longer any particular type of language has existed, the greater will have been the number of its dialects & their divergence from the parent stock, not only by desuetude but by new inventive additions, some of a higher, some of a simpler kind. New wants arise & new terms are invented by the same volition by which the first words were coined—not always by a secondary law of inflection or addition.

Languages [p. 56]

When new conditions arise, new wants & sentiments, new scientific discoveries, new refinements of honour, depth of religious sentiment, abstract ideas, terms which could not have been understood in a rude stage of society, which if at hand w.[d]

have been rejected are for the first time invented & added to a simpler tongue.

Suppose it possible that some man of genius, born before his powers c.[d] be appreciated by the rest, to feel the want of such terms, to foreknow them, still if isolated, he c.[d] not give them—a stranger of more advanced race might land [&] wish to inculcate them.

If the volition of the Deity alone can be supposed to raise the animal Kingdom gradually from irrational to rational & that such a transmutation is a supernatural exertion of [p. 57] power, if it be a power which adds new causes, new laws in harmony with those previously in force yet unwitnessed before on Earth, why sh.[d] we refuse to grant the probability of such supernatural agency or of such a miracle?

If the power be exerted consistently with some law of continuity it may nevertheless be equally supernatural, a power from without, of that same incomprehensible & transcendental nature as the power which added the laws of vitality to those which previously governed brute matter & then super-added sensation, instinct etc., etc. Reason—To a being who has no relation to time, there can be no difference whether all these phenomena are produced . . .

May 23 [1860] [p. 58]

Instantaneously or gradually, but according to our ideas, breach of continuity & violation of the ordinary laws of nature imply supernatural agency.

Unless we are prepared to say that Natural Selection could have created the first living types, we cannot suppose the power capable of adding sight to the blind, instinct & intelligence to mere sensation, & reason & moral attributes to mere animal instinct. If we believe that the originally created types were less complex & advanced than those now flourishing, if we adopt any law of progress, we immediately require a greater intensity of the intervention or manifestation of the First Cause as Time rolls on.

[p. 59] Instead of referring to the days of a remote ancestry for the exertion of powers of the highest order, we must ascribe those miraculous interventions to the most recent epochs.

Supernatural Intervention

It may be asked what do we gain (if compelled to admit that the Creative Power must interfere) by disproving the reality of species, why not grant it as equally probable that there have been breaks in the series, additions not connected lineally with an ancestral archetype? Which is most of a miracle? How is it more comprehensible, to suppose the highest intelligence to grow out of the irrational brute or to come more directly from the hands of the Creator?

In reply we must candidly concede that the latter hypothesis is *a priori* as comprehensible, perhaps more so than the other, but it will be found that both are so [p. 60] entirely beyond the reach of the human understanding that we get no nearer to uplifting the veil of mystery which shrouds the origin & source of the laws of Nature, whether of things inanimate or vital, material or spiritual.

We have a humbler task assigned to us & that is to try & discover in what way the machinery of the terrestrial world, as observable to our senses, is made the instrument of introducing new forms & powers & qualities & attributes into the earthly theatre. How are they ushered upon the stage? If we cannot positively ascertain this, with what hypothesis can we reconcile the greatest number of well established facts derived from a vast lapse of time? [p. 61] Suppose a question to have arisen as to the birth of a prophet, etc. (see above p. [47]) some may argue that so universal a belief could not be referred to the mere enthusiasm & fanaticism of the prophet's followers, & why they may ask, sh.d not an unusual intellectual & spiritual phenomenon be accompanied by some deviation from the everyday course of Nature. At length scientific enquirers are sent to the scene of action to determine, before it is too late, into the truth of the alleged occurrences & marvels. Curiosity, being excited by controversy, [a] philosophical commission is appointed—they report that they are satisfied, etc. They visit the birthplace of the legislator & prophet & report that all the stories of the earth having trembled at his nativity—

His parents were only of ordinary ability & some of their children, the brothers & sisters of the prophet, quite commonplace, of less intellectual & moral character than their parents—

[p. 62] Would this report render the event less of a miracle or make it more unaccountable than ever? Whence came the spiritual power, the spark of heavenly fire & how? If several such leaps be made, from what humble origin so far as its mere animal genealogy be concerned, may not the ancestors of the prophet have come?

If we sh.[d] not believe that a giant 9 f.[t] in height was born in the ordinary way of the dwarf only 2 f.[t] high, why imagine that in intellectual stature so great a difference sh.[d] exist between one individual & others of his nature, unless some particular circums.[s] had attended his birth? Such a departure from the usual law of inheritance, by which like produces like, may have been attended by other causes of difference.

It may be that a similar advance was made before from a rude to a higher stage of intellectual, social & moral perfection & improvement & that, after retrograding somewhat, new steps in advance are realized & at length the capabilities of a whole race are elevated. All this leaves the progressive scheme, [p. 63] whether called development or creation by Law, equally supernatural & we cannot, by proving the genealogical link, render it less miraculous or more intelligible.

If we ascribe this progress to Natural Selection, we simply deify that Power & it no longer bears any true analogy to the selection which the Breeder exercises in choosing or rejecting varieties offered to him by Nature or by the variety-creating Power.

It becomes the union of a secondary Law with the divine First Cause—by which that law has been instituted. It is to confound the working of a secondary cause, the operation of which we can analyse & observe, with phenomena which we can only refer to the divine First Cause by which the Universe was made.

It will be objected that if you refer varieties to a divine source, you must attribute those which fail, monstrosities, abortions, those which meet with an untimely end, [p. 64] to the same; useless organs, rudimentary ones, immoral, vicious individuals, evil, insanity—but here we plunge into depths which we cannot fathom. We must not, because of this, hesitate to ascribe organic laws to the design of a Supreme Being & the order which reigns, because there are disturbances, or the good to imperfection, because of the existence of evil.

There is some exalted power far above us, or one comprehensive operation in the development of the world, of matter & of spirit, of life & intellect, the constant intervention of whose creative & sustaining power will be more & more felt, the further progress is made in unravelling the laws & secondary cause of the terrestrial system. If we make sex a continuity of creative power thro' the instrumentality of ordinary generation whereby the succession of animal & vegetable forms has been introduced into [p. 65] the earth, we shall have made a step, very slight indeed in a metaphysical point of view, but great in reference to Natural History & Geology. It may not do much in enlarging our mental philosophy but it cannot fail to advance our facility in interpretation of Nature.

To know what is to be assigned to the unknowable is the great secret of every important step in natural & psychological enquiries. The attempt to solve the problem of the manner in which species are introduced in succession into the world's stage need not be condemned as overambitious any more than to investigate the phenomena of ordinary generation of individuals, but as we are now just where Pythagoras & Plato were in the comprehension of this so far as the causation goes, so we shall doubtless be in regard to species & the enigma is probably less perplexing.

Separation of Man [p. 66]

They who sh.[d] select a particular language of some highly civilized nation, & sh.[d] say it is so superior that it must have had a distinct origin from all the rest, w.[d] soon find certain lines in common wh. no coincidence, no rare chance c.[d] be suppos.[d] to sanction, it w.[d] not be millions, but billions of billions to one ag.[t] such laws of structure & certain irregularities & coincidences.

It w.[d] be better to deny the whole filiation of other languages to invent some arbitrary law by which nations independently of each other, & by similarity of mental constitution struck out, & by identity of the organs of speech, hit on similar expressions & sounds, but to grant the connection in general, the continuity of growth, the formation of dialects & language & then, because some one tongue we find wider apart from the rest (with a great gap & yet with some very marked [p. 67] connecting links) to select this as made by a distinct set of laws w.[d] be an unphilosophical proceeding.

All the arguments derived from the structural affinity of the different species of a genus & genera of a family, or order & orders of a class, fall to the ground unless we can apply them to the human family & its different races. It is far better to believe with Agassiz & Owen & to follow Linnaeus, that species are all originally distinct & not connected by the hidden bond of descent, than to grant the hypothesis & not apply the rule to Man.

In rejecting this they are logical & consistent—whether we look to structure or instincts, or intelligence, or osteology, or habits, or the action of epidemics & disease, some actually common to Man & the inferior animals, to such laws as that most perish [p. 68] before attaining the adult state, or that race wars against race, or the differences of individual capacity, albinos, migration, fossilization; in nothing can we separate them. We must yield this or nothing. For if Man is not linked by the bond of descent then neither are the quadrumana with each other. — has said that the distance between the Orang & Lemur is greater than between Man & the Orang. As to their being, in the one case, many extant transitional links which are wanting in the Case of Man & the higher apes, we ought to expect such a gap having as yet never searched into the history of Man or of the highest & most anthropomorphic of the Quadrumana. That thousands of extinct [p. 69] genera of Quadrumana have prevailed in 10 ? successive Tertiary Periods, we may infer. Whole assemblages are as yet unknown. From the Eocene to the Recent Period how many forms must have served & yet which probably are indelibly written in the great flourished of which the memorials may be but scantily pre-book of Nature. They & hundreds of extinct races of Man, some inferior to any to which History or Tradition may relate—We have not looked into the volume of the tropics. That we sh.[d] have already some glimpses from extra-tropical documents is more strange than our ignorance of these tree-frequenting creatures, confined as our researches have been to regions least congenial to them & how scanty our enquiries into these extra-tropical monuments!

Instantaneous Creation [p. 70]

If there be some law of succession in time, we have only to substitute for a million years a million[th] fraction of a second & the result would to a human being be instantaneous, while to a being to whom Time can have no relation it may be the same as the longer period.

In the one case we might be unable to stretch our imagination to comprehend the time required, in the other it may be instantaneous but both w.[d] be the same to the Superior Being.

If the longer time be required, if Man & the higher animals are developed slowly & by an indefinite number of generations, it is the same & no more intelligible. The metamorphosis belies the embryo, & adult states of the same individual seem to point to there being no necessity why the whole sh.[d] not proceed at a more rapid rate.

Isolation of Man

[Isolation of man] implies great amount of extinction & no animal so extirpating as Man or one who would press so hard on the nearest allied forms. His most savage wars are against his own race, tribe against tribe, [p. 71] the weakest yielding.

May 30, 1860

Geological Society

Falconer says the Grays' Thurrock rhinoceros is not R. hemitoechus. Clacton one is R. hemitoechus.

All the cave animals of England, except the Machairodus of Torquay of Kent's Hole are in the Gower Peninsula, S. Wales—plus El. antiquus & Rh. hemitoechus—Hippop.^{s} major with the above—No mixture of these three with mammoth & tichorhine rhinoceros.

1100 reindeer horns Cervus Guettardi? or Tarandus priscus? in one cave (Bacon's cave).

A few other mammals only in same, wolf? equus? bos?

One cave inhabited by hyaenas. No streams but small brooks & few of these in some caves & yet the [p. 72] marine shells in the substratum of all the cave deposits are recent—3 Littorines—rudis, littorea & another—being qy. if more than 12 species—Prestwich says we must take in all the shells of the raised beach, elevated 30 ft., in order to estimate—but qy. as to the conchol.^{l} fauna.

Austen[24] remarks that the Pagham mollusca, contemporary with E. antiquus, are not British & imply more genial climate. The erratics are over these & then gravel with mammoth.

Brighton Eleph.^{t} remains—(qy. as to hippopot.^{s} in Brighton breccia).

Prestwich says the hippopot.^{s} of St. Roch, Amiens occurs with El. primigenius. Falconer does not seem to have seen [p. 73] El. antiquus at Amiens & Abbeville but did he see their collections?

El. meridionalis Rh. etruscus	occur under the glacial deposits in Norfolk mud cliffs
2 El. Primigenius Rh. tichorhinus El. antiquus 1 Rh. hemitoechus Hipp. major	contemporary with— Ursus speleus Hyaena spelea Felis spelea Bos Equus Meles taxus

So many species coexisted with 1 & 2 that all were contemporary.

Mr. S. W. King[25] remarks that El. antiquus may have overlapped El. primigenius but this does not prove that both were contemporary properly so called.

Prestwich thinks that the [p. 74] Welsh evidence proves that

El. antiquus & R. hemitoechus lived *after* the glacial period. Now this is contradicted—

1.st by the Loess with El. primigenius mammoth, (Arctomys ——) of Aix, Rh. tichorhinus all being newer than the El. antiquus of valley of the Oise [France] & absence from Loess of El. antiquus.
2.dly by Pagham erratics.
3.dly Norfolk Mud Cliffs.
4.thly by absence of Cyrena from the mammoth beds.
5. by absence of erratics from El. antiquus beds of Grays' Thurrock.
6. by cold fauna of N. Germany accompanying El. primigenius, e.g., Lagomis, Mus, Lemming & another Bos moschatus. Reindeer.
7. by absence of hippopot.s from ordinary mammoth beds & from Loess.

June 2, 1860 [p. 75]

Andrew Murray on Darwin[26]

The objectors seize on the bear converted into a whale, the descent of all from one prototype & showing this to be extravagant, all bond of descent is denied & the explanation of many distinct sets of phenomena, otherwise unexplained, is neglected.[27]

Extract from Mallet's MS Report
to Royal Society, June 2, 1860[28]

On visiting the Temple of Serapis at Pozzuoli where the notoriety it has acquired on this point, & the daily attention given to it, presented the best chance of indication, no evidence whatever could be [p. 76] found, of change of level. The '*gardien*' of the place, however, on being questioned as to whether he had observed any change of level at once directed our attention to the base of one of the worm-eaten columns and stoutly affirmed that the level of the water, which was then standing at *a* (Fig. 117) had directly, after the shock of December, fallen to *b*, equivalent to a rise of the Temple of 7 inches, but that since that time the water had gradually returned to its former level, i.e. the land had sunk again. He denied that the difference [p. 77] could be due to variability in the sea level. The utmost limit of disturbance by wind or tide within the sheltered walks of the ruins, being according to his stated experience, far within seven inches.

I could not find that any man of science in Naples had ascertained what these limits of aqueous disturbance were & on my return to Naples (from the interior) took the occasion of a severe gale of wind inshore (the Garbino from the S.W.) & at the presumed time of high water, to visit the Temple again in company with Sig.r Guiscardi,[29] when I found the water rather above the sill of the entrance iron gate [p. 78] and fully 22 inches above the level of the 5.th Febr.y and it had been nearly 3 inches higher about two hours previously.

It is obvious therefore that any deductions whatsoever as to levels, whether of elevation or of depression based upon the tidal level of the Mediterranean on this coast, cannot be depended upon within the limits of 18 inches or 2 feet at the very least & several of the speculations as to minute oscillations of level of the Temple of Serapis, so based, must be received with doubt.

Impressed with this fact, in which I found that Prof. Capacci[30] & [p. 79] Sig.r Guiscardi coincided with me and with the extreme value to physical science in this unstable region of possessing some definite and unimpeachable standard of level, I addressed a formal letter to the Gov.t suggesting the importance of having an accurate line of levelling run thro' to Naples from the sill of the front door of St. Peter's at Rome, which may be presumed at present an invariable datum point—and the difference of level marked upon beach marks ? at & around Naples. (See Appendix No. 4.)[31]

While the limits of error as to level deduced from the sea affect all minute questions of rise & fall [p. 80] of Serapis, they do not touch the great change of level as evidenced by the celebrated columns, but they appear to be sufficient to destroy the force of the conclusions of Niccolini & others as to oscillating changes of level of small extent.

The evidence of elevation of the whole building since its original construction appears to me irrefragable, but not so that upon which the supposition of its subsidence, first after its erection—& previous to its elevation, are based. The argument for subsidence rests upon the improbability that the level of the floor of the building was originally designed & constructed below that of [p. 81] the mean tide of the Mediterranean. Now it appears to be that the probability runs just the other way. Archeologists appear to have settled that the so-called Temple of Serapis was not a Temple at all—but a public Bath—a conclusion that forces itself upon the mind of any untheoretical observer of the general architectural structure of the place. If a bath, nothing was so probable as that its level should have been so fixed with reference to the sea, that sea water would run in on command, the baths in a place where there appears to have been no fresh water but that of the thermal spring. The objection to this, that there would [p. 82] then be no drainage for the waste water of the baths, being met by the fact that the dry & porous subsoil consisting of 12 to 20 feet of tufa, lapilli & scoriae would soak away any amount of water if simply discharged into a pit sunk in it below the level of the baths, a method of drainage actually practised from a remote age to the present day—a considerable district of Paris at present discharging the whole of its sewage into such a *'puits d'absorption'*.

The land at the existing level of the terrace called La Starza —upon which the Temple was [p. 83] built, is in rapid & constant process of marine degradation at present, so much so that unless artificial means be soon taken to prevent its inroads, the sea will in another half century probably have swept away the whole Temple (so called).

It therefore was probably very much more inland when first constructed and was probably built either in some natural depression of 10 or 12 feet below the sea level or in one excavated to that depth by a race whose burrowing tendencies are revealed by many of their buildings in all directions around. If much inland there was doubtless a [p. 84] sufficient mass of material between it & the sea to be watertight—but if as more & more of it became removed—the sea water percolated the bank incessantly at the seaward side, it could no longer be kept out from the building & the place would have been abandoned as untenable.

The water of the sea w.[d] then stand permanently at a level with the highest line of testaceous perforations of the Limestone columns—say about 20 feet above the level of the present floor—*assuming that the general level of La Starza, was then about 8 feet* [p. 85] *under what it now is,* and that the floor was originally founded 12 feet below the sea level.

The channels or ducts that had before brought the sea water to the baths, w.[d] also bring the young testaceae, & preserve sufficient change for their healthy existence. If then the land, bearing the so-called Temple upon it, were gradually elevated about 8 feet, resting at about its present level, we have sufficient to account for the phenomenon observed, without having recourse to several successive depressions & elevations.

To the view here advanced as offering the simplest & most probable [p. 86] solution of the Serapis problem, it may be objected that the adjoining ruins of the temples of Neptune & of the Nymphs are some feet under water & that the arches of the so-called mole of Pozzuoli are covered above the level of the springing. The levels of these two latter temples will not accord either with the presumed depression nor elevation of Serapis—and may hence be made to argue as much against as for the oscillatory view, and as to the arches & general structure of the so called mole—now deeply immersed, I am satisfied that it [p. 87] never was built on dry land and for other purposes. It would be a work of no small difficulty to construct these piers & arches in the open seaway where their remains now stand. With all the aids that modern engineering afford, and without the diving bell, we may safely affirm that they never could have been built in open sea-water. They further are of dimensions & construction that no Roman or any other architect would have adopted for a marine mole. [p. 88] How then came they [to be] immersed as they are? It appears to me that they and the nicherent tufaceous land that sustains them, & these Temples are now & have long been in gradual process of insensible land slip downward & seaward by the continual removal of the loose material from the foot of the submarine talus which the soundings prove to be outside them in the Roadstead, and hence unequal subsidence, but always greatest where nearest the seashore—and this view is strongly corroborated by the fact that all the standing columns at Serapis lean some inches out of plumb [p. 89] to seaward & that the whole floor of [the] place is waved & uneven and with a general out of level slope to seaward also, as tho' the whole mass stood upon a base of loose, soft material that was gradually settling and going seaward from the effects of sub-littoral erosion. This seems also to be the solution of the cases of the Roman roads under water between Pozzuoli and Baiae & the Lucrini Lake. Moreover, if Serapis had been ever depressed to the extent required—then this marine mole must have been equally so, but it [p. 90] is quite obvious to an enquiring eye that were the arches upon the piers, as now standing, depressed but a few feet, so as to receive the full stroke of the waves in storms or the entire impulse of the moving superficial columns of the sea, they would have been overthrown long ago. They only stand because they never yet were under water, wholly.

[Page 91 is blank.]

Baron Anca[32] to Falconer, Palermo, March 1860 [p. 92]

In cave [of] Winds of Bay of Palermo & another N. Sicily—

Herbivores:	Carnivores:
Hippopotam.s 2 species?	Canis
Elephas	Ursus
Equus	Hyaena
Bos	Felis
Cervus	& small carnivora.

With Major Von Benningsen-Förder of Berlin[33]

Limia marly & muddy	lehm on loess on limus
H. V. Meyer Elephas?	helix, succinea oblata Marne, loess-marne
Mammoth rarely	sable bryozoa cf. chalk

V.B. thinks that the countries round the Baltic (& the land of Baltic?) were elevat.^{d} very high during Glacial Period & then cov.^{d} with glaciers. The erratics then journeyed South to N. Germany. Meanwhile the mammoth flourished in Siberia where there were no contemporary glaciers. When at last the country subsided the mountains sank more in proport.^{n} just as they had risen more proportionally, so that the glaciers [p. 93] melted back leaving giant cauldrons, *osars* & lakes, & erratics. (Hence no marine shells). According to this view the plains of N. Germany may be the diluvium of the retreating glacial torrents.

South of the borders of the upraised glacial land, V. Benningsen-Förder thinks that there was sea—why? See 8? Berlin Geol. Gesch.^{t} Zeitung.[34]

Dr. A. Fleming in 1852 published—Orthis crenistria, carbonif.^{s} sp. See Quart G.S. Journ.^{1}, 1852, p. 190, Vol. 9—list of salt range fossils.[35] These carbonif.^{s} fossils, so called by DeVerneuil[36] & Davidson are in the same rock, almost the same slates, as the Ceratite & Orthoceratite. Here therefore is a link between Paleozoic & Trias.

The Ceratite is not C. nodosus as in the same slab of limestone.

The salt is a diff.^{t} form & below the ceratite & orthoceratite limes.^{t}.

June 5, 1860 [p. 94]

Owen's Lecture in *Times,* May 19, 1860: Marsupials of Australia[37]

Sloths (in S. America) Armadilloes, Anteaters, Platyrhini, monkeys, Llamas, Peccaris, Fossil Megatherium, Glyptodon, Glossotheres, Macrauchenias.[38]

Australia

Diprotodon, huge size.

Nototherium, big.

The post-pliocene & pliocene mammals of same region are of same families.

Brutes

"The mysteries of their future" [39]

With Prestwich: Ilford

It was in a pit south of Ilford on the road to Barking that we saw in 18 [blank] a mass of reconstructed London Clay like L.C., filling a great hollow scooped out of older fluviatile drift (gravel, sand, etc.), showing what a series of rivers may be in the great plain of the Thames. [p. 95] Prestwich admits that the tributaries & streams wandering over the valley of Thames have cut and filled channels such as *a a*—[See Fig. 14.]

Fig. 14

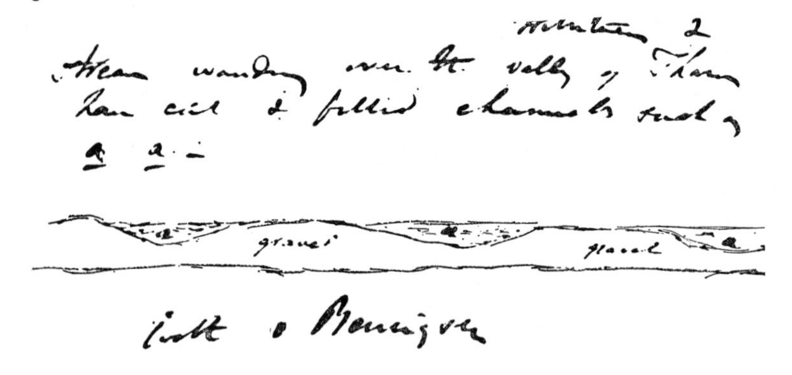

With Von Benningsen-Förder

Le loess, at 2000 ft. French, in the Alps near Appenzal & nearly as high at Coire, also near 1400 ft. at Kaiserstuhls, upheaval of Alps raised the loess—uplifted it higher than in the Hartz.

Near the Hartzburg 1000? ft.

The mammoth & musk ox of Kreutzberg was anterior to the map of Benningsen.

Not only anterior to that, but also to the 3 diluvial epochs, i.e. sand, marl & loess.

The mammoth is in the loess, but the cold of the Scandinavian glaciers killed the tertiary animals.

The arctic shells of the Clyde are later than the loess.

[p. 96] Textiliariae, Rotaliae in the loess from Basle, Kaiserstuhl, Weissenburgh near Mayence, near Bonn Siebengebirge at Dieppe, Proville—also Coblentz, Cologne, Liège, Landst[uhl], Dusseldorf.

This Cretaceous debris in the chalk came from the north but the Bryozoa are not in the Rhenish loess.

The Thonmergel or till is a fluviatile format.^{n} wh. was formed when Scandinavia rose.

Till of England [is] a mixture of London clay & chalk. Von Benningsen-[Förder] seems to identify the loess by infusoria of chalk & by bryozoa [also] of chalk, but these he does not find in the loess of the Rhine.

June 6, 1860

With Huxley

[See Fig. 15.] [p. 97] The spermatozoa pass slowly up the fallopian tubes & the ova, after breaking out of the bag wh.

Fig. 15

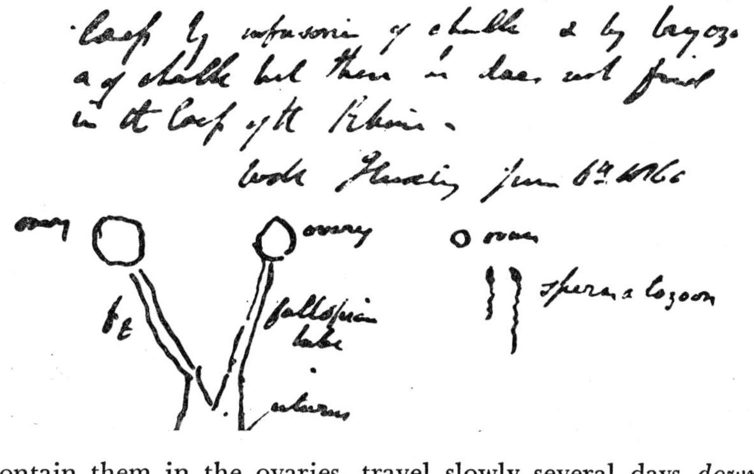

contain them in the ovaries, travel slowly several days *down* the fallopian tube. Hence a difference in the *apparent* periods of gestation. If the spermatazoon finds an ovum in the womb it may at once impregnate it. It pierces the skin of the egg & entering becomes intimately mixed with it. Several may enter one egg.

[Page 98 is blank.]

Von Benningsen-Förder [p. 99]

The separation of Engl.[d] & Continent was posterior to the 3 stages of the loess, which last is made up [of] tertiary materials, the fine sand of loess being derived from Brown Coal formation.

Mica in loess

Norfolk	Lehm I	Loess lehm
	Marne II	Loess marne
	Sable III	
	Till IV	in deepest parts

The osars, says B.[Benningsen]-Förder, were more modern. But ice & floating ice in all the three periods.

June 12, 1860

Development

Assuming that the succession of Life implies the development of a plan of progressive advance in organization, has the progress been effected continuously, or discontinuously, or if there has been [p. 100] a law of continuity, has the bond of descent been the mode in which that continuity has been kept up (sustained). In imagining this we have the advantage of knowing as a fact that in form & instincts & various properties the descendants of the same parents may vary, may come to constitute races, that races may differ (in the animal kingdom) in instinct & intelligence. So that this kind of bond, if no limit can be found except by an arbitrary hypothesis to the modifiability of species, is, if not a *vera causa,* allied at least to one.

But no other, hitherto suggested, has the same advantage &

if imagined, it must partake in some respects of the consequences which belong to ordinary generation. It must assume a similar near affinity, as shown by rudimentary organs, the archetype & antetype.

Progression

It may be regarded with the more favour, especially when it includes Man, because it is in no small degree the reluctant testimony borne [verso p. 100] by those most opposed to Transmutation to a gradational advance approaching most nearly to that which favours the developm.[t] of simpler to more complex forms of structure & instinct, & to the highest or culminating point in Man. The chief objection to this view is avowedly a moral one because it leads, say the anti-transmutationists, to the lowering of Man. It may be equally grand so far as the Creator's manifest.[n] of power is concerned, but it links the inferior animals to the rational ones.

Hence no school w.[d] by inclination be so adverse to admit embryological progress as coincident with chronological as those who are disinclined to the Lamarckian law of continuity by genealogical descent. Their testimony has the force of unwilling admission. It is true that they are most of them believers in catastrophe & gaps in the series, but by no means all of them. It is also true that some like Hugh Miller[40] are predisposed to place Man as the last because of its agreement with Genesis, but the majority are uninfluenced by this tendency. There are [p. 101] other temptat.[s] such as that no system but that of develop.[t] has ever yet held out a prospect of our being able to comprehend in one grand generalization the work of creation from the monad to Man & even comprehend the develop.[t] of plants & animals.

Languages

We may reason on the probable diminution of the number of languages as civilization increases, as nations become larger, we may show that languages in France & England have diminished in a few centuries.

All this does not require me to decide whether I believe that there will ultimately be only *one* language.

We may trace the Aryan tongues back to one & believe that all the existing languages are affiliated without deciding whether there were originally one, two or more languages invented separately by tribes of men who acquired, as they were developed, ideas enough to require speech.

June 13 [1860]

Geological Society [p. 102]

Dr. Falconer showed us Rhinoceros hemitoechus with shells attached wh. he said were littorina. The marine shells believed by Prestwich to be contemporary with the beach, but G. Jeffreys[41] says the sea-shells in Paviland were *over* all the fossil bones of El. primigenius, Ursus, Hyaena, etc.

Buckland however shows that the shells seen by him were eatable species, *Rel. Dil.*, p. 86, etc.[42]

The marine shells at bottom are compared by Falconer to the marine bed of S. Ciro & there is a singular resemblance. In Wales also a hippopot.[s].

Falconer says the elephant behind the Happisboro' forest bed is E. meridionalis, but Rev. Gunn[43] says he has recently obtain.[d] also E. antiquus & thinks that there may be three levels. [See Fig. 16.]

The various Elephants may overlap but the order of their flourishing may be as above indicated.

Murchison[44] says the hippopotamus was found in the Brighton Elephant Bed. If so it outlasted the Selsea Glacial Period.

This would favour the idea of a part of the Glacial Period having been anterior to the Hippopot. & if so it may be also to El. antiquus.

But Falconer argues that because El. antiquus & Rh. hemitoechus & E. primigenius & [Rhinoceros] tichorhinus are both contemporary with Meles taxus [p. 104] & Felis spelaea, Ursus spelaeus, Hyaena spelaea, & other extinct & fossil car-

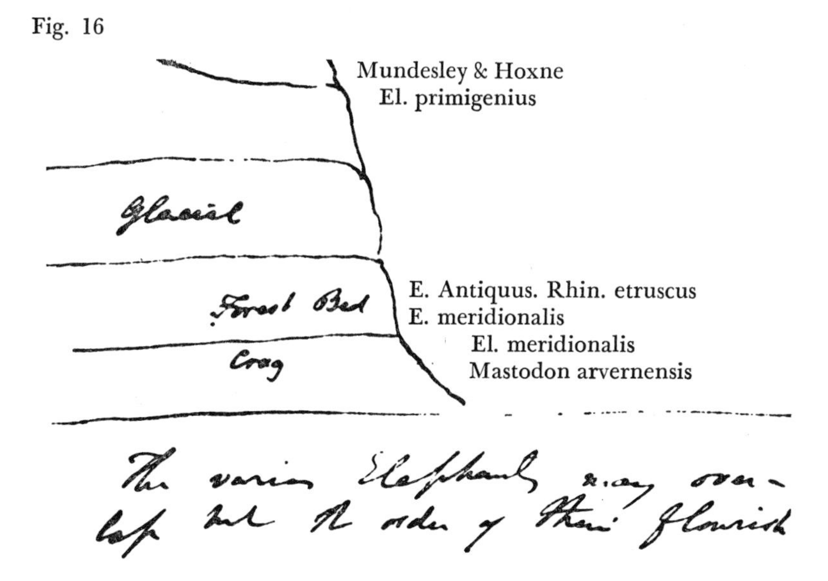

Fig. 16

nivora, ruminantia?, equine? & rodent mammals, therefore both & all belong to one great period.

Mr. Jeffreys says that in Ireland he found two sea beaches on same level, one with glacial and the other with recent species of sea-shells, agreeing with living species. Therefore sea beaches may be confounded wh. may be of different ages.

Prestwich is much puzzled in regard to Glacial Period & admits that it may be complex & not all of one period.

The Welsh case of Cefn mentioned by Prestwich as a case of northern [p. 105] drift is not glacial, says Murchison, but local drift & therefore does not bear on the caves.

According to Prestwich we have [see Fig. 17].

The 2 sets of mammalia of Elephantine & Rhinocere groups in caves & in old alluvia of Thames—& boulder clay 400 ft. high in each case.

The Cyrena of valley of Thames helps us conchologically. [p. 106] If Falconer c.[d] connect the glacial Beds with E. meridionalis & sinking down of Forest Bed of Happisboro', he w.[d] make it very distant from E. primigenius & Bos moschatus, nearer the heat than the ice.

With the 1100 reindeer were many of bear & some of Bos?

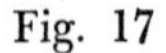

Fig. 17

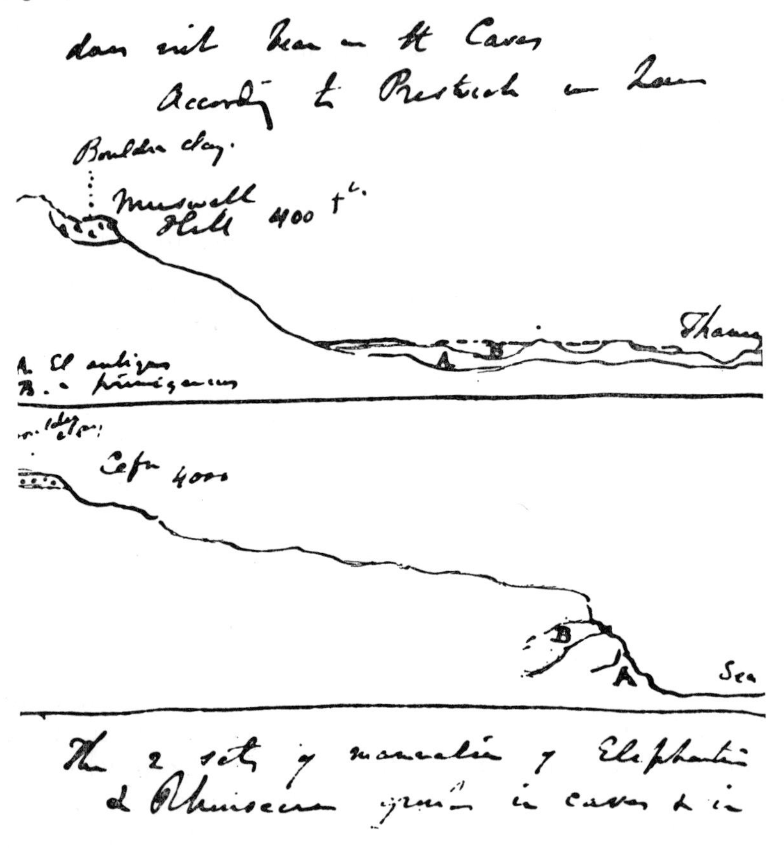

Much migrating of mammalia as of crag shells which disapp.[d] in glacial period & returned.

Glacial Period

The greatest of all the difficulties of this period is connected with the possibility of submergence beneath the sea over vast tracts & no marine shells left as monuments—Glen Roy., U.S. of America S. of Great Lakes, the country round Manchester (& Macclesfield where Prestwich at last found Turritella, much of the Irish Drift.

Geikie[45] finds proofs around Edinburgh of submergence & striae, yet no marine shells like the Glacial beds of the [p. 107] Clyde.

There are three cases of S. Welsh Glamorgansh[ire] caves.

1.st With E. antiquus & Rh. hemitoechus
2.dly " " primigenius R. tichorhinus
3.dly " 1100 reindeer, some bears but no Eleph.t or Rhinoc.s

Mr. King[46]

[Mr. King] says the Donacia indicates climate like our present one—that insects are very sensitive to climate & the range of these beetles small & restricted.

Desnoyers[47]

Footprints in gypsum show wonderful amount of our ignorance of the birds, reptiles & vertebrates of periods of which we thought we knew most.

June 15, 1860

Extract from Letter of C. Lyell to Darwin [p. 108]

Your comparison of Selection to the architect—variations to the stones, is what I deduced from some passages but cannot accept. The architect who plans beforehand & executes his thoughts & invents the Corinthian and other styles of architecture & then by means of machinery, living & inanimate, cranes & horses, & even sometimes intelligent men (former allowed some discretion & power of choosing), such an architect must not be confounded in his functions with [p. 109] the humble office of the most sagacious of breeders. As I have said all along it is the deification of Natural Selection.

Hereditary Descent

That ancestors who have led a very immoral life should have virtuous descendants is not thought unnatural nor that clever parents sh.[d] have stupid offspring. If the coming into the world of good men sh.[d] be thro' vicious parents other anomalies may also be permitted.

Northern Drift

If there has been great submergence beneath the sea of the glacial period without any marine shells having been left as a general rule in the drift formed during that submergence, it may explain much of the difficulty.

June 15, 1860 [p. 110]

By rejecting the bond of descent we lose the power of explaining many phenomena—embryological—rudimentary organs—geological succession—non-appearance of new creations of distinct species while determinate forms are becoming extinct—& we do not gain, as the anti-transmutationists would fain do, the severance of Man from the chain of being which links him with the lower animals. The law of continuity of progress from the fish to the gyrencephala, & from them to the humblest type of humanity, once embraced, what great gain is it to substitute some imaginary law in place of that which is capable of making permanent [p. 111] varieties, which can evolve a negro out of a white Man or a European out of a bushman.

A law being once granted, creation by law subject to numerous conditions in the contemporary state of the organic & inorganic world, at the period of each new creation & above all connected with another law of progressive development in Time, what effort is it, after all this is conceded, to embrace an hypothesis in harmony with a known law of reproduction in place of a dissimilar & unknown one?

In the present state of science it can only be a balance of probabilities in favour of one hypothesis over another [as to] which is the least arbitrary. The consequences so [p. 112] far as the dignity of Man is concerned are not seriously at stake even if a philosopher had any right to regard them as a makeweight.

Few have so little faith in the maxim, that whatever is is best, or so low a conception of the deity, or the wisdom wh. has made & which governs the universe, as to doubt that whenever we arrive at the discovery of the truth, the *modus operandi* by wh. the whole series of phenomena has been evolved will manifest the grandest power & intelligence.

It is merely in its relation to human destiny that the question is agitating, anxious, perplexing. But, granting progression, [p. 113] the intimate link is established without lineal descent—which last solves in some degree the deep meaning & mystery of the close alliance of the human & the animal structure & instinct.

The theory of progression, apart from transmutation, points to the successive appearance of the more advanced species & naturally to more advanced races, that the humblest developments of instincts preceded the higher or those approaching nearer to human reason. If this be true the more powerful manifestation of human intellect ought still more to be the work of Time inasmuch as here we have transmission, not merely by hereditary descent but also by instruction. But if the savage [p. 114] preceded the civilized, there will have been a time when the first man of exalted genius was born of parents of humbler capacity, or say who had never enjoyed an opportunity of displaying high faculties.

That moment was the greatest event which we can conceive in the annals of the planet—a greater one than w.[d] be the appearance of a bushman preceded by no human antetype. Just as the creation of a gyrencephalous mammifer, if by a direct act of creation w.[d] exceed the power displayed by a coral succeeding a plant, or the first living thing w.[d] transcend the origination of a planet without life.

The discovery that each individual is derived from a germ cell, microscopically minute, in which is stored [p. 115] up many habits of its immediate parents, good qualities perhaps, or bad, moral or physical, passions, diseases, healthy constitution or the reverse, malformation, beauty, ugliness & not only peculiarities of the parents but of remote ancestors, of a nation or a race, the effects of climates which are no longer in the same region, perhaps have vanished from the earth—if all this can be stored up in this minute atom—thousandths of an inch in circumference, what wonder may we not be prepared for?

June 15, 1860

With Mr. Jeffreys[48]

Fusus despectus & F. contrarius same, Buccinum undatum carinated & smooth insides & some gigantic.

Mytilus modiola, enormous Scotch variety as compared to English.

Buccinum Dalei found recent 30 yrs. ago.

These varieties have a certain constancy of type in spite of existence of intermediates.

Varieties with Ditto [i.e. Mr. Jeffreys] [p. 116]

Mr. Cumming's[49] collection has done harm because none of the young preserved, & intermediate varieties eschewed.

Dealers dislike mongrels. Typical forms with names alone sell.

Buccinum acuminatum rare & admitted by Jeffreys as distinct from B. undatum tho' he showed a long var. of latter. One var. of B. undatum is quite volutiform—another smooth except in some of the young whorls.

In Buccinum Dalei the striation of the recent [individuals is] different & much finer than in the Crag specimens, [and they] might well be thought a [different] species by some naturalists.

Letter to Morris[50]

1. Megaceros hibernica
   Cervus Tarandus
   Bubalus moschatus
   Elephas primigenius
   Rhinoceros tichorhinus

} Post pliocene

2. Boulder clay

3. Elephas antiquus Rhinoceros hemitoechus Hippopot.^{s} major Cyrena fluminalis	Pleistocene or beds of passage from post-plio-

[p. 117]

Mastodon arvernensis El. meridionalis	Crag
El. " " antiquus Rhinoc.^{s} Etruscus	Post-pliocene Forest bed

El. antiquus
Hippopot. major
Rhin. hemitoechus

June 19, 1860

Letter to Darwin

Did I refer you to a passage which I told Huxley I objected to in his review, in the West.^{n}, p. 546. 'Matter & Force are the two names of the one artist who fashions the living as well [p. 118] as the lifeless.' [51] This I presume is a specimen of the reaction against what you call Paley[52] & Co. who in search of some higher power outside of the machinery called Nature by us, & capable, it is thought, of adding new causes to it from time to time, such as vital powers introduced into our inorganic planet & then instinct & then reason, have instead of deifying Matter & Force or Natural Selection likened the Unknown Cause to the Mind & Soul of Man, this being the loftiest conception they could form, they have striven to intensify Man's [p. 119] intellectual & moral attributes & seem to be far more philosophical in so doing & to keep nearest to analogy. For the free will of Man which, however inconsistent with belief in constant laws, you must admit or give up your source of all knowledge & ignore the constitution of your own mind, must, I think, have some counterpart in the Deity or First Cause according to the highest conception I can make of Him or It. Volition or Free Will in Man is a new cause which in the time of the Deinosaurians had no action, did not interfere with the course of vital [p. 120] action in the globe. The idea of such an anomaly being ever allowed to play such pranks as the breeder has played & to sport with God's creatures & the laws of reproduction, so as to perpetuate pouter pigeons & other monstrosities, would have been scouted by a philosopher of the Wealden Period, if you can suppose some human reasoner, disembodied or not yet in corporeal existence, to have studied the organic phenomena as displayed so long before the human era. If you say that in such [p. 121] matters we are beyond our depth, this objection naturally arises when too much is attributed to any secondary cause whether it be such a 'unity in duality' as Huxley's 'one artist', or your 'Selection.' But perhaps I misunderstood Huxley.

What you say of the gestation of the hound is very remarkable. How comes it that in the human race there is such regularity? Why have we not had years ago many specimens of the fertile hybrids called leporines between the rabbit & hare in the Zool. gardens? Do speak to the [p. 122] Council. Lewes[53] talks of a thousand having been sold from the Angoulême stock. It is the greatest fact yet asserted. Can it be true? It would help me to believe Pallas' theory of the multiple origin of dogs from jackal, wolf, & c., which I suppose I must reluctantly embrace. Perhaps variation in time of gestation is a consequence of such multiple origin whereas if Man belongs to a species which came from one original stock, (or one area) & is not a mixture of several species [p. 123] this might explain why the Negro & European have the same period of gestation. (Have they?)

Has Hooker written to you about the absence of peculiar forms in extra-arctic Greenland & his explanation? It confirms my notion that the glacial period tho' it may have required half a million of years was a brief episode in the last geological epoch. Not above 1 or 2 per cent of difference in the shells—so that the species-making power had not time to produce new plants.

Diluvium [p. 124]

E. de Beaumont[54] thinks that the Diluvium in Scandinavia is the oldest, called also diluv.^{m} des plateaux to wh. also belongs the forms called Limon de Picardie.

The *detrit.*[s] *des Plateaux* cover Picardie (chalk) also extending to La Beauce as far as Orleans & Tours.

It overlies the *depot de St. Pris* wh. is contem[poraneous].[8] with the Val d'Arno.

Next to the plateau came the opening of the actual valley such as the Seine, the Eure, Soineau were filled with a more modern deposit called Diluvium, Alluvium or Diluv.[m] de Valée. [p. 125] which changes in every great river. It varies in Moselle, Seine, Rhine.

Loess

E. de Beaumont thought that it came from Scandinavia.

But he alludes to the loess such as I found (brick earth) in the chalk plateau above Menchecourt. Difficult to say what the age may be.

June 20 [1860]

Hare, Lepus Timidum

Caesar in his commentaries speaks of ancient Britons, Lib. V., ch. 12, not eating hare, nor gallinum nor anseram.[55]

With Col. Strachey[56]

The Kunkur of the Ganges is only in the N.W. provinces & may arise from the decomposit. of land shells which are numerous on the surface of the Gangetic plains & yet seem to disappear in [p. 126] the dense deposits of clay which are seen in the Gangetic mud cut through recently by railways near Calcutta & the Hoogley.

A red clay underlies this blue-mud which last is unstratified (loess & *Nil-erde*?). The rivers as you ascend the Ganges cut thro' the modern alluvial clay down into the red clay which underlies all & high up may be seen to repose immediately upon the old slate.

Large land shells (Balani) on the surface & a layer seen at slight depth.

Clay in the (sunderbund?), also in the delta, depth of sections?
[p. 127] St. Cassian beds, work by Freiherr v. Richthofen.[57]

Predazzo, S. Cassian & Seisser Alp, South Tyrol. St. Cassian beds are upper Triassic, but not uppermost.

With Falconer

Rhinoceros megarhinus, Christol is the name proposed by Falconer for the Ilford & Grays' Thurrock rhinoceros also found at ——.

Rhin. Hemitoechus in Welsh caves.

Clacton Rhin.

Elephas antiquus, Gunn[58] now says is not with El. meridionalis.

Baron Anca has shown Lartet one Eleph. from Palermo cave which turns out to be El. Africanus!

Hence may not Africa have been united with Sicily & the African hippopot.[s] & Elephant have ranged there? [p. 128]

Cropthorne

Rhin.[s] of Montpellier

Clacton species, R. hemitoechus

Rhin.[s] of Cortesi[59] at Milan, wh. Cuvier called R. leptorhinus, has *no* bony septum

Rhinoceros Etruscus (wh. accompanied El. meridionalis at Happisboro) has a bony septum, the Florence specimen wh. is millions of years older than R. hemitoechus (wh. also in V. d'Arno) Miocene? plants with R. Etruscus in V. d'Arno.

Much Sari [p. 129][60]

El. Primigenius & antiquus in Tuff with amphegre [?].

[p. 130]

Mastodon arvernus
El. meridionalis
R. Etruscus } Val d'Arno

A. El antiquus
El primigenius } Kirkdale, rat hole, tunnel not high eno' for deer

Grays' Thurrock rhinoceros at Much Sari seems to be with El. primigenius.

In Clacton when El. primigenius occurs Cyrena trigenuti occurs also—& El. antiquus.

Pliocene Rhinoceros

Rh.^{s} tichorhinus—Fisch & Waldheim, Cuvier

" antiquitatis Blum[enbach]

Erith Crayford & all the glacial deposits, Kingsland.

[p. 131]

Rhin.^{s} megarhinus
Grays' Thurrock
Ilford } no caves
oldest animal

A. { The tunnel-like channel at Kirkdale shows the washing in of the two kinds of Elephants, milk molars of each ticketed in Buckland's museum. & made out by Falconer.

[p. 132]

Rh. hemitoechus { Clacton
Northamptonsh.
Durdham Down
Gower caves in one set
with El. antiquus

R. megarhinus, Christol or the typical R. leptorhinus of Cuvier, Grays' Thurrock—Ilford.

Rh.^{s} Etruscus, Norfolk Cliffs

Upper erratics
Lower boulder clay
[illegible] clay
submerged forest
Eleph.^{t} bed, crag? meridionalis
Chalk

[p. 133] For Falconer on Rhinoceros see Quart. Journ. G. S. No. 61, p. 585, Vol. 14?, June 1857.[61]

Mr. Gunn says that beneath the cliff at Bacton there was dug out [of] the hard marl of 'the Elephant bed' an Elephant's jaw

—after removing part of the gravel which covered it he found the teeth to resemble as much Loxodon meridionalis as E. antiquus.

14 laminae in tooth, doubtful if more, but even if that number the tooth is smaller than many that have only 12—intermediate perhaps between E. meridionalis & E. antiquus & are there not others between the latter & E. primigenius.

Perhaps an illust.^{n} of Darwin's theory. If the jaw in question be E. meridionalis then every well authent.^{d} spec.^{n} of that is from the hard pan of the Elephant bed & every specimen, well authenticated, of E. antiquus is from the submarine forest—& it is possible (as you [discussion of this point continued on p. 137]

[p. 134] Falconer's paper, Quart. Journ. 1856, 1857 [62] On Mastodon & Eleph.^{t}

Chartres Eleph.^{t} ment.^{d} & R. leptorhinus now called, qy. what? but certainly not tichorhinus.

It may be Etruscus or megarhinus, not hemitoechus nor tichorhinus.

2.^{d} Oct. 1856, Falconer's note book on Chartres.

Cefn in Denbighshire, N. Wales, the warm climate must have come on with African animals after the northern drift.

Capt. Thomas Brinelwy,[63] Denbighshire, w.^{d} show Cefn cave. Hippopot.^{s} of Leeds must have been posterior to the Glacial Period.

Kirchberg in Wurtemberg

Rh.^{s} with El. [illegible]

Loess w.^{d} indicate by [its] animals, the upper or northern animals of the Caves.

June 22 [1860] [p. 135]

Dr. Falconer considers that after the boulder-clay submergence there was a warm climate fit for El. antiquus, Rhin.^{s} hemitoechus & hippot.^{s} (probably major), that the Elep.^{t} & Rhin.^{s} above mentioned which occur in certain caves in S. Wales gave place after a time to the Siberian mammoth & Rhino.^{s}, the same Cave Bear & Cave Hyaena & Wolf & Beaver & many other quad.[s], extinct & living, continuing during both elephantine periods & finally the reindeer & musk buffalo & Megaceros representing a colder period.

The Loess of the Rhine, judging by the quadrupeds, resemble this more modern cave period. If so, the intercalated warm period was between the northern drift & the Loess.

The Cyrena of Grays lived on the Norfolk coast *before*? the boulder formation. Is Grays' Thurrock after the latter? Its rhinoceros is R. megarhinus, Christol.

[p. 136] Wolf' & badger down at the bottom of all [?] in the first warm period with El. antiquus & R. hemitoechus.

A living? arvicola with Rh. hemitoechus &

Wealden Denudation

p. 203—*Geologist,* Mackie's theory[64]–take it with me

G. Austen Quart. Journal 1851 Tabular view—[65]

[p. 137] (from p. 133) Falconer hinted that the hard rock of the Elephant bed does belong to an older formation than the submarine forest.

The forest bed is now uncovered at Hasbor' [Happisboro].

According to the above the El. antiquus lived before the boulder clay period of Norfolk—also after it, says Falconer, in Wales—[&] at Pagham before, for under the P.[agham] erratics. [p. 138] [See Fig. 18.]

[p. 139] Gaudin has given some of Falconer's Rhinoceros evidence in a paper, which he sent me, on Val d'Arno?[66] Qy. as to interval between lower & upper boulder form.[t] of erratics in Norfolk Cliffs?

June 24, 1860

With Falconer [p. 140]

El antiquus & Rh. megarhinus in a form.[n] both together between Florence & Pisa.

Write to Owen as to Macacus pliocenus. [See Fig. 19.]

Fig. 18

Fig. 19

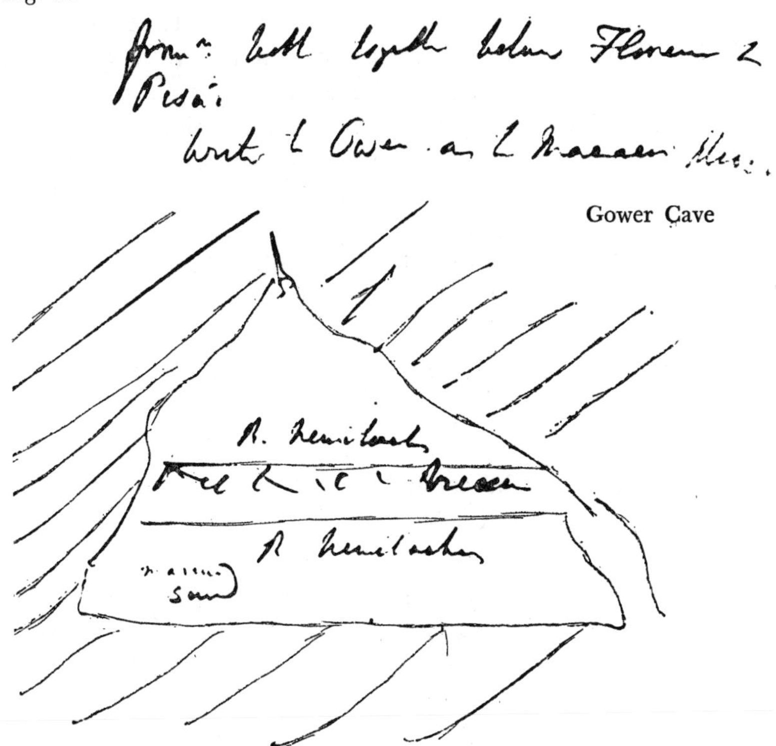

June 25 [1860] [p. 141]

Dawson's Letter: Modern Causes & Darwin's Theory

J. W. Dawson says that Lyell could point to changes in progress which support his theory of modern causes, such as elevation & subsidence, volcanic ejections, river-sediments, now actually in progress—whereas Darwin has to invoke new & unheard of powers & to produce new units of existence. As if the Modern Cause theory required Chemists to admit that new elements were developed out of others, or physically, the many different properties of matter had arisen out of some one original property.

Given weight that malleability, ductility, conversion into a solid, or [p. 142] fluid, sh.[d] be successively added.

The question is whether much is not going on in the organic world. Extinction which widens the breach between existing varieties, 2dly, increase of variation—migration & spreading accompanied by the fact that the wide geograph.[l] distrib.[n] is attended with increased variability—the formation of races.

The develop.[t] of new elements in chemistry out of a few simpler preexisting ones, of magnetism, elasticity, ductility, formability [?] to mere weight & amorphous condition, crystalline form added. New properties superinduced on the primeval matter.

Darwin's original germ-cell, the antetype of all vegetable & animal existence, is as if we could "First matter see [p. 143] undrest—

> He caught her naked all alone
> Ere yet one rag of form was on—" [67]

But in truth such is the germ cell out of which every individual is developed, the addition of sensation, complex structure—organs, instinct, mind, form—all these are watched from the embryo to the adult.

We cannot expect to find an adequate cause, as in the inorganic world, or to find the material & the vital, the inorganic & the spiritual, to admit of exact comparison.

Natural Selection is inadequate but the Variety-making Power may perhaps be adequate. It may be no secondary Cause but the Cause of Causes. If it be progression, if it can add mind to instinct, it is far beyond parallel in the inanimate world. Volcanoes, for example, if they had really first begun in the Tertiary epoch, as some have supposed, would not be a more exalted Cause, superadded to antecedent ones, but one equal in grade.

Creation [p. 144]

If the Variety-making Power be not the direct presence or intervention of the Cause of Causes itself, it is the nearest to it, of any of which Man is permitted to witness the effects.

In any case where the variety is an advance, not simply an adaptation of the offspring to new conditions which may be a move either forward or backward, to a higher & more complex, in place of a simpler & inferior grade of structure, but where it is a product to which nothing equal had antecedently been seen on earth, this is a manifestation of power of supernatural force & wisdom & design not easily reconcilable with our ordinary notions of secondary causes or of delegated powers. It may be a slight advance & require a multiplication [p. 145] of such units to effect a decidedly higher grade whether of organism or instinct or mental or moral attributes, still if it be an onward step it can only come from some force greater than that of Nature as it preexisted, some power external to Nature.

It is unnecessary to confound this with ordinary causes, because generation & Natural Selection & many other secondary causes may be part of the machinery by which it works. If you refer it to the First Cause (it is objected) why not do so to the development of a wise man out of a wayward child, why not attribute abortion, degraded beings, retrograde steps to the direct influence of the First Cause. Why are there failures, imperfections, monstrosities, bad constitutions. Are these direct interferences? Here we meet with the question of the existence of Evil, of the power, as the Manicheans have it, of another [p. 146] Cause capable of marring the creation. But they who believe in progressive development have all the same difficulties to contend with. They are common to the theory of the reality of species.

If there were only two races of Man & one is superior in intellect to the other & not contemporaneous in origin, then the argument for intervention of a First Cause, in creating the higher of the two races, is stronger than that which favours the creation of two species of the Horse, or feline or canine species.

If we could see a new plant created, a new satellite added to our terrestrial system, we could not be sure that we witnessed so great an act of power as by the birth of a human being of more genius than any one previously existing, or suppose that we witness a step in the process, to a being to whom time has no relation, the summing up of the units may imply an equal demonstration of power.

Hippopotamus

Never seen fossil, says von Baer, in a thesis written 30 yrs. ago,[68] in the plains of [p. 147] northern Germany. Falconer.

([K. E. von Baer] *De reliquiis animalium mammalium in Regno Borussico inventis*—Prussia) Col. Wood, Struthall, Swansea[69]

[Synonyms for the Pleistocene Bison]

Bos priscus [?]
Bison priscus [Owen]
Bos ferus Linn.
Bos priscus of Bojanus [Urus priscus, Bojanus]
Bos (Bison) priscus generally
Bison Europaeus—Owen
Aurochs [European wild ox, *Bos primigenius, Bojanus*]
Bos (urus) primigenius Goldfuss
Zubr [the wild bison of the Caucasus]
[p. 148]

Eleph. primigenius Bos Bison Bos priscus	found together in upper Val d'Arno newer than all the older mammalian format.[s]

Three hundred hippopot.[s] from Val d'Arno—a great lake.

The Grays' Thurrock Eleph.[t] & Rhinoceros (R. leptorhinus—syn. R. megarhinus Christol) have been found together with El. primigenius at Rome.

But qy. if mixture in the tufac[eou].[s] deposits (at Ponte Molle?)

Falconer admits that he once found a spec.[n] of E. meridionalis, a tooth which must have been older, also mixed with the above.

2 Glacial Periods

May there not, says Falconer, have been 2 periods of submergence?

The difficulty arises from our inability as geologists to learn from [p. 149] collectors the exact position of fossil bones. 2.[dly] the confusion of several species of Elephants & of Rhinoceros.

Sicilian Cave

The Cave of San Teodoro looking N.E. at Acqua Dolce at foot of San Fratello being 65 metres (or above 200 English ft.) above sea-level, 1041 m. from the sea & yet containing marine shells, deposit attests changes of level of land & sea. Therefore the presence of extinct mammalia, as well as recent, natural. The same change, to wh. the upheaval has been due, has caused sinking of land between Sicily & Africa, the land by which African Eleph.t (& hippopot.s?) crossed.

Yet 180 flint implements mixed with these remains of fossil animals which proves that man existed when geography of Europe was very different. Perhaps Etna only beginning then?

[p. 150] Two new grottoes or caves discov.d by Baron Anca in 1859.

One at Mondello at the northern extremity of Mont Gallo called Grotto Perciata because it is open on two sides—near Palermo, 2 leagues N.W. of the town. Mont Gallo is of hippurite limestone, *calcaire magnesien* of Hoffman[70], at its base is the Pliocene form.n & recent conglomerate. The cave 30 metres (100 ft.) above the sea-level & in one part 49 m.

Land shells & sea shells long known by naturalists to be in this grotto.

A	0.50	0–50 sandy land & sea shells
B	0–10	cindry (tufaceous, calcareous?) compact with land shells fragments of bone & flints.
C	0–40	Cervus, Sus, Equus, Lepus, Toad, Bird, Patella vulgata, Helix aspersa, Bulimus decollatus
D	0–50	Murex & Fusus in reddish loam

The cave of S. Teodoro, see p. 149, penetrates 70 metres into its interior of the mountain, e.g. is broad in the middle. The rock is hippurite [p. 151] limest.

S. Teodoro [Fossils] Named by Lartet

Spotted Hyaena
Bear like large one of Alps
Wolf—Fox—
Porcupine
Rabbit (lepus cuniculus)
Elephas antiquus
Elephas africanus?
Hippopotamus species
Sus scrofa
Equus asinus
Bos middle size & a smaller species
Toad large species
Bird undetermined
Coprolites of Hyaena 130 in number
Stone implements, phonolite & trachyte
Oysters
Cardium edule
Helix aspersa

Elephant Bed,[71] Gunn [p. 152]

For more than 40 miles along the coast runs the Eleph.[t] bed from Kessingland to Yarmouth several ft. thick—contains huge vertebra of whale & El. meridionalis.

But now at Bacton, El. indicus turns up.

Gunn thinks that as at Kew there is a pan of gravel in the river & wood in islands.

N.B. at Corton best section (between Lowestoft & Yarmouth or 3 m. from Gorleston Pier)—at Whiton, near Bacton, El. primigenius [is] 14 ft. from top.

[Page 153 is blank.]

June 26, 1860 [p. 154]

Dr. Falconer says there is no true Hippopot.[s] of the living section, Tetraprotodon, in the Miocene of the Sewalik, only in the Pliocene of India.

The two Merycopotamus of Sewalik [belong to the] older type. None in America except that—Leidy's comes so near that he thought of calling it Myricopotamus, H. v. Meyer. So that one might have imagined that the continent of rivers par excellence w.[d] have had hippopot.[i] in it, come down from Nebraska? times.

Oxford British Association Meeting

H. Rogers[72] says that Silliman, B. Jun.[r],[73] analyzed some corals brought from reefs by Dana & found carb.[e] of magnesia as well as carb.[e] [p. 155] of lime & from hence may be derived much of the dolomite instead of its being injected. It may, says Jukes,[74] be segregated, first having been disseminated thro' the mass, then attracted to points.

Dr. Wright,[75] Avicula contorta beds diff.[t] from—; C. Moore[76] says that the above agrees with the St. Cassian formation of the continent.

{ 14 genera of shells in Avicula contorta [bed]
fishes Hybodus acrobus, (Placodus rept.[d])
Squal[u]s—raia—
Avicula contorta bed, intermediate bed between Trias & Lias.

Estheria

Agassiz s.[d] that the small Acrodes of the bone bed & the Kupferschiessen Acrodes both belong to a different genus Lophodus.

Sauricths & Hybodes, says Mr. Brodie[77], [p. 156] common to bone bed & lias

Passage beds between both, say some—

Wright concludes that the Avicula contorta & bone bed are *trias*, not as he first thought lias.[78]

The shells are specifically distinct from the true lias, confirming Sir P. Egerton's original view—

[p. 157]

boulder clay
mytilus } F
cyclas

boulder clay
shell bed
Forest bed

[p. 158] [unlabeled incomplete diagram]

[p. 159] See Rose of Swaffham in *Geologist* of April, on divis.[n] of upper beds in Suff.[k] & Norf.[k] [79]

He does not [think that] limes.[t] [is as rich] in Ca. as upper boulder clay.

Oxford Elephants

El. primigenius alone found in gravel deep down in which the so-called Lickey quartz rock, thought by Phillips[80] to be the Trent valley quartz, abounds.

Above this elephant gravel is a newer one containing fluviatile shells—J. Phillips.

C. Moore[81] [p. 160]

In triassic drift, or materials chiefly triassic, but in part, 5 per cent? carbonif.[s] & oolitic, in a fissure of mountain limestone. Avicula contorta, Sauricthys Acrodes now called ——, Placodus, —— Hybodus, Thecodontosaurus, Microlestes & a mammal 6 times as large, 19 teeth of Microlestes, [Is Spalacotherium the larger mammal? Another mammal? 3 vertebrae.

Ludlow Elephant

In gravel at Woferton, 4 miles S. of Ludlow, Rob.[t] Lightbody[82] announces the finding of Eleph.[t] remains 4 ft. from surface of gravel, first discovery of such in those parts—tho' Mr. Knight[83] of Henley s.[d] to have found some foss. mammals on his property.

July 1, 1860

Zoological Gardens

The Capucin monkey of Brazil, Cebus apella, which used to break nuts with a stone smoothing away first from the floor any straw or sand which [p. 161] might be there, is dead. He used to put the nut on the ground & with one hand almost encircled it so that when struck it sh.[d] not fly off, also avoided crushing it, only breaking the shell. I have often remarked all this except the clearing away of the straw & dirt, but two keepers assure me he habitually did this also.

Leporines

Saw 2 supposed intermediates between Rabbit & Hare & the offspring of one with a she rabbit.

The period of gestation differs. The rabbits linger blind after birth 8 or 10 days instead of two? The hare has 1, 2 & 3, the rabbit 6, 8 & more in one litter. The hare does not burrow like the rabbit.

The English hare differs from the Irish. Both in the gardens. The Lepus variabilis of Scotland differs again. Perhaps not 3 true species.

July 1, 1860 [p. 162]

Fossil Elephant in Ireland

Jenks[84] tells me of a new or 3.[d] locality discovered in a railway cutting—where [they have found a fossil elephant].

Maltese Mole

. . . gigantic found in cave. Spratt.[85]

L.[d] Ducie[86] told me of Eleph.[t] & hippopot.[s] in cave in Malta.

Wealden

Phillips thinks no part of it cretaceous.

Transgressive stratification—same as overlapping? Phillips. [See Fig. 20.]

Fig. 20

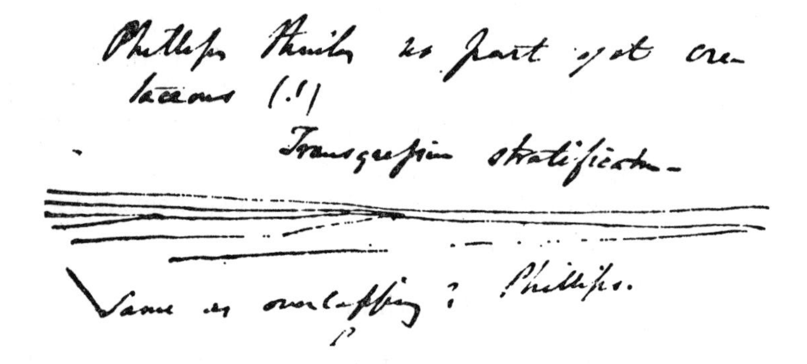

Saurians

Phillips thinks that not only Megalosaurus but Cetiosaurus were probably land quadrupeds.

Magnesia in Rocks

. . . may be derived from corals, for Dana brought from reefs corals in wh. [p. 163] B. Silliman found by analysis carb. of magnesia. Muriate of magnesia in sea water.

Elephas Antiquus

So much allied to E. Indicus or the living one that when a young tooth was brought by Baron Anca from Sicilian cave to Dr. Falconer he could not say which it was, but says he must have a rigid comparison. In the Sicilian caves E. antiquus & E. Africanus are now both found & with the latter is the living spotted hyaena. This goes well to confirm Lartet's views as to a modern fauna having peopled Europe from Africa.

Virgil

"At[que] genus immortale manet" *Georg.* Lib. 4. 1, 208.

Agassiz on Pallasian Theory

Agassiz believes in the amalgamation of several distinct species in our common fowls. *Classif.*[n] p. 55.[87]

August 28, 1860

Letter [from Lyell] to Darwin [written] from Rudolstadt[88] [p. 164]

The grand argument from absence of mammalia & batrachians in Oceanic islands is probably felt to be strong by Owen as he has not ventured to impugn it and therefore declined to bring it into notice. It is a very telling argument & I think just & true, tho' if I had been helping Owen & the Bishop[89] I fancy I could have suggested objections, but I have not thought them out yet, & have not therefore troubled you with them, thinking I sh.[d] be able to [p. 165] answer them myself.

I remember however it struck me that the putting it, even as possible, that the Galapagos or Cape de Verde could have been joined to the nearest Continent was dangerous, as in that case (p. 398)[90] they must have retained some of the smaller mammalia. Also if atolls be remnants of sunk continents, would not some of the old continental land quadrupeds have survived or have been slowly modified into amphibious & volant species.

Also, if, as I believe, the Canaries & Madeiras have been islands ever since the Miocene period (& the Azores), would not the theory of Nat. Selection have time to modify any vertebrate [p. 166] animal into the lowest of the Lyencephala in such a lapse of ages. Or if not, does it not give us a sort of scale for the vast geological time required for such a transmutation. No Galapagos saurian having risen to a mammal. Perhaps you will say that this would be a greater puzzle to the out-&-out progressionists who may be converted by you to transmutation. Bronn,[91] e.g., for you do not believe in that constant advance, as an ever working tendency, & allow for stationary & retrograde movements.

You would prefer, I conceive, to derive Eocene mammalia [p. 167] from the Microlestes of the Trias rather [than] from any reptilian of however high a grade, e.g., a deinosaurian. You w.[d] rather conceive a bird to be turned into a mammal than derive the latter from a reptile. The Microlestes, or its nearest living analogues (marsupials), w.[d] in its embryology go through the bird-state or likeness before it became even a non-placental mammifer?

But if an island be very ancient the bats would be modified into mammalia of other genera & orders long before any bird could be turned into a mammal. Here comes the question, how old is the oldest isolated Oceanic island? The miocene *littoral* shells of Grand Canary & of Porto Santo [p. 168] give a miocene date to those lands, already in all probability separated from the Continent as now. But there are no bats there.

New Zealand is so large it probably existed from Eocene times as land? ? Why did not bats getting there diverge into a dominant mammalian volant fauna & some lose their wings & become non-volant mammalia? Or why did not seals & marine mammalia turn terrestrial in their habits if they had such a geological period as the post-miocene for nat.[l] selection to work in. The seals swarmed in Madeira before Man drove them away. This difficulty strikes me as greater inasmuch as the rate of change in mammalia is more rapid than in [p. 169] inferior grades & there have been in Europe several changes of mammalia since the Upper Miocene period. The adaptation of a seal, which is already half terrestrial, requires less time than the conversion of Hearn's bear into a whale which they are never tired of quizzing you for.[92] I shall be glad if you will tell me whether these objections are of any value. Probably you have not thought the evidence of the antiquity of any islands good for much. Yet it seems necessary to grant a good deal of time to cause so many endemic shells in Madeira & Porto Santo & land shells would change much more slowly than mammalia. Still I grant that the conversion of a Phoca into a terrestrial quaruped is very different from one [p. 170] Helix into another, but ought not the seals to have begun in ancient islands to keep somewhat longer out of water. These are perhaps unfair, imaginary difficulties & ought not to weigh against an hypothesis which explains so much as does transmutation in the case of absence of mammalia & frogs from remote islands while they swarm in islands which we know have been united in newer pliocene times.

If there was time in Galapagos to convert a marine Amblyrhincus into a land species or vice versa, why not a Phoca into a land species or other Cetaceae into terrestrial. Did the land [p. 171] Amblyrhincus come first into being & of what S. American Saurian was he or his aquatic congener a modification? If all these questions apply equally against independent acts of Creation they will not be very damaging. But when the free agency of the First Cause is allowed to come into play, it may be said that no mammalia were wanted because it was foreseen that Man would come with his attendant train of useful mammals, taking even the camel with success into the Canaries.

St. Acheul, Amiens, Letter to Horner, 1860 [p. 172]

Alphonse Faure[93] says he found in the white sand above the gravel with hatchets a grade of crystals of hyaline quartz cemented together & posteriorly to the hatchets.

Also crystals of quartz adhering to a hatchet of S.[t] Acheul—July 1860.

Extracts from "The Christian Doctrine of Prayer," an Essay by James Freeman Clarke, Boston 1859 [94]

p. 88 [89] "According to them [men of science] God now does everything in accordance with law, & nothing, in any strict sense, as a free Being. Or, in other words, they view the operation of the Divine laws in such a way as to exclude the Divine freedom." [95]

Note [p.] 89. Mr. Combe[96] in his Constitut.ⁿ of Man ". . . excludes all that we have called Special or particular Providence" [97]

p. 92. ". . . the errors of almost every theory or system consist rather in what [p. 173] it rejects than in what it asserts, rather in its negations than in its positions." [98]

[p.] 93. "Is there any reason why these views sh.ᵈ not be united? Are they necessarily inconsistent with each other?" [99] i.e. God working by general laws for the good of the whole & sometimes sacrificing individuals for the whole as Combe says, etc. etc.

[p.] 94 Plutarch[100]

[p.] 95. ". . . we maintain that we may believe in God acting through natural laws for the general good of races, & also in God acting supernaturally in the sphere of Freedom for the special needs of individuals. When it is objected, that all such supernatural action implies a miracle, & that the days of miracles have ceased, we must ask in turn for the true meaning of a miracle. If a miracle means a violation or suspension of the laws of nature, then there w.ᵈ be a real contradiction between [p. 174] the Supernatural & the Natural, & belief in the one w.ᵈ so far nullify belief in the other. But if, as the better theologians of all schools maintain, a miracle is no violation of a law of nature, but the coming in of a new force from a higher sphere, which, while the old force, or law, works on, controls it according to the special need, then there is no such contradiction. We may believe at the same time in God's natural & supernatural action. We may believe in general and special providence. We may believe in the natural laws & also in the answer to Prayer. For, according to this view, miracles have not ceased, & never will cease until the God of Christianity abandons the world, & until living faith is no more found on [p. 175] the earth." [101]

[p.] 96. "If we suppose, accordingly, that God steadily maintains the order of the universe & the laws of nature, but that besides this he continually sends new & special influences into the world of matter & of mind to meet the rising exigencies of the hour & that this is no afterthought, but part of the great plan of the universe from the beginning, the conflict between the Natural & the Supernatural falls to the ground. That God does influence the world from a realm of freedom by ever-new creative activity, no less from a realm of law, is not only asserted by Christian faith & needed by the human heart, but is demanded by the deepest philosophy." [102]

[p.] 88. "This metaphysical objection to prayer is an objection lying against free will [p. 176] altogether; & by whatever argument we defend the freedom of the will, by the same argument we may defend Prayer, so far as this objection is concerned." [103]

[p] 97. [98] "For human freedom is in the strictest sense a force wh.[ch] acts within nature, but from above nature. It is surrounded by laws, & limited externally by the laws of organization & circumstances, but it cannot itself be brought under law. Every act of freedom is a new creation & wholly inexplicable. The moment you explain it as resulting from any thing already in existence, you deny Freedom, & introduce Necessity. The moment you make outward motives to be the cause of our actions & not merely the occasions, you deny, so far, human freedom. If this view be true, then every human being is himself an illustrat.[n] of the coexistence & harmony of the Natural & the Supernatural. Part of his life is [p. 177] natural, resulting from organic tendencies, determin.[ed] by external motives, & another part is supernatural, the reaction of the free will & the power of choice. If man, therefore, himself can act in the world at the same time in a sphere of freedom & of Law, shall we deny a like capacity to God, & limit his activity to the support of existing laws? Much rather must the adequate view of the Deity suppose in Him the perfect harmony & absolute synthesis of law & freedom,—an infinite, creative activity, for ever combined with an unchanging support of the never-failing laws of the universe.

"But if, [after all], we cannot fathom the depth of this mystery, . . ." [104]

[p.] 251. "They told pleasant truths in a pleasant way, & of unpleasant ones they said nothing. These men were never found with the minority, struggling to open men's eyes to prevailing abuses; always in the self-complacent [p. 178] majority, willing to let everything remain as it was."

". . . never to win popular applause or fame." [105]

". . . religious instincts" lead to prayer.

[p.] 219 ". . . nebulae can be seen by L.[d] Rosse's telescope,[106] the light of wh. has been millions of years in travelling the depths of space,—consequently nebulae which we see as they were millions of years ago." [107]

Beyond what we know with certainty about the soul comes a region of belief, & further out still a region of opinion, & still further one of speculation.

[p.] 235. ". . . evil tends to strengthen & educate the soul." [108]

[p.] 261 "He does not see God in nature, he only sees a dead machine, ingeniously put together long ago, & left by its maker to grind out what results it may with its iron mechanism. He sees no God in events, but only a hard necessity . . ." [109]

Man—Material Link [p. 179]

Sedgwick has compared the human mind not to a sheet of white paper, as Locke said, but to a sheet prepared chemically to receive photographic impressions. But the difference between the mode in which it is prep.^{d} for 2 individuals, even men of the same race, may be such that it may be more unlike in degree of int.[elligence], in kind of intellect, moral tendencies, spiritual aspirations, constit.^{n}—these two allied species of the same genus. Yet the link which connects the two individ.^{s} may be that of normal generation in the same race. It may not be very distant even in time & time may produce greater contrasts.

Coburg, September 8, 1860

C. Lyell to C. Darwin

Your argument from absence of mammalia in islands, excepting those which we know to have been joined to main-land in Pliocene times, is a favourite one of mine when I wish [p. 180] to prove that y[ou]r bond of descent is required for the coming in of new species. The older the island, the stronger y[ou]r case against special creation without a closely allied antetype, as Wallace would say.

The land shells of Madeira and the Canaries are so endemic as to show that none were united with each other or Africa. I believe all were separate from Upper Miocene times; if not, small African species of quadrupeds would have been retained.

But the antiquity of islands void of mammals checks the hypothesis of the easy adaptability of one species of a genus [p. 181] to new conditions. On a new volcanic island before all the best places are seized upon, seals or walruses, manatees, dolphins & other Cetacea, when pressed hard for food, w.[d] go up the rivers or if amphibious devour the eggs or young of land-birds. Bats, though not primates, but here Lissencephala ought to get the better in some stations of marine birds if higher classes tend to become dominant.

The long reign of the Dinornis family in New Zealand implies that for ages there had been room for mammalia if bats & rodentia & aquatic placentals being at hand had been convertible even into allied structures fitted for land habits. And why bats [p. 182] & rodents should have peopled all Australia without having been developed into something higher in the placental line, seeing that the representatives of the latter [which] were imported from Europe, can run wild there, is strange, for why sh.[d] it take more time for a bat to work up into a lemur than for a Myrmecobius to improve into a Thylacinus. As to the land shells of Madeira & Porto Santo, 95 per cent of them have remained absolutely unchanged for a period sufficiently long to allow the form & size of the islands to [p. 183] be altered materially by the waves of the Atlantic, long enough for some species which [were] once very common to become almost extinct—in a few [instances] quite lost. The only speculation I have been able to make is this that by far the majority, 9/10.[ths] perhaps, are immutable &, when changes come, must die rather than yield. The hereditary power has got too strong a set, after a million of years, in one way. It cannot go in any other. But there are some eminently metamorphic species in the Madeiras & these are producing the allied species & sub-species of Lowe in each island & after a long period there might be found, in spite of extinctions, [p. 184] just as many true species of Helices in the Archipelago as now, supposing Man would let it alone. So I suspect it has always been. Most species immutable & true to the death, as I maintained in the *Principles* that all were, but a few of the whole [are] plastic & becoming the centers of new genera & ultimately higher groups as you maintain. So it seems to me to hold good in the Tertiary series. Some shells, once abundant in the Mediterranean formations & in the Crag, still survive unaltered though excessively persecuted & rare. [p. 185] Your antagonists are driven to great straits when they try to set up any machinery other than that of normal generation. The greater we can show the role of extinction, the worse is their case because the reactivating power ought to keep pace, if as Bronn pretends the creation is always growing richer & more varied, which however I doubt, since Eocene days.

According to the progressive theory, why sh.ᵈ there be a living platypus or ornithorhynchus, for we have nothing synthetical or elementary in the coal strata, & if they sh.ᵈ be found there, how have they escaped [p. 186] being altered, improved, and specialized in 30 periods.

Ammonites in uppermost part of the Maestricht, of a cretaceous type, if not, species of Texas are against the doctrine of wide gaps between Maestricht & Thanet sands.

What you say of the power of preoccupancy is good—so, as to Atolls, but if they subsided very slowly, the absence of volant & amphibious forms of reptile & mammifer in such a region proves rate of transmutation slow, even as compared to revolutions in physical geography.

Bonn September 18, 1860

Copy of Letter, [C. Lyell] to C. Darwin [p. 187]

I found your letter here & thank you for intending to send me Asa Gray's last article which I will return without delay, as also the M.S. on dogs. Agassiz in some one of his Essays leans to Pallas' views, which always struck me as very rash in him who is for the sanctity of species, whereas if any thing could show Nature's entire indifference in regard to the violation of such supposed sanctity, it would be that those now existing should be of multiple origin.

You need not be afraid of my starting any theory of successive creation of types. I have never had any leaning towards a compromise [p. 188] and thought Carpenter[110] very inconsistent in going so very far with you & then concluding that there was probably in the case of Man an exception, whereas there would be an end of all the morphol.[1] & embryological arguments, if such a concession was made.

But what I asked, or intended to ask, in my last letter was different—perhaps you meant to reply that you had no fixed opinion on the subject—but I wish to be sure. Can we assume as at all probable that all mammalia came from one original stock instead of several distinct mammalian types, each developed by small & successive [p. 189] modifications out of lower ornithic, reptilian, or perhaps monotrematous prototypes. It would greatly simplify matters if single & exclusive areas could be assigned or even speculated upon, on some even slight data, or if single periods could be proposed as those of the first coming in of mammalia. But as I understand your views this is not very probable. If for example we found before the Triassic Microlestes, some monotremes in Permian or Carbonif.[s] beds & fixed on them as the probable starting point of mammalia, what influence would the development of a mammal in Asia or Africa have in preventing [p. 190] some other line of monotremes from improving into a totally different genus of mammal in Australia or America & supposing the mammal first formed, say in Asia, to be extinguished, what is there in your system to cause us to despair of the higher or placental grade from ever being evolved, because after a geological period the earlier formed mammalia died out.

I have always expected to find mammals intermediate between Stonesfield & Eocene, but I also looked for them before Stonesfield. When therefore they turned up both in Trias [p. 191] & in Purbeck, it agreed with my anticipations. The Triassic surprised some people more than the Purbeck, for they held the doctrine, propounded by you in p. 316, once begun the group is continued till it dies out.[111]

I wish you could give the slightest reason why it should not begin more than once in more than one place. I incline to think it has not, but why? According to the principle of selection, why when once in any one quarter of the globe, at any one period, the step in advance has been taken, are the inferior types elsewhere to be checked, & not to presume to work up into any Gnus of corresponding grade & class? A frog has a [p. 192] vast deal of one type in him, even the double occipital condyles, to say nothing of his limbs & digits & the absence of a tail, tho' he is low in the reptilian scale, (but it is true he does not as yet go far back in time), but take some Triassic reptile or bird, I cannot see why they are never to be modified into something higher, because a Microlestes has entered on the stage. Two independent chains of development would not be discordant with your machinery of selection?

I am not aware of any fossil bats or rodents in Australian caves, so the small antiquity of rodentia may be a reason for their not having given origin [p. 193] as yet to Gyrencephala & does there not seem some connection between the low grade of Lyssencephala as the only occupants of that vast Australian continent.

Can you not refer me to some book or paper where I can find how many plants may have been extirpated in St. Helena & may I not presume that as many or more endemic insects have been annihilated there besides some land-shells?

As to Australia, I cannot help thinking that the Marsupials must in the long run (if Man were away) be best fitted for that continent & that after European or Asiatic Gyrencephala had been introduced & then left to themselves, they w.[d] during some 4 years drought, after a century or much more, have all perished & given way before their pouched rivals, who as Owen has somewhere suggested, w.[d] have [p. 194] migrated with their young in the pouch wholly indifferent to the long leaps of the mother. This w.[d] not apply to bats, but perhaps you may be able to object that some of the native placental rodents be no better off than a lion or an elephant in a 4 years' drought?

In your chapter on grafting you might, it struck me, have alluded to those wonderful cases of nat.[l] grafting recorded in the fir-woods of Germany. I think I could find a reference to it, where a totally leafless stump keeps alive by its roots, borrowing from the roots of adjoining fir-trees till the bark closes over. It answers the question why Nature has given such a power & was it for the mere sake of human gardeners?

I am not going to press the [p. 195a] objection about the non-variation, adaptation & advance of bats & rodents, so as to destroy the weight of y.[r] argument respecting absence of other mammalia in islands, as I cannot conceive such absence not being due to the want of migratory powers of mammalia, & that alone w.[d] not do, unless the coming in of new mammals was by inheritance from pre-existing types. But it is better to show that one does not overlook objections.

I think I asked you once before whether the admission of the multiple origin of dogs w.[d] not impair one's faith in the single origin of the human race, & as you did not reply, you perhaps meant that [p. 195b] you had nothing to hazard in the way of speculation. You remember a passage in Agassiz about negros being black where, the Chimpanzee is black & yellow in Borneo? where the orang is yellow, or something to that effect. I could hunt out the passage if you do not remember it. As he does not connect Man genealogically with any anthropomorph, he says this without any hesitation.

To shorten correspondence I will imagine the best answer I can for you & beg you to try & strengthen it.

It is so many millions to one against the right variation offering itself in the [p. 195c] right place & time that many geol.[1] periods must pass away before the first mammal is formed & as an equal lapse of ages w.[d] occur before a 2.[d] offered, the first will in the meantime have spread over the globe & branched out into many genera, wh. being of high grade w.[d] keep down any incipient mammals, just as *homo sapiens* when once evolved w.[d] take good care that no other anthropomorphs sh.[d] improve into rationalism & compete with him for the good things of this world. No island has existed for so long a time as that which separated the Lower Miocene [p. 195d] from the human epoch, therefore the chances have been far too many against transmutation of any creature of an inferior class into one of a higher class on an island, more especially as bats & rodents w.[d] help to check any signs of an invasion of the prerogatives of the dominant class, they being already there to represent the vertebrate aristocracy.

INDEX [p. 196]

1. Amiens flints section ideal.

2. Binney, marine drift.
Symonds, Cropthorne.

3. Amiens, Prestwich, on rolled worked flints.
—Cyrena of Abbeville of human period?

4. Species, their origin by blood relationship, not to be mixed with question as to one original type, or 2.[dly], the Nature of the Creative power.

6. & 7. The subject in *Principles*, how treated—extinction & reproductive power always active.

8. *Origin of Man* if admitted by intervention of First Cause draws with it creation of other spec.[s].

9. Grandest creative acts last.

10. I adopted limited, Darwin unlimited, variability.

11. Independent phenomena explained by last.

" Alternate generation will not help.

" Simple organisms, why abounding now.

" Progress, arguments in favour of.

13. Transmutation—man of superior powers born of ordinary parents (p. 81).

18. Deification of Matter. Creation by a power external to Nature.

19. Language may be historically & continuously connected, yet distinct.

" Free agency of Man—Originating Power.

[p. 197]

24. First Cause, self-governed by Laws, yet free.

25. Brachiopoda, Davidson's Tables & results.

26. Creation, imperfection to be admitted.

27. Negro & White, common parentage of Races.

28. Question of Descent, more modest one.

32. Species, Nature made him & broke the die.

33. If we could observe long eno', many problems of races w.[d] be solved.

34. Life defined as power to resist laws of dead matter.

35. Northern drift older than Welsh cave animals?
37. Microlestes
" Bossuet. The nearer apes approach man the better.
38. Accommodation of Creative power to laws previously established.
47. Prophet supposed to have originated not in ordinary way.
50. Increasing intensity of miracle working power.
55. Languages, divergence of types, & origin.
63. Deification, secondary causes.
65. This mystery as great as ever even if descent unbroken be proved between species.
67. Man must be included if Darwin right.
73.} 103.} Falconer on Meles taxus contemporary with E. antiquus.
128. Falconer on Rhinoceros & p. 132.
135. " on change of mammalia with phases of glacial period.
136. Wolf & badger as old as E. antiquus & Rh. hemitoechus.

Notes

1. Joseph Prestwich (1812–96), geologist.

2. Edmond Hébert, "Sur les poches diluviennes à ossements fossiles d'Auvers (Seine et Oise)," *Bull. Soc. Geol. Paris,* 1848–49, *6,* 604–05. Edmond Hébert (1812–90), French geologist, in 1857 was appointed professor of geology at the Sorbonne.

3. Edward W. Binney, "Sketch of the Drift Deposits of Manchester and its neighbourhood," *Mem. Manchester Phil. Soc.,* 1848, *8,* 195–234.

Edward William Binney (1812–81), lawyer and geologist, in 1838 had helped to found the Manchester Geological Society. In 1853, he was elected a fellow of the Geological Society of London and, in 1856, a fellow of the Royal Society.

4. William S. Symonds, "On some phenomena connected with the drifts of the Severn, Avon, Wye and Usk," *Edinburgh New Phil. J.,* 1861, *14,* 281–86.

William Samuel Symonds (1818–87) in 1845, was appointed rector of Pendock, Worcestershire, where he pursued geology and archaeology.

5. Hugh E. Strickland, "An account of land and freshwater shells found associated with the bones of land quadrupeds beneath diluvial gravel at Cropthorne in Worcestershire," *Proc. Geol. Soc. London,* 1833–38, *2,* 111–12.

Hugh Edwin Strickland (1811–53), geologist, had studied geology under the Rev. William Buckland at Oxford University. After a period of travel to Asia Minor, Italy, and Switzerland, he assisted Sir Roderick Murchison with his study of Silurian formations.

6. Charles Lyell, *Principles of Geology,* 3 vols., London, Murray, 1830–33, vol. 2, 1832, chs. 8–11, pp. 123–84; Charles Lyell, *Principles of Geology,* 9th ed., London, Murray, 1853, chs. 141–42, pp. 677–700.

7. Lyell, *Principles,* 1st ed., 1830–33, *2,* 183–84.

8. Dominique A. Godron, "De l'espèce et des races dans les

êtres organisés du monde," *Mem. Soc. Sci. Nancy,* 1847, pp. 182, 239–88. Dominique Alexandre Godron (1807–80), botanist.

9. Blaise Pascal (1623–62), French mathematician, scientist and religious philosopher.

10. Thomas Davidson (1817–85), *A Monograph of the British Fossil Brachiopoda,* vol. 1. *Tertiary, Cretaceous, Oolitic and Liassic Species . . . ,* London, Palaeontographical Society, 1851–55.

11. Charles Lyell, *A Manual of Elementary Geology,* 5th ed., London, Murray, 1855, p. 99:

> It appears, that from the remotest periods there has been ever a coming in of new organic forms, and an extinction of those which pre-existed on the earth; some species having endured for a longer, others for a shorter time; while none have ever reappeared after once dying out. The law which has governed the creation and extinction of species seems to be expressed in the verse of the poet,—
>
> Natura il fece, e poi ruppe la stampa. Ariosto.
> Nature made him and then broke the die.

12. Hippolyte Cloquet (1787–1840). Cloquet defined life in a paraphrase of Bichat's famous statement that life is the sum of the forces which resist death. He wrote that what characterizes life is "la faculté de resister jusqu'à un certain point aux lois générales de la nature." See Hippolyte Cloquet, *Traité complet de l'anatomie de l'homme comparée dans ses points les plus importants à celle des animaux, et considérée sous le double rapport de l'histologie et de la morphologie,* Paris, Dondey-Dupré & Baillière, 1825–27, 216 pp.

13. Lyell had found the reference to Cloquet in Isidore Geoffroy Saint-Hilaire, *Histoire naturelle générale des regnes organiques, principalement étudié chez l'homme et les animaux,* 3 vols., Paris, Victor Masson, 1854–62, *2,* 101.

14. George Louis Duvernoy (1777–85), French anatomist and zoologist. Lyell is referring to a quotation from Duvernoy's article, "Vie," in *Dictionnaire des Sciences Naturelles,* 60 vols. & 12 vol. atlas, Paris, Le Normant, 1816–30.

15. Pierre Jean Marie Flourens (1794–1867), French physiologist and professor at the Collège de France. Lyell is referring to the following passage:

> Depuis qu'il y a des physiologistes qui écrivent, il y a des physi-

ologistes qui cherchent à définir la vie. Quelqu'un d'entre eux y a-t-il jamais réussi? . . . Il faut dire de la vie et de toutes les forces de la vie ce que la Fontaine a dit de *l'impression:*

L'impression se fait: le moyen, je l'ignore;
On ne l'apprend qu'au sein de la divinité

Pierre J. M. Flourens, *De la longévité humaine et de la quantité de vie sur le globe,* Paris, Garnier, 1854, p. 187, quoted in I. Geoffroy Saint-Hilaire, 1854–62, *2,* 102.

16. David Milne, afterward Milne-Home (1805–90).

17. Samuel Peickworth Woodward (1821–65) was assistant in the department of geology and mineralogy at the British Museum and a leading authority on Mollusca.

18. Charles Moore (1815–81), geologist, was a bookseller at Ilminster, Somersetshire, until 1853, when he gave up business, moved to Bath, and devoted himself to geology. In 1861 he discovered the English formation corresponding to the Rhaetic beds on the continent.

19. Hugh Falconer (1808–65), palaeontologist and botanist, was in India from 1830 to 1842 and again from 1847 to 1855. He described the vertebrate fossils of the Sivalik hills. In 1860 he showed that Elephas antiquus and Rhinoceros hemitoechus had formed part of the cave fauna of England and that the Bovey Tracey lignite was of Miocene age.

20. Charles Lyell, *Supplement to the fifth edition of a Manual of Elementary Geology,* 2d ed., London, Murray, 1857, pp. 17–18. Lyell discusses the mammalian fossils recently found in the Purbeck beds and refers to woodcuts of Microlestes teeth in Lyell, *Manual,* 1855, p. 343.

21. This may be a reference to transformist views expressed in one of Isidore Geoffroy Saint-Hilaire's numerous memoirs on the classification of mammals and on fossil man. Richard Owen would be almost certain to disagree with him.

22. Jacques Bénigne Bossuet (1627–1704), Bishop of Meaux. We have not been able to identify the source of the reference in Bossuet's writings.

23. In a discussion of definitions of life, Isidore Geoffroy Saint-Hilaire wrote that contemporary definitions were but modifications of that given by Georg Ernst Stahl a century and a half earlier: "The preservation of the corruptible mixture of which our body

has been formed, that is life." He attributes to the followers of Georges Cuvier the following version: "La vie est la faculté de résister aux lois générales de la nature." He quotes Cuvier in a footnote to *Histoire naturelle générale,* 1854–62, 73–74:

> Notre propre corps, et plusieurs autres qui ont avec lui des rapports de forme et de structure plus ou moins marqués paraissent résister pendant un certain temps, aux lois qui gouvernent les corps bruts, et même agir sur tout ce qui les environne d'une manière entièrement contraire à ces lois, nous employons le nom de vie et de force vitale pour désigner ces exceptions, au moins apparentes aux lois générales [Georges Cuvier, *l'Anatomie comparée,* 1800, *1,* 1–2].

24. Robert Alfred Cloyne Godwin-Austen (1808–84), geologist and country gentleman, had, like Lyell, become interested in geology while a student at Oxford (B.A., 1830), where he came under the influence of the Rev. William Buckland.

25. The Rev. Samuel William King (1821–68).

26. Andrew Murray (1812–78), naturalist, was educated for the law and practiced law at Edinburgh, but his interest in natural history gradually prevailed. In 1857, he was appointed professor of natural science in New College, Edinburgh, and in 1860, gave up the legal profession and moved to London.

27. Lyell is referring to an example cited by Darwin in the *Origin of Species,* London, Murray, 1859, which suggested to him the power of natural selection to change the characteristics of a species. Darwin wrote, p. 184:

> In North America the black bear was seen by Hearne swimming for hours with widely open mouth, thus catching, like a whale, insects in the water. Even in so extreme a case as this, if the supply of insects were constant, and if better adapted competitors did not already exist in the country, I can see no difficulty in a race of bears being rendered by natural selection, more and more aquatic in their structure and habits, with larger and larger mouths, till a creature was produced as monstrous as a whale.

The example provoked so much ridicule from Darwin's critics that he omitted it from the second printing of the first edition.

28. Robert Mallet (1810–81), civil engineer and scientist, was sent by the Royal Society of London to Italy to study the effects of the earthquake at Naples, December 16, 1857. His report, which Lyell examined in manuscript, was later published in part. See

Robert Mallet, "Report to the Royal Society of the Expedition into the Kingdom of Naples to investigate the circumstances of the Earthquake of the 16th December 1857," *Proc. Roy. Soc. London,* 1859–60, *10,* 486–94.

29. Guglielmo Guiscardi (1821–85), geologist of Naples.

30. We have not been able to identify Professor Capacci. He was presumably a professor at the University of Naples.

31. This appears to be a reference to appendices to Mallet's manuscript report.

32. We have not been able to identify Baron Anca further. He published a paper on Sicilian caves. Baron F. Anca, "Note sur deux nouvelles grottes ossifères découvertes en Sicile en 1859," *Bull. Geol. Soc. France,* 1859–60, *17,* 680–95.

33. Rudolf von Benningsen-Förder was a German naturalist whom Lyell probably had met during his visit to Berlin in March 1855.

34. Rudolf von Benningsen-Förder, "Beitrage zur Niveaubestimmung der drei nordischen Diluvialmeere," *Deutsch. Geol. Gesell. Zeitschr.,* 1857, *9,* 457–63.

35. Fleming lists Productus costatus Sow., Productus Flemingii Sow., and Orthis crenistria Phill. as "well known Carboniferous forms." Andrew Fleming, "On the Salt Range of the Punjaub," *Quart. J. Geol. Soc.,* 1853, *9,* 189–200, p. 190.
Dr. Andrew Fleming was the son of the Rev. Dr. John Fleming (1785–1857), Scottish naturalist, and went to India as a military surgeon.

36. Philippe Eduard Poulletier de Verneuil (1805–73), French palaeontologist.

37. "Palaeontology," *The Times* (London), May 19, 1860, p. 11:

> Professor Owen has lately concluded a course of lectures on Palaeontology at the Government School of Mines, Jermyn Street. The chief subject of this discourse was the extinct quadrupeds whose remains have been recently discovered in the caverns of Australia, and in the auriferous and other tertiary deposits of that country.

38. The genera listed are South American forms. Owen compared the resemblance between the gigantic fossil and smaller living representatives of genera peculiar to South America with the re-

semblance between the fossil and living species peculiar to Australia.

39. We have been unable to trace the source of this quotation.

40. Hugh Miller (1802–56), Scottish geologist and author.

41. John Gwyn Jeffreys (1809–55), conchologist.

42. The Rev. William Buckland found the following species of shells in the Paviland cave: Buccinum undatum, Turbo littoreus, Patella vulgata, Trochus crassus, Nerita littoralis. He wrote: "These are all common on the adjacent shore, and the animals that inhabit them are all eatable." See William Buckland, *Reliquiae Diluvianae; or observations on the organic remains contained in caves, fissures, and diluvial gravel, and on other geological phenomena attesting the action of an universal deluge,* London, Murray, 1823, 7, 303 pp., p. 85.

43. The Rev. John Gunn (1801–90), rector of Irstead and vicar of Barton, was born at the village of Irstead on Barton Broad, Norfolk, the son of the Rev. William Gunn (1750–1841), literary scholar and historian. He was educated at the Norwich Grammar School, where his schoolmates included George Borrow, the author, John Lindley, the botanist, and James Brooke, later Rajah of Borneo, and at Exeter College, Oxford where he was graduated in 1824. In 1829 he was ordained and appointed to the livings of Irstead and Barton and in 1830 he married Harriette Turner, a daughter of Dawson Turner, the banker and antiquary of Yarmouth. Throughout his life, Gunn was an enthusiastic geologist and collected especially the fossil bones of the Forest Bed of Cromer and the Norwich Crag. See John Gunn, *Memorials of John Gunn . . . being some account of the Cromer Forest Bed and its fossil Mammalia . . . from the M.S. notes of the late John Gunn, with a memoir of the author.* Ed. Horace B. Woodward and E. T. Newton, Norwich, Nudd, 1891, 7, 120 pp.

44. Sir Roderick I. Murchison (1792–1871), geologist, described the Silurian rocks of Great Britain. From 1855 to 1871 he was director-general of the Geological Survey of Great Britain.

45. Archibald Geikie (1835–1924), geologist, in 1855 had joined the Scottish branch of the Geological Survey of Great Britain. He began surveys in Haddington, Midlothian, and Fife and was led to study the volcanic history of Great Britain during the Tertiary period.

46. The Rev. Samuel William King (1821–68).

47. Jules Pierre François Stanislas Desnoyers (1800–87), "Sur les empreintes de pas animaux dans le gypse des environs de Paris, et particulièrement de la vallée de Monmorency," *Bull. Soc. Geol. Paris,* 1858–59, *16,* 936–44. Desnoyers described the footprints of many animals whose fossil remains had not been discovered or described.

48. This was probably John Gwyn Jeffreys (1809–85).

49. Joseph George Cumming (1812–68), geologist, was appointed warden and professor of classical literature and geology at Queen's College, Birmingham, in 1858.

50. John Morris (1810–86), English geologist, was professor of geology at University College London, London.

51. Thomas H. Huxley, "Art. VIII, Darwin on the Origin of Species," *West. Rev.* (Amer. ed.), 1860, *73,* 295–310, p. 297: "We know that the phenomena of vitality are not something apart from other physical phenomena, but one with them; and matter and force are the two names of the one artist who fashions the living as well as the lifeless."

52. William Paley (1743–1805), Archdeacon of Carlisle, wrote the *Evidences of Christianity* and was the most influential writer, in the English language, on natural theology, or the attempt to adduce evidence for the existence of God from the presence of design in nature.

53. Of the hare and rabbit, George H. Lewes, *Studies in Animal Life,* New York, Harper, 1860, p. 121, wrote: "Nevertheless between species so distinct as these, a new hybrid race has been reared by M. Rony of Angoulême who each year sends to market upward of a thousand of his *Leporides* as he calls them. His object was primarily commercial, not scientific."

George Henry Lewes (1817–78), English philosopher, naturalist, and literary critic, was the husband of Marian Evans (George Eliot).

54. Jean Baptiste A. L. L. Élie de Beaumont, "Sur les preuves de la grande étendue qu'ont embrassée les courants diluviens," *Bull. Soc. Phil. Paris,* 1843, pp. 105–9, p. 109.

Jean Baptiste Armand Louis Leonce Élie de Beaumont (1798–1874), French geologist, was professor of geology in the Collège de France. He was best known for his catastrophic theory of the simultaneous elevation of mountain chains.

55. Of the inhabitants of Britain, Julius Caesar wrote: "Leporem et gallinam, et anserem gustare, fas non putant," *Commentariorum de bello Gallico,* bk. *5,* ch. 12.

56. Colonel Richard (later Sir Richard) Strachey (1817–1908), in 1836, went to India as a second lieutenant in the Bombay engineers. In 1839, he was transferred to the Bengal engineers and, in 1843, was appointed executive engineer on the Ganges Canal. In 1848 he traveled through the Himalayas into Tibet. In the passes entering Tibet, he observed a great series of Palaeozoic beds with Jurassic and Tertiary beds overlying them. He later wrote on the geography and geology of the Himalayan region. He was promoted Lieutenant Colonel on July 2, 1860.

57. Ferdinand P. W. von Richthofen, *Geognostische Beschreibung der umgegend von Predazzo, Sanct Cassian und der Seisser Alpe in Sug-Tyrol,* Gotha, Perthes, 1860, 327 pp.
Baron Ferdinand Paul Wilhelm von Richthofen (1833–1905) was a German geographer and geologist.

58. The Rev. John Gunn (1801–90). See note 43, above.

59. Guiseppe Cortesi, *Sugli scheletri d'un Rhinoceronte Africano ed una Belena ed altre ossa di grandi Quadrupedi e Cetacei disotterate ne' Colli Piacentini,* Milan, 1808, 26 pp.

60. We have not been able to identify Much Sari, which was evidently a village in England.

61. Hugh Falconer, "On the species of Mastodon and Elephant occurring in the Fossil state in Great Britain. Part I. Mastodon," *Quart. J. Geol. Soc. London,* 1857, *13,* 307–60, p. 308:

> But it is deserving of remark, that the fossil genera and species which are in the most unsatisfactory and unsettled state, as to definition and nomenclature, are not those that are the rarest, but often the reverse. Take *Mastodon* or *Rhinoceros* for example, in which the array and confusion of specific names are signally perplexing.

Lyell's reference is wrong.

62. Ibid.

63. We have not been able to identify Captain Thomas Brinelwy.

64. Samuel J. Mackie, "Geological Localities.—No. I. Folkestone," *Geol.,* 1860, *3,* 41–45, 81–90, 121–31, 201–07, 281–84, 321–27, 353–57, 393–96.
Samuel Joseph Mackie was editor of *The Geologist.* Mackie

argued that the Weald had been denuded by tides at a time when the area between the north and south Downs of Kent and Sussex had been a funnel-shaped arm of the sea terminating the English Channel.

65. "Tabular view of equivalent portions of the Upper Tertiary series of accumulations," in R. A. C. Austen, "On the Superficial Accumulations of the Coasts of the English Channel and the changes they indicate," *Quart. J. Geol. Soc.*, 1851, *7*, 118–36; table opposite p. 136.

66. Charles T. Gaudin, "Modifications apportées par M. Falconer à la faune du val d'Arno," *Bull. Soc. Vaud. Lausanne*, 1859, *6*, 130–31. Charles Théophile Gaudin (1822–66) was a Swiss geologist.

67. We have not been able to identify this quotation.

68. Lyell's citation seems to have been written from a rather hazy memory. The work he had in mind was Karl Ernst von Baer, *De fossilibus mammalium reliquis in Prussia adjacentibusque regionibus repertis*, Regiomontani, 1823, 38 pp.

69. We have not been able to identify Colonel Wood further.

70. Friedrich Hoffmann (1797–1836).

71. This was the so-called forest bed exposed in the cliffs of the Norfolk coast between Happisburgh and Cromer. The bed was "chiefly composed of vegetable matter, with scattered cones of the Scotch and spruce firs, and many other recent plants, and with bones of the elephant and of other extinct and living species of mammalia." It was the chief object of the Rev. John Gunn's studies. See Charles Lyell, *Elements of Geology*, 6th ed., London, Murray, 1865, p. 160.

72. Henry Darwin Rogers (1808–66), geologist, a native of Philadelphia, in 1835, was appointed professor of geology at the University of Pennsylvania. In 1835, he undertook the geological survey of New Jersey and from 1836 to 1842, worked on the geological survey of Pennsylvania. When the state legislature refused further appropriations for the latter survey in 1842, Rogers continued it at his own expense until 1847. His report was published in 1858. In 1855, he was appointed Regius professor of natural history at the University of Glasgow, where he remained until his premature death.

73. Benjamin Silliman, Jr. (1816–85), chemist, was professor of

chemistry at Yale University. See B. Silliman, Jr., "On the chemical composition of the calcareous corals," *Amer. J. Sci.*, 1846, *1*, 189–99.

74. Joseph Beete Jukes (1811–69), geologist and director of the Irish division of the Geological Survey.

75. Dr. Thomas Wright (1809–84), physician and geologist, formed a large collection of Jurassic fossils, especially rich in cephalopods. In 1855, he began a systematic description and classification of the sea urchins and starfish. He had undertaken this work together with Edward Forbes, but Forbes's death in 1854 ended their collaboration.

In 1860, Wright showed that the Avicula contorta beds, which in England lie at the base of the Lias, contain a fauna corresponding to that of the Upper St. Cassian beds and the *Kössener-Schichten* formations grouped with the Trias by continental geologists. Since the grouping of the Avicula contorta beds with the Lias had been based on their position and lithological characters instead of their fossils, Wright felt justified in altering their classification. Lyell, in his *Manual*, 1855, pp. 337–38, had placed the English bone bed in the Trias, following Sir Philip Egerton and Louis Agassiz in this.

76. Charles Moore (1815–81), geologist. See note 18, above. Charles Moore, "On the Zones of the Lower Lias and the Avicula contorta Zone," *Quart. J. Geol. Soc.*, 1861, *17*, 483–516.

77. Peter Bellinger Brodie (1815–97), geologist.

78. Thomas Wright, "On the Zone of Avicula contorta and the Lower Lias of the South of England," *Quart. J. Geol. Soc.*, 1860, *16*, 374–411.

79. Caleb Burrell Rose (1790–1872), geologist, had begun medical practice in 1816 at Swaffham, Norfolk, where he remained until 1859, when he retired to Great Yarmouth. His collection of fossils was deposited in the Norwich Museum. See Caleb B. Rose, "On the Divisions of the Drift in Norfolk and Suffolk," *Geol.*, 1860, *3*, 137–41.

80. John Phillips, "The neighbourhood of Oxford and its geology," *Oxford Essays*, London, Parker, 1855, pp. 192–212, p. 197.

> Oxford stands upon a bed of gravel, which rests upon the thick blue clay of Otmoor. . . . It is a late tertiary (or pleistocene) deposit, due to the action of water when the land was at a lower level than it is now. By this action, great quantities of rolled

stones have been brought from the Lickey hills and scattered over the region of the Cherwell, and the sides of the valley of the Thames. It contains the bones of quadrupeds, but not the remains of men. Among these quadrupeds are the fossil elephant.

81. Charles Moore (1815–81), geologist. See note 18, above.

82. Robert Lightbody appears to have been an amateur geologist. In 1863, he published a single short geological paper, but we have not been able to identify him otherwise.

83. We have not been able to identify Mr. Knight of Henley.

84. We have not been able to identify Mr. Jenks.

85. This was apparently a personal communication from Captain Thomas Abel Brimage Spratt (1811–88), who was surveying for the admiralty in the Mediterranean. Spratt had written a brief account of the geology of Malta. See Commander Thomas A. B. Spratt, *On the geology of Malta and Gozo,* 1852, 26 pp.

86. Henry George Francis Moreton, second Earl of Ducie (1802–53), had his seat at Tortworth, Gloucestershire, where he was known as a breeder of shorthorn cattle and a leading agriculturist.

87. Louis Agassiz, *An essay on Classification,* London, Longman, Brown, Green, Longmans, & Roberts, 1859, pp. 81–82:

> When domesticated animals and cultivated plants are mentioned as furnishing evidence of the mutability of species, the circumstance is constantly overlooked, or passed over in silence, that the first point to be established respecting them, in order to justify any inference from them against the fixity of species, would be to show that each of them has originated from one common stock which, far from being the case, is flatly contradicted by the positive knowledge we have that the varieties of several of them at least are owing to the entire amalgamation of different species.*

88. From July 6, 1860–September 23, 1860, Lyell was traveling on the continent in France, Belgium, and Germany. During the period August 2–8, 1860, he was at Rudolstadt, Germany.

89. Samuel Wilberforce (1805–73), who was from 1845 to 1869 Bishop of Oxford. Bishop Wilberforce had, with the help of Richard Owen, written a review of Darwin's *Origin of Species* for the *Quarterly Review* (1860, *108,* 225–64), in which he attacked and ridiculed Darwin's theory.

* Our fowls for instance.

90. Charles Darwin, *Origin of Species,* 1859, pp. 398–99:

The inhabitants of the Cape de Verde Islands are related to those of Africa, like those of the Galapagos to America. I believe this grand fact can receive no sort of explanation on the ordinary view of independent creation; whereas on the view here maintained, it is obvious that the Galapagos Islands would be likely to receive colonists, whether by occasional means of transport or by formerly continuous land, from America; . . . and that such colonists would be liable to modifications; the principle of inheritance still betraying their original birthplace.

91. Heinrich Georg Bronn (1800–62).

92. Darwin, *Origin of Species,* 1859, p. 184. See note 27 above.

93. Jean Alphonse Faure (1815–90), French geologist and conchologist.

94. James Freeman Clarke, *The Christian Doctrine of Prayer. An essay,* 3d ed., Boston, Walker, Wise, 1859, 313 pp.

95. Ibid., pp. 88–89.

96. George Combe (1788–1858), phrenologist, was born at Edinburgh and after attending the University of Edinburgh was articled to study law. In 1812, he was admitted as a writer to the Signet and practiced very successfully until about 1836 when he decided to retire from business and devote himself to writing. He had early become interested in phrenology. In 1819, he published *Essays on Phrenology, or an Inquiry into the principles and utility of the system of Drs. Gall and Spurzheim and into the objections made against it,* Edinburgh, Longman, Hurst, Rees, Orme and Brown, 1819, 392 pp., and, in 1828, the first edition of *The Constitution of Man considered in relation to External Objects,* Edinburgh, Anderson, 1828, 319 pp.

97. Clarke, *Christian Doctrine of Prayer,* 1859, pp. 89–90.

98. Ibid., p. 92.

99. Ibid., p. 93.

100. Ibid., p. 94.

101. Ibid., pp. 95–96.

102. Ibid., pp. 96–97.

103. Ibid., p. 88.

104. Ibid., pp. 97–98.

105. Ibid., p. 251.

106. William Parsons, third Earl of Rosse (1800–67), in 1845, at Parsonstown, Ireland, built a very large reflecting telescope with a tube six feet in diameter and a focal length of fifty-four feet.

107. Clarke, *Christian Doctrine of Prayer,* 1859, p. 219.

108. "We can understand a little the mystery of evil when we see that evil tends to strengthen and educate the soul" (ibid., p. 235).

109. Ibid., p. 261.

110. William Benjamin Carpenter (1813–85), physiologist, had reviewed Darwin's *Origin of Species* in the *Brit. & For. Med.-Chir. Rev.,* 1860, *25,* 367–404.

111. Charles Darwin, *Origin of Species,* 1859, p. 316:

> Groups of species, that is, genera and families, follow the same general rules in their appearance and disappearance as do single species, changing more or less quickly, and in greater or lesser degree. A group does not reappear after it has once disappeared; or its existence, as long as it lasts, is continuous. I am aware that there are some apparent exceptions to this rule, but the exceptions are surprisingly few, . . . and the rule strictly accords with my theory.

Scientific Journal
No. VII

[p. 1] no entry
[p. 2] Candia [p.] 52. Gratiolet & entries [p.] 74.

September 25, 1860

Copy of Letter to C. Darwin [p. 3]

I return the M.S. on dogs which I think excellent. The case you make out seems very strong not only of crosses from distinct living species having blended into the dog, but in favour of diff.[t] savage races having domesticated different canine types, wolves, jackals, etc. by domestication in accordance with the hypothesis of Pallas, having eliminated the dislike to cross with other species as well as the tendency of such crosses to sterility.

All this helps the doctrine of the several so blendible wild species [p. 4] having themselves come down from a common remote progenitor & I suppose you will say in reply to Quart. Review that the only reason that pigeons have not so mixed is that Man had not the same motive to cross them with other species distinct from the Rock pigeon.

I think the subject too important to bear shortening. As to the antiquity of the Dingo I have turned up to your two references in Quart. G. S. Journal & they by no means bear out your inference. The fossils occurred in a cave in the basalt & were certainly posterior, probably very long so, & there is no evidence as caves are gradually choked up [p. 5] that the Dingo was not one of the last creatures whose bones were introduced. I asked Falconer yesterday & he said, "let him take care he does not get himself into the same scrape which Owen did about Strzlecki's Australian Mastodon,[1] I read the papers about Dingo & thought they had no proof."

Falconer knows of no fossil, rodent or bat in Australia. He says Strzlecki's Mastodon was from S. America.

It seems strange that the question whether any Newfoundland dogs are semi-web-footed or not sh.d be so doubtful, also gestation of jackal. On these and many points y.r book will draw forth a number of facts for the 2.d Ed.n.

Some one sh.d go at once to [p. 6] Angoulême & see into that question of the Leporine. If true, it w.d take away the antecedent improbability of Pallas' hypothesis of dogs when domesticated losing sterility, etc. Bartlett[2] says at Zool. Gardens that we must wait till next spring for good experiments. He ought to have as many as possible to start with.

The chapter on dogs makes me wish your book soon out. That & the pigeons & the Tables of large & small genera w.d alone make a useful beginning.

You conclude yr. letter received this morning by doubting whether I care for so speculative a way of dealing with the question.[3] It is just what I wanted & not more conjectural than [p. 7] my letter, much of which I w.d not, of course, touch or venture on in print, nor wish shown to third persons, more especially as one may alter one's speculations the day after. I have referred to the pp. you allude to of *Origin* with much profit, tho' I remembered them all, but they contain such a condensation of matter that they require to be often re-read.

I have been too deeply interested & taken up with the Dogs to have done with Asa Gray,[4] but see enough to wish it to appear in Annals of N.H. & cannot but think they w.d print it for their own sake & not require you to pay. Tell them that the addition of A. Gray's name w.d give a wholly new value to it instead of its appearing anonymously. I want to buy it for one. [p. 8] I think you have understood my point & the idea that if the original type of Mammalia had been last & the reptilian had been greatly raised in grade, they w.d have produced some other great class as high perhaps or higher, but not the existing Mammalia, is a grand notion & believing as I do in the infinite capacity of the creative power inherent in the organic world, worked out by variation & natural selection, I do not think it an extravagant special[iz]ation at all. You might say the same if some Monotreme, instead of a reptile, was improved into a mammal, by a series of changes independent of those which gave rise to the first mammal. Possibly birds may be an instance of [p. 9] such a class, though I fear that would not do, because however low the Monotreme, it must perhaps rank typically before birds & birds therefore must have preceded in order of development.

I am glad you reminded me of the N. Guinea Marsupials. I had imagined that there was a larger admixture of placentals. I sh.[d] think still the Australian genera & species were so well fitted for the extraordinary droughts that they w.[d] get the better of the Dingo had he run wild as in Juan Fernandez. The Brazilian case does not tell for much as that country has in the Marsupial line turned out nothing but opossums.

The case of nat.[l] grafting stated by Göppert (Uberwallung der Tannenstiche) [p. 10] Bonn 1842), where silver fir borrows from roots of Pinus abies is the more analogous to grafting as being one conifer borrowing from another.[5] Plants of other divisions w.[d] do the same probably if near enough to allied trees, which owing to rotation may not be often at hand. If the genus were remote, it w.[d] fail.

Do you not think, when Man was less advanced 50,000 years or generations back & tribes more isolated, there must have been rather more races than now tho' of course at a remoter period much fewer, & originally only one, & that one lower intellectually than any one now existing. In proportion as more powerful & more cosmopolite [p. 11] races arose they w.[d] exterminate inferior ones & also there w.[d] be more mixture & a check to divergence into species or the centripetal force of hybridity (excuse the use of the phrase) would come into play.

Falconer has been holding forth today on the drift, Mastodons & Elephants not coming in chronologically as they sh.[d] do, according to y[r]. views, but then one sees the new Maltese dwarf intermediate between El. antiquus & E. meridionalis & Anca's new Silician cave elephant, a modification of the living Indian one leaning towards antiquus, & when one thinks that Falconer can distinguish all American varieties of Mammals from all European fossil [p. 12] species, I confess I attach little value to the objection.

September 27, 1860

[C. Lyell] to C. Darwin

I am haunted with a kind of misgiving, which bye & bye I shall be able to express more clearly—that the multiple origin of the dog will furnish an argument for the multiple origin of a Mammal or of Man. Do try to consider it that way, for I incline to go far with Hooker (& with you?) in believing that whatever is true in domestication is essentially possible in Nature also, & I am not sure that you confine the multiple origin of the dog to Man's selection.

September 30, 1860

[C. Lyell] to C. Darwin [p. 13]

I expect that when the nondiversification of rodents, bats, manatees, seals, etc., on remote Miocene islands is fully worked out, it may merely end in satisfying you & me that the time required for change is longer than was supposed, or without such facts, demonstrable.

It will require renewed enquiry into the antiquity of such islands, also considerations as to the time before first bats & rodents arrived, also the force of preoccupancy—also the arrival of the same species of bats & rodents again & again, acting like European colonists into U.S., checking the formation of a new race.

Have you ever speculated in print on turtles changing to land tortoises [p. 14] in remote islands like Galapagos?

From Falunian (Upper Miocene) to Recent is only a part, tho' a large part (2 3.[d]) of one geol.[1] period measured by change of mollusca.

If as a rule a few species even of large genera vary, then if bats & rodents migrate into an island, the chances may be a thousand to one that these species are among the unpliant, unvarying ones, & if as you say we knew nothing of the law which governs the exceptional cases, we cannot conjecture whether insular conditions would favour such divergence if when the thousandth chance did turn up in some island & an insular species was ready to sport & improve & w.[d] have done so on land [p. 15] far from sea. Perhaps the chief use to make of my difficulty is this, & it accords with my old notion that species as a rule, or the majority of them are immutable, & have always been so.

Asa Gray is afraid of argument from imperfection of record.

Falconer observes that the Apteryx has strong clavicles & other bird-like characters without wings. I said, the more useless some of these parts not yet suppressed, the more likely to come down from an ancestor who had need & use of them or to be nascent organs advancing towards perfection, for a posterity which w.[d] enjoy wings. He only laughed as if the whole was a joke, yet I got him to admit that the hypothesis of limited [p. 16] modifiability was quite as arbitrary as yours.

By the way, that reminds me that the keel in the middle of the breastbone is wholly wanting in some (or all?) of the Dinornis family, tho' I think a little remains in the Apteryx & I am almost sure that H. v. Meyer[6] has lately found this keel in the Pterodactyl.

The long duration & numbers of the Ammonites from trias to chalk & then their extinction, tho' high in the scale, is certainly striking. I presume from the myriads of Sepia bones, which sometimes strew the Jutland coast, that naked Cuttle-fish now play their [p. 17] part & are higher in the scale, though not higher than Belemnites who accompanied Ammonites. The vast number & size of Hippurites & their sudden appearance & disappearance, the only extinct order of Mollusca, is also very striking. It would not make much impression here as we have few, but in S. of France, Italy, Sicily & all round Mediterranean.

What you say as to my difficulty is I suspect an explanation up to a certain point, but I hope to put the whole more clearly soon. Dogs of multiple origin & leporines would greatly weaken the objection to regarding the negro as of a different species, in the same sense that the Prairie wolf [p. 18] & common wolf may be. I should like a good naturalist to give me a list of reputed species in Mammalia, not more remote than negro & white man. Would they not be many?

Instead of Selection I sh.[d] have said, Variation & Nat. Selection. My only objection is not to the term, but to your assigning to it more work than it can do & not carefully guarding against confounding it with the Creative power to which "variation" & something far higher than mere variation, viz. the capacity of ascending in the scale of being must belong. Most likely you would have chosen some term less worthy of [p. 19] Deification, for "Selection" you had an excellent technical reason.

Letter from [Albert] Toilliez of Mons[7] to C. Lyell, September 28, 1860

I beg to announce to you that the day before yesterday I positively made out that the superficial bed of pebbles at Spiennes is above the Hesbayan *limon,* & very different from it; & consequently from the pebbly deposit at the base of this *limon,* which sometimes contains in this country remains of Rhinoceros & elephant. This is to be seen in a deep lane, at the base of the hill & on the path near which I pointed out to you a fragment of the Maestrichien of Dumont,[8] after we had passed [p. 20] the plank at the mill. I don't comprehend how I had not before remarked it before your visit, nor how we did not see it then, but I was thinking of other things & the heat was very great when we were there. I have also made out that the superficial zone with pebbles rests indifferently on the *limon* & on a yellowish brown sand with fine black grains of *glauconie* which I think I can refer to the Ypresien. In my opinion all the surface of the soil where the Spiennes pebbles are found, presents merely remains of ancient workshops for the manufacture of arms in silex. I hope to be able to prove this in a memoir on which I am going to work vigorously & of which I will send you a copy.[9]

Albert Toilliez

October 5, 1860 [p. 21]

With Prof. Ramsay

Glaciers carried blocks from Alps to Jura—& this leads to doubt whether many of the glacial striae of N. America, where the furrows go up & down hills, were made by terrestrial or subaerial ice.

If such a mass of ice as that which went from Alps to Valde Travers, why not much more in Britain & elsewhere.

Glaciers in the Schwartzwald, Welsh glaciers.

The subsidence must probably have included South Wales, Drift traced to S. Wales, Erratics there.

Varieties

The geograph.[l] vars. of some species on the American & European sides of the Atlantic, Vanessa Atalanta, Pteris aquilina, remarkable. See p. 700, 9.th Ed. *Principles*.[10] The time required by 9/10.th of the living species to accommodate themselves to the new circums.[s] will never be granted. Not that a red Indian might [p. 22] not be developed into something as good or better than a White Man but as it w.[d] take 10,000 or more years & he is not allowed as many centuries, he must be improved off the face of the earth & he is therefore not transmutable for any practical purpose of salvation in this world.

In no other way can any theory of develop.[t] or transmut.[n] be reconciled with the stationary condit.[n] of species in general, & the fact that they are dying out instead of becoming altered when their existence is at stake.

October 6, 1860

Letter to Darwin

Apteryx

The coracoid [bone] large, wh. in birds connected with organs of flight, why not reduced as the clavicles are to some extent & in the ostrich being separate bones not united in [the] merrythought.

The coracoid in reptiles largely developed, but this connected with fore limbs & this w.[d] not apply to Apteryx.

Messina [p. 23]

Hippopot.s, species undeterminable, seen by Falconer from the lignite Miocene of Sequine's in Museum. A Sus also, qy. age, this w.d be the oldest known hippopot.s.

Norfolk Coast

Pinus abies, doubted by R. Brown[11] latterly, says Falconer.[12]

Macrauchenia

Huxley[13] has received it from Bolivian Andes 20,000? f.t high, charged with copper, a synthetic type, post-pliocene in Patagonia.

Cwm Drift [p. 24]

In Cwm cave, Trimmer[14] made northern drift *posterior* to the cave animals.

With Huxley

Machairodus with Leidy's Nebraska fossils. This a newer pliocene or still newer European type.

Asa Gray on Darwin

Two hypotheses divide the scientific world, very unequally, upon the origin of the existing diversity of the plants & animals which surround us. One assumes that the actual kinds are primordial, the other that they are derivative. One, that all kinds originated [p. 25] supernaturally, and directly as such, and have continued unchanged in the order of Nature. The other, that the present kinds appeared in some sort of genealogical connection with other and earlier kinds that they became what they now are in the course of time, and in the order of Nature.

October 14, [1860]

Miocene Hippopotamus

The hippopotamus of the lignite of Gravatelli, Messina specimen in Museum of Univ.^{y} examin.^{d} by Dr. Falconer, consists of two incisors or tusks, perhaps new species. Ponzi[15] when he saw the drawing said he had seen some from tertiary of Italy (Marerna?) in which many fossil leaves. The Sewalik Hexaprotodus is also a Miocene hippopot.^{s} with the normal number of teeth.

October 14, 1860

Staring[16]—Haarlem [p. 26]

Beyrich[17] & V. Dechen[18] have in their map made the Lower Rhenish & Rhenish Hessian Brown Coal too ancient. It comes in, Dumont thought, between Bolderian & Diestien. (Falunian?)

Diestien

Sable a lignite du Limburg
Bolderien Nymphean (qy. marine bolderian?) —Lower Rhenish & Rhenish-Hessian Brown Coal
Nucula loam Lower Rupalian
Sable de Klein Spaura
Argiles de Lethur Toregreen sup.^{r}

Sable de Lethea Taigrien Inferum de Dumon[t]

The Diestien is the sand-dunes of the Crag Noir or Pliocene sea. Between Berchem & Wildrych near Antwerp. [p. 27] The Bolderian has been lately found well characterized by Ours Dufrénoy,[19] immediately overlaid by the Crag Noir with Terebratula grandis, & Nyst[20] has shown me (Staring) the fossils, another proof that Diestien which is found quite near that place is the same as the Crag Noir.

But how did these Diestien dunes keep their external shape

during all the pleistocene & hodiernal periods (asks Staring) p. 6.[21]

Hennesey, *Athenaeum,* October 20, 1860, p. 516 [22]

On James' change in earth's axis, Henslow, same number on St. Acheul,[23] not contin.[d] in Lat. 27. N.[o]

[Bunsen, *Egypt's Place*]

Egypt's place in universal Hist.[y], Ch. C. J. Bunsen, V.1.[24]

The Egyptians of Asiatic origin & having affinity to the Semitic or Aramaic stock. Preface 8.

[p. 28] The Icelandic language was the old Norse from wh. the Swedish & Danish branched off, p. 10.[25]

The Hieroglyphic language of Egypt between the Indo-Germanic & the Semitic.

Vol. 4, p. 14.[26]

33 centuries between Menes & Alexander.
Upper & Lower Egypt, 2 kingdoms.
Hieroglyphics & with characters in time of Menes.

[p.] 18 The contemporary records of the Egyptian language go back 4000 or 5000 yrs. B.C.[27]

[p. 29] p. 21, 4000 yrs before [Christ] a great empire with organic members (upper & lower Egypt?) of very ancient type, a peculiar written character, & national art, & sciences, we must admit that it required 1000.[s] of yrs. to bring these to maturity in the rich valley of the Nile.[28]

It is a sober conclus.[n] that it requires 20,000 yrs. to explain the beginning of the develop.[t] of Man.

p. 431, Vol. 4,[29] Tradition of the flood,
Creation 20,000 B.C.
[p.] 480, Flood 10,000[30]

November 13, [1860]

Copies of Letters from Mr. Smith of Jordanhill [31] [p. 30]

My dear Sir Charles, I have come to the conclusion that the

canoe beds are marine & belong to the period you suppose, in which I found, about 20 feet above sea level at Port Rush, 100 species all recent British & in which Dr. Landsborough found at Largs 70 species d.° [ditto] See Geol. Soc. Proceed.gs III, 445.[32]

I at first supposed that the beds at Glasgow, stratified sand were lacustrine & the shells found in a canoe being edible might have been brought from the sea, but assuming the sea [p. 31] at its present level a navigable communication could not have existed; we must therefore bring the last elevation into the human period—and to come nearer to the relative age of the embedded remains, the flint arrows & hatchets recently exhumed in France are excessively rude, the stone celt found in the canoe at Glasgow exhibits a great advance in the art of forming stone implements. I happen to have it lying before me at this moment & have traced its outline on the annexed slip of paper. It is of dark coloured [p. 32] bloodstone or jasper. The artist has first of all brought this stone to its present shape with a smooth & true but unpolished surface. He has then traced faint lines *c.c.* leaving the centre part B. unpolished, either to save labour or to prevent the haft from sliding, the portions not covered by the haft are as highly polished as my seal.

The polishing must have been done by the hand & not a wheel, for it is not bounded exactly by the faint lines. [p. 33] I have reread Mr. Jamieson's paper.[33] I am glad to find so industrious an observer at work on the superficial beds & find nothing to controvert.

I observe what you say as to the rarity of marine remains except in the lower beds. I am inclined to attribute this to the slow but gradual solution of the shelly matter. The glacial shells seem to owe their preservation to the clay, in or under which they are found, but even here we find them in every stage of decay, often mere casts with a shelly powder [p. 34] round them, like the enclosed.

November 14, 1860

Smith of Jordanhill

Your understanding of my account of the Cumbra dike is quite correct as well as the relative rate of wasting of the sea at its present level & at its former height. [See Fig. 21.]

Fig. 21

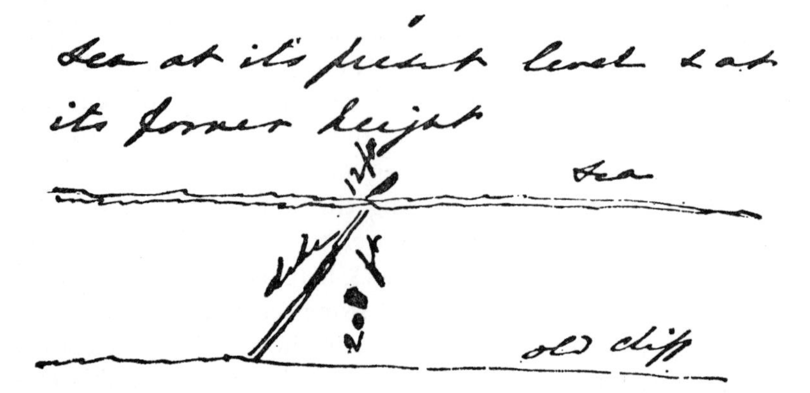

Hills red sandstone

If I have exaggerated anything, it is the proportion of 13 feet caused by the sea at its present levels & that 1/16 is too small [p. 35] a proportion, however, it was taken from actual measurements.

This cliff & terraces represent a period of stationary levels of which 32,000 years is a portion, possibly a small portion. Yet it is but a minimum of geological time. From everything I have observed it is posterior to the glacial epoch.

If the rate, at which solid sandstone yields to the battering of the sea, impelling sand & stones against it, could be ascertained, we would come nearer the mark. Opposite the Cumbra dike are old [p. 36] mill-stones set up as a breakwater. I remember them nearly 70 years, & can see no difference in their thickness. Tried in this way, I have never been able to account for the effect produced in so short a period as 100,000 years. Two inches in a century would give 125,000, but when I look at old pieces of at least that age, I can see no such loss of substance.

With regard to the Mammoth I know but three instances, two described by Mr. Bald in the [p. 37] *Memoirs of the Wernerian Soc.,* Vol. IV, p. 58.[34] One of these found in the Till or boulder clay, the other in clay of a light brown colour associated with marine shells, from the locality, the shells must belong to the glacial period. Near that another Elephant was found certainly in the Till. Dr. Scouler[35] visited the locality before the bones were altogether exhumed & assured me this was the case.

The boulder clay, whatever its origin, belongs to the preceding or glacial epoch, (in my paper, Geol. [Soc.] Journ. v.[ol.] VI, p. 386, you will find proofs, & of course the mammoth [belongs] to the same epoch [p. 38], 2.d or 3.d of Jamieson).[36]

Some of our peat certainly belongs to this stone period, not only from stone hatchets & arrow heads being found in it, but a small structure or cattle pen was found embedded in peat formed by stakes bearing the marks of having been pointed by stone hatchets.

The conclusion I have been led to is that we have evidence of movements both of elevation & depression in the human period, but the natural tendency of Man to improvement forbids the supposition of his being contented [p. 39] with stone implements for geological epochs. In a late paper to the archeological society here, I chose the subject of pre-historical antiquities showing the progress from the rudest implements of the stone period to the highly finished celts described in my last. I did the same with bronze implements, the earliest bronzes being rude copies of stone celts only thinner & sharper, then with gradually improving devices for fixing the handles, last of all ornamental mouldings.

The opposite movements of elevation & depression produce [p. 40] apparently contradictory phenomena. Thus we have the wearing away of hundreds of feet of solid rock & at the same time rocks constantly washed by the sea with the glacial scratching unobliterated. This points to a comparatively late subsidence.

November 24, 1860

[C. Lyell to C. Darwin]

My dear Darwin,

In former editions, in speculating on age of Uddevalla & still more elevated beds of glacial period in Norway & Sweden, I have [p. 41] assumed an average rate of rise as 2½ feet in a century. See *Manual* p. 119. 125 feet takes 5000 years to rise or to sink. I wish to keep to my old standard or estimate unless shown to be objectionable as applied to countries like Greenland & Sweden in full movement, a mere guess at an average rate, at p. 120 I give 27,500 years for 700 ft. of upheaval in recent times. But I expressly guard myself in same page by saying I make no allowances for pauses or for oscillations of level (of the minor kind).

What I want to do in the new [p. 42] edition is to make another conjecture & to allow for the pauses. When the amount is reversed, as if (which I believe with Trimmer, Ramsay, Jamieson & others) Scotland & Wales have moved first down & then up again 2200 feet or more, then I conceive the chief pauses would be before the downward was converted into an upward movement. Successive sea beaches & terraces & inland cliffs mark long pauses. Smith of Jordanhill thinks he has a cliff which took more than 100,000 years to cut, subsequently to the glacial re-elevation in Scotland.

Now I propose to conjecture [p. 43] keeping on the safe side & not exposing myself to the charge of exaggerating the probable time, that 4 parts of Europe are stationary for one which moves at the rate of 2½ feet in a hundred years, or that the stationary area exceeds that in motion as 4 to 1 & 4 expresses the period of rest or the pauses where 1, in a given region, expresses the movement. Therefore if Scotland has first gone down 2250 feet after the period of land glaciers, as I believe, this would take 90,000 years & the re-elevation rather more because it went up higher than its present level, say 100,000 years, but then I must [p. 44] give 400,000 additional (in round numbers) for the intervening periods of rest.

Thus the last oscillation is about 50 feet in Scotland in each direction, since the 2 vast movements of 2250 [feet] down & up. But when we try to estimate the time required for the 2 movements, we encounter, as Smith of Jordanhill observes, 2000 years since the Romans built the Pictish wall & find that we do not get back to the close of the last oscillation.

Then I regard this 2000 years (and we know not how many

1000 before) as a part of the great [p. 45] excess of stationary condition.

I also consider the successive beaches cut at Glasgow in the canoe-bearing sands, all before the era of the Roman Wall, as portions of the same excess of 4 to 1.

I am aware that the North Cape moves, or is said to move, 5 or 6 feet in a century & D. Forbes[37] thinks that Chile has gone up from 40 to 60 feet in 350 years, but then north of Africa, he thinks there has been a subsiding & about Africa, a stationary one as proved by Indian tumuli & mummies buried near the old shore level & elevated since.

But I look on S. America [p. 46] as rather exceptional. It would be balanced by other more inactive regions of oscillation. The average in Scandinavia would not, I think, exceed 2½ feet [per century]. I am however much more in doubt as to the comparative areas of rest contrasted with those of movement. I once made them as 9 to 1. If I could stimulate the geographer to make objections it would do good.

Perhaps the contemporaneous inland oscillations, especially in mountain chains, may be greater than the sea-coast ones. But we may leave those speculations out for the present.

There may be scarcely [p. 47] any areas in an absolute state of rest, but we cannot yet take account of minute quantities. I wish you to reflect on my principle, & if possible say if 4 to 1 is a reasonable conjecture.

Large plateaux of denudation & inland sea cliffs, are monuments of immense pauses.

1.st	Period of rest of unknown duration before the extreme cold & glaciers of Scotland		0 ? years
2.dly	Period of gradual submergence 2250 feet.	90,000	
3.dly	" of reelevation 2500 feet,	100,000	
4.thly	Intercalated pauses up to the close of period of reelevation	400,000	
		590,000	

During this period, the whole of the [p. 48] glacial period & the present establishment of provinces of species has occurred—the mammalian fauna greatly changed, but the shells very little.

The last oscillation of Scotland of about 50 feet in each direction w.[d] take 4000 years & for pauses 16,000 more which would give 20,000 years. The Glasgow canoes & polished celts come into this brief era, but the Somme valley flint implements probably into some part of the great period of reelevation, during which in Scotland the erratic boulder clay was getting denuded & my Forfar gravel beds manufactured.

Chas. Lyell

November 30, 1860

Letter to Darwin [p. 49]

You must have thought so much, when on your coral-reef-areas of subsidence & volcanic areas of elevation, that I think you [a] high authority in speculation of that kind & if you think 2½ feet per century a good allowance (as I believe it to be & above an average) & if you agree with me that between two great movements in an opposite direction there would generally be a long pause, I am satisfied; even if we have no means even of guessing at the probable duration of such pauses. Suppose the passage of a great subterranean [p. 50] lake of liquid lava into a crystalline rock to be the cause of secular upheaval—the heat gradually diminishing & the rock e.g. granite, expanding, it is likely that the rate of upheaval tho' it may be equable & persistent for tens of thousands of years, like the heat of thermal springs, will, when on the decline, decrease very gradually & one cannot expect the reheating of the same mass, or of one in the same part of the earth's crust & the melting thereof, to take place just when the cooling had ceased. The chances are that before the [p. 51] reversal of the movement there would be a very long interval. But you speak of astronomical causes. Did you ever read of any as yet suggested, which seemed in the least degree feasible, & one cannot doubt the connection in S. America of volcanicity & movements of upheaval & depression? I understood, however, the force of your objection to my reasoning as to movements, from space to time. If I were to judge by sea-coast evidence alone I should think a tenth rather than a fifth was all that is in motion. But I cannot help suspecting that the Alps & other mountain chains move more than [p. 52] lower countries. The Alps seem so often to have risen & sunk considerably. This perhaps strengthens your objection as showing that you cannot apply the rule of one area to another, as certain spaces are repeatedly moved, while others may continue at rest for one geological period after another. Spratt's proof of Candia or Crete nearly 100 miles long having gone down at one end—Greek cities being submerged & up at the other, the old docks being upraised high & dry, & in the intermediate space no movement, agrees with many facts [p. 53] to indicate that the Mediterranean is one of the areas of unequal movements in the last 2000 years.[38] Smith of Jordanhill shows great oscillation of rock of Gibraltar since [the period of] existing Mediterranean shells.

On the other hand there are parts of Norway stationary where beaches with recent shells have been upheaved 700 feet. I doubt there having been periods of general upheaval or general downward movement in Europe in Glacial Period.

As to Scotland having been as high above the sea as now, before the great submergence, all that I know of the Grampians, which are covered by unstratified till [p. 54] & old moraines, under these more modern gravels which Jamieson has correctly described at all heights, lead me to the belief that Jamieson is right, pp. 368–369, in believing that glaciers or land-ice in Scotland & Ireland acted powerfully before the land sank.[39] You may say, however, that the land need not have been as high as now, but I think it more likely that it was, as the Scottish Mountains are not high enough to have had glaciers on them, unless there was a great volume of land & that would almost imply height. [p. 55] I shall take your hint & be satisfied in claiming, as a minimum, 100,000 years for the downward, & as much for the upward movement, & then give as my reason for inferring a very long stationary intervening period, that so much of Europe, even where there has been pleistocene & post-pliocene upheaval & depression as e.g. Scotland, part of Norway & parts of Sicily, have been at rest for last 2000 years.

You will have no objection, I presume, to my going so far without pretending to define by centuries the length of said interval.

My idea is that subterranean [p. 56] upheaval, like volcanos, shifts its theatre of action & by degrees visits every country, & if there are Silurian regions of horizontal strata, they too can be proved to have been submerged & denuded, pressed & upheaved. As active volcanos occupy a small area so did extinct ones & so do deep seated volcanic agencies or those causing upward & downward movements. Rest therefore always has been, as it is now, the character of the surface generally, & change of level exceptional, & inland sea-cliffs, like those of Greece or the Morea, show very long [p. 57] pauses. The quantity of pleistocene & post-pliocene movement makes me rather doubt the continents & oceanic basins remaining as constant as you suppose, through several successive geological periods. From the Miocene to our era I can imagine it. But not from the Cretaceous to our times.[40]

December 17, 1860

Letter, King,[41] July 18, 1860, Burnt

Remains of hyaena & Equus fossils etc., found by him in the bed of the modern excavated valley (river bed) at Hoxne.

Ditto, July—Burnt

In Mundesley cliff no laminated clays in the stratified drift sands in his section so far as he remembers tho' occurring farther N.

Rev. S. W. King, October 3, 1860 [p. 58] [42]

Between Cromer & E. Runton "Forest bed, blue clay, with bones & scales abundant of fish, (large) scapular of mammoth, seeds & elytra of beetles, all blending into & perfectly continuous with red marine shelly crag & pan."

Blue clay rests on crag at Runton & crag as chalk seen at low tide. p. 3.

3 Elephant's molars from blue clay & black gravel of Forest bed.

Black lignite of forest bed rests on rich red clay in vertical laminae under horizontal forest bed. This is King's "lower crag". Middle crag, Forest bed & marine shells, Red crag.

Upper crag, laminated sand & clay series over wh. Till.

[p. 59] Forest bed has yielded Hasbro' bed (at or below Cromer & Runton?) cones of Pinus abies, antler of Roe buck, coloured elytra of beetles, seeds.

Runton fresh water [bed] is not, as he at first thought, the equivalent of, or continuous with Cromer Forest bed, the latter ranging now with shelly marine crag.

From Same, July 3, 1860 [43]

p. 2. Elephas antiquus has been found below the glacial beds but no good case of its occurring above.

[p.] 3. Mastodon arvernensis not found on Norfolk coast.

Same, November 24, 1859

p. 2. In Mundesley as in Hoxne, Valvata planorbis, cyclas.

[p.] 3. Fluviatile stratif.^{n} of part of Mundesley

[p.] 4. Lignite bed in clay at Mundesley, is [also] at Hoxne. Silt above peaty mud in both.

King's Letters Continued [p. 60]

p. 5. Unio in Mundesley peaty mud.

Same, November 25, 1859

p. 2. No interstratif.^{n} at Mundesley as stated in my paper on Norfolk cliffs.

[p.] 3. Icebergs incompatible with Donacia & Mundesley fish.

Ditto, June 9, 1859 [44]

Whether Diatomaceae abundant in the Mundesley peaty bed.

Same, January 11, 1860 [45]

[p.] 2 Paludina lenta? in Mundesley beds.
" Paludina marginata was found by Strickland in Cropthorne beds & Stutton.
p. 3 Diatoms abound in Mundesley beds.

Ditto, December 31, 1859 [46]

p. 2 Gunn says Runton beds under till.

Ditto, November 16, 1860 [47]

p. 2 Vertical strata underlying forest bed, a local nodule in blue forest bed.

December 13, 1860 [p. 61] [48]

p. 2 Paludina marginata in Gunn's Mundesley specimens.
p. 4 Cyclas amnica var. fig.[d] in *Manual* [49] like Grays—new name suggested.
p. 6 Le Verrier's calculat.[ns] of secular variations of distance of tropics from equator from 19.[°]42[′]to 27.[°]30[′] in 50,000 yrs. & from 27.[°]30[′] to 23.[°]27[′] in another 50,000.[50]

January 4, 1861

Progression

If the transmutationists had as implicit faith in progression in time, or development from the simplest organism to Man, as some of their chief opponents, there can be no doubt that their confidence in their own system would be increased.

January 11, 1861

Cretaceous [p. 62]

E. Roberts on deep sea Wallich's[51] Ophivesma & Globigerina. *Geologist,* vol. 4.[52]

Carboniferous

Salter lecture on Coal. *ib.* p. 6 (1861, to be continued.)[53]

Post-Pliocene

Mackie? on flint implements, *ib.* p. 19.[54]

January 15, 1861

Cambrian

Taconic of Emmons,[55] primordial of Barrande.
See Marcou's[56] paper Tracts—
Billings[57] told Salter in a letter that he had discov.[d] 60 or more species, chiefly primordial in Canada. See Report Aug.[t] 1860 trilobites of Barrande primordial.[58]
Also a report by Frank Bradley[59] on fossils of Keesville rocks read at American Assoc.[n] [for Advancement of Science], Montreal, 1860, in wh. primordial forms.

British Cambrian

Salter showed me primordial Dikelocephalus & other trilobites from Lingula [p. 63] beds & above these Tremadoc slates with intermediate forms between Cambrian & Llandeilo, but new species. At the top only of the Tremadoc Graptolites.
No Graptolites below this. None in the primordial, the species at top of Tremadoc. Same as a Llandeilo one.
Salter supposes the downward passage of Potsdam, as describ.[d] by D. D. Owen,[60] into magnesian limest.[e] & other beds, to correspond to Emmons' Taconic rocks.

Silurian Trilobites

Whether Annelids—They differ from crustacea in having no legs. No living crustacea with tripartite thorax & cephalic portion.

Crustacea do not eat mud like worms. Trilobites did so. Salter.

Post-Pliocene [p. 64]

Mr. Busk says that the angle of the jaw alluded to by Malaise is a very variable character not to be depended on.[61]

He says that in wild races, savage tribes, there is as remarkable a uniformity of type of skull as in wild animals. But in civilized man such intermixture & crossings & such influences brought to bear that they vary exceedingly.

Newer Pliocene

Mr. Bristoe[62] tells me that in the Eocene at Bracklesham are seen trunks of trees of the New.[r] Plio. rooted. In same old terrestrial surface the pholas has bored. Shells of warm latitudes & colder in the New-plio. of Pagham & Selsea. An entire elephant's [p. 65] skeleton of same age as the trees lately dug up & now at Chichester, seen by Waterhouse [&] Hawkins.[63]

Eocene

Bristoe says there is an old coast-ice formation with angular flints & trap, in the middle of Eocene, contemporaneous & cov.[d] by Eocene at Pagham & Bracklesham! (qy.)

Cambrian

Scolithus of Rossh[ire] & N. of Scotland agrees with N. American, says Salter. Worms eat mud, so did trilobites, & extract nourishment out of them & then reject it. Many of the old rocks full of rejected worm-excrement or earth.

March 1, 1861

Development [p. 66]

They who advocate this are the most influential promoters of Darwin's views, for if there has been general progress it becomes more probable that Man has been progressive from a humble beginning to the highest present development. And if instead of starting with exalted endowments, which some races have perpetuated, but from which others have degenerated & deteriorated, we suppose the original pair to have been like some of the rudest races with few abstract ideas, many instincts like those of brutes, a scanty language, unable to count as far as the digits of one hand & that at the end of time of thousands of years, some one of the descendants of such a pair was a being of the highest genius & much exalted moral power yet possessed by [p. 67] any one of the human race, we are forced to grant such a deviation from the original standard, effected simply in the way of ordinary generation, that all divergence of instinct or organization, such as is usually deem[ed] sufficient to constitute specific & generic distinctions, shrinks into insignificance in comparison.

We shall witness character of a higher order & seeming to men to require some intervention of a power transcending any one previously manifested on the globe, there w.[d] be exhibited by the varieties of the species of a genus or the genera of an order in the lower mammalia.

This difficulty may be met [p. 68] by assuming that we have no proof that Man did not start in the most perfect form morally & intellectually, but in assuming this the advocate of developm.[t] deserts his usual line of inference & the rude implements of very ancient tribes seem to be at variance with a primeval perfection. It is a chapter of the earth's history that we are only beginning to open.

March 3, 1861

If we ascribe this coming in of the higher states of being to inheritance, or to the variety-making power, or to Natural Selection, we immediately deify those laws, they instantly cease to be secondary causes, they begin to partake of the infinite & omnipotent, to usurp the place & attributes of the Cause of Causes by which [p. 69] all the physical, moral & spiritual forces of the universe, as known to us, were called into being & yet how have those wonderful changes, compared to which the variations of instinct & organization within the range of a species, or even of a genus or an order in the organic world, shrink into insignificance, how have they been introduced, not by any interference with what we call the usual course of Nature—the nativity of a superior race or of the highest [p. 70] intellectual or moral exemplification of that race, has been attended by one prodigy, has been ushered into being, not only with "the earth trembling" or "the front of heaven being full of fiery shapes," but even without any [even] the slightest deviation from those ordinary laws of generative reproduction, which are common alike to the higher & lower animals & even to the greater part of the animal [Kingdom] but even to the vegetable kingdom. Such are the unavoidable consequences of admitting the doctrine of progression if it be pushed as far as it is by many of the opponents of transmutation.

March 19, 1861 [p. 71]

Brain of Animals

Owen says the brain of the humming bird is larger in proportion to the size of the creature than is the brain of Man—also in some field mice.

It seems, says he, as if in these diminutive animals a certain quantity of brain was required in order to enable them to play the part required by their grade in the animal kingdom.

Man's brain is out of all proport.[n], greater in absolute quantity.

Quadrumana

Huxley says the distinctness of the hind limbs from the fore was known to Blumenbach[64] & taught by Cuvier.[65]

The difference in this respect between Man & higher Apes is a good family distinct.[n], not class or sub-class. [p. 72] Geoffroy S.[t] Hilaire[66] wishes to make 3 kingdoms, Animal, Vegetable &

Human because of moral & intellect.[l] develop.[t] of Man. But in such cases the inferior races & individ.[s] of human family not sufficiently regarded.

Trias

Ceratadon, a fish of Keuper [beds], one found by Plieninger,[67] occurs with humerus of Dicynodon in England (Oldham[68] plates) showing that the Nagham beds or Demorah format.[s] are also Trias. This format.[n], one with Cycads oolitic & between the Glossopteris Browniana wh. occurs in Bardwan & in Australia. Oldham has found Dicynodon [p. 73] & long snouted Labyrinthodon in the Glossopteris Browniana beds in India, Bengal.

East V.[a] coal may be of same age [as] upper part of Trias.

Oestheria [Estheria] has a wide range from Lias? to Trias? T. R. Jones[69]—Labyrinthodon [is found] from Keuper? to Coal, says Huxley, therefore these alone cannot decide age of Indian beds, but Ceratodon [has a] very restricted range. Plants show absence of dicotyledons, very few if any cycads, & in this respect Richmond, V.[a], differs from the Bird-wan & Nagpur coal of the Demurah age, because Zamia [is] in E. V.[a] beds, but Heer thought E. V.[a] triassic—Stangeria, [p. 74] some think the Dicynodon beds may be Permian.

April 9, 1861

Gratiolet[70]

—tells me that in the more advanced races & more cultivated, the sutures of parts of the skull are longer in consolidating than in negro & allow of more expansion at a later age. This seems a great explan.[n] of indefinite improveability, may explain why some slow children end in being cleverer than quicker ones. If also Natural Selection acts in multiplying those whose sutures are longest in ossifying, this may help to explain progress by hereditary transmission.

September 1861

Letter from Mr. Jamieson[71] of Ellon
to Cha.[s] Darwin [p. 75]

Dear Sir, I returned a few days ago from a trip to Lochaber where I spent a fortnight and now hasten to present you with some of the results of my visit and I may at once state that all I saw tended to impress upon me the conviction that these parallel roads have been formed along the margin of freshwater lakes and finding the marks of ice action so plain over the whole district I cannot help thinking that Agassiz hit upon the true solution of the problem when he pronounced these lines to be the effect of glacier-lakes. [p. 76] I attentively examined the entrance to Loch Treig and found both sides of the gorge to present the clearest evidence of most intense glacial action and that to heights of many hundred feet above the lake, rounded rocks, scores, flutings & perched blocks abound and all these phenomena are most conclusively seen to have been effected by a great volume of ice flowing down the valley, now occupied by the Lake and issuing out by this gorge into Glen Spean. Mr. Milne[72] states that the glacial markings here point S.W., this is quite true as regards the east side [p. 77] of the gorge, but on the opposite or W. side they point S.E. [See Fig. 22.] Mr. Milne says he could find no transverse scratches on the rocks of Glen Spean. I was however more fortunate and found them on the north side of the Spean opposite the entrance to Loch Treig, as Agassiz had correctly described. The profusion [p. 78] of blocks on this side of Glen Spean over an immense expanse is very striking—in fact the whole ground about the mouth of Loch Treig is a perfect study of glacial action. I could see no line of erosion due to water action on either side of the entrance and have no doubt that the rounding you noticed was simply the glacier action which is there very conspicuous & striking. I indeed noticed some scores & furrows on this level of the shelf. I should also mention that the protrusion of a glacier from Loch Treig across Glen Spean would, by resting on the north side of that valley, block up the outlet at the head of Glen Glaster so that this Glen Glaster outlet does not [p. 79]

Fig. 22

(as Mr. Chambers[73] supposed) present any great difficulty on the glacial theory.

I went up to the head of Loch Treig and explored a wild glen which runs from the S.W. corner of the Lake to Spean bridge & called the Larig or Larig Leachach.

There is no trace of any line round the head of Loch Treig, nor in fact round any part of it altho' Sir Lauder Dick[74] says it is visible on the W. side. There are some indications of alluvial detritus here & there towards the lower end of the lobes, but a marked absence of such at the upper end & instead I found hillocks of what seems to be [p. 80] Moraine matter abounding in blocks. There is an absence of all deltas at the mouths of the streams that come into the head of the lake at levels above the present upper limit of its waters which leads me to think that the upper end of the lake must have been occupied by the glacier during the time when the waters in Glen Spean stood at the level of the lowermost line of Glen Roy. In traversing the Larig glen I found much evidence of glacier action but not lines like those of Glen Roy.

The only place where I [p. 81] observed any wearing away of solid rock due to the action of the waters that had formed these Glen Roy lines was at the head of Glen Roy or rather what has been called the head of Lower Glen Roy, where there is a sort of pass caused by a set of rocky eminences in the hollow of the glen.

Here also there had been previous glacial action so that it is not easy to determine how much is properly due to the effect of water. This pass is a notable place for wind (as I can testify, having been nearly blown off my feet there), which coming up the long hollow of the glen [p. 82] rushes over this high narrow pass with immense force, and when the water stood at the levels of the two uppermost lines of Glen Roy it would lash against these rocks with considerable force during gales from the S.W.

On the face of the hill marked Tom Brahn in your map, the shelves are more broadly & rudely marked than any where else that I saw. This broad marking however consists of a protrusion of loose matter and I could find no noticeable erosion of rock along any of the lines. The middle line crosses a mass of solid rock *in situ* here, [p. 83] which I examined with the view of detecting traces of such erosion but the angles were quite similar to such as are usually to be found on weathered rocks not exposed to the flow of water.

At the West end of Loch Laggan there is clear evidence of the water having formerly stood at a height of probably 40–50 feet above their present level, there being a great bank of alluvial matter to that height which seems to have been the delta of the Gulban river. Along the shores of Loch Laggan however there is no clear trace of any line that I could see, most of it indeed is so [p. 84] encumbered with block[s] & rocks that the mark of a line could scarcely be expected. At the E. end there is also evidence of the lake having formerly stood at a higher level than at present and that to such a height as would, I have no doubt, cause its waters to flow eastward by Makoul towards Spey at the pass of Makoul (or Muckul as it has been written by some), or rather a little to the Eastward of it there is clear evidence of a large stream of water having flowed out towards the basin of the Spey, although I should describe it somewhat differently from Mr. Milne. [p. 85] There has no wall-sided gorge been cut, there is a ravine due to a series of rocky eminences occupying the middle of the valley. Here also there had been previous glacial action, but the subsequent effects of water flowing eastward are quite unmistakeable. The glacial scoring has been washed out & smooth winding curves with a tendency to *form potholes* are seen, whilst quantities of well rounded pebbles like cocoa nuts or cannon balls lie on the water worn surfaces, and heaps of shingle in the recesses and sheltered spots to the East of the rock masses. This shingle is *intensely* water worn & has been used in some [p. 86] places for making the road. I satisfied myself that this could not have been caused by tidal action but that it was due to a current flowing downward and Eastward. In the rocky narrows a little above the height of the summit level where tidal action would also infallibly have left its traces I found stony detritus but presenting none of the water rolled complexion of that which I have described. As there is no clear line along the E. end (or indeed almost any where else) of L. Laggan I doubt much whether this Makoul outlet has ever been properly connected [p. 87] by levelling with the lower line of Glen Roy, although I think it is highly probable that it does coincide with the level of that line. The outlet at the head of Glen Glaster *does* coincide with the middle line of the Glen Roy & here also there appears to have been an outflow of water. The outfalls at the head of Glen Gluoy & Loch Spey I was unable to visit. In looking up Glen Roy from the Gap, I was much struck by the extreme neatness & precision of the lines, which seemed to me very unlike what might be expected from the shore of a lake subject to tidal action. Very different from the appearance of true old coastlines [p. 88] which I had seen on the West coast of Argyleshire last summer & also from those depicted by Mr. Bravais.[75] But what seemed to me even more important evidence in this respect was the wonderfully fine preservation of the deltas at the mouth of some of the streams near the head of Glen Roy. These deltas have the appearance of being lodged in the waters of a placid lake, even in a stagnant pool, so undisturbed is the outline of some of them. This seems inexplicable to me had the lake been an arm of the sea subject to the flux & reflux of the tides.

Again the delta at the mouth of the Turret is out of all proportion [p. 89] too large for the size of the stream. I estimated the waters of the Roy as 3 or 4 times greater than those of the Turret at their point of junction—whereas the Turret delta is 2 or 3 times bigger than the Roy delta. This I think can be explained only by supposing the Turret delta to be partly due to the outflow from Loch Gluoy. The glacier streams that had blocked up the mouth of Glen Spean have probably been those issuing from Glen Arkarg & the other glens at the mouth of the Caledonian Canal. At Fort Augustus the traces of glacial action are very noticeable.

In Glen Spean to N.E. of Bridge [p. 90] of Roy the scores run E. & W. but further up they sweep beautifully round into the mouth of Glen Treig, opposite which they run N. & S. The glacial markings on the syenitic granite on the N. side of Glen Spean opposite the entrance to Loch Treig are amongst the finest specimens of ice-work I have seen, this with the heaps of moraine matter & the perfect wilderness of boulders made me stare with astonishment how any one, after Agassiz had drawn attention to all this, could go on the ground & yet deny that there had been any glaciers here! I do not suppose there is any [p. 91] place in Britain where the traces of a great ice stream are more complete.

There are however some facts connected with the former glaciation of Lochaber very remarkable but they refer to a period anterior to the formation of the parallel lines, and although very extraordinary they are consistent & harmonize with what I have seen in other parts of Scotland. They indicate however a climate more like that of the *poles* than any thing else we have at present on the face of the globe & the state of things which they disclose seems to me one of the most curious & inexplicable in the whole range of the geological record. [p. 92] I have written the above notes very hastily but shall be happy to answer any queries you may wish to put. I may mention that I had most villainous weather, which prevented me making so long excursions as I otherwise might have done and it was only by going doggedly to work with a waterproof & umbrella that I could get any thing done at all.

I am,
Thos. F. Jamieson

Ellon, Aberdeens.[h]
3.[d] Sept. 1861

Sept. 14, 1861

Letter of C. Lyell to T. F. Jamieson of Ellon

Stratif.[d] drift in Forfarsh. 800 f.[t] & more high. Moel Tryfan, shells [of] Norwich [p. 93] crag in Chillesford?

If we submerge Perthsh. 1200 ft. & raise it more than lowest shelf of Glen Roy, we have still Jamieson, difficulty of no foraminifera or deep sea corals in mud or gravel.

Glen Roy shells valuable as showing:

1.st The non-invasion of the sea to height of lowest shelf 950? f.[t] in post glacial period.
2.[ly] No local revolution in physical geography & surface levels within Lochaber—

Were Grampian boulders carried to the Sidlaw [hills] as Alpine erratics to the Jura by glaciers? [76]

September 22, 1861

Letter from C.L. to Dr. Hooker [p. 94]

I was very glad to have yr. letter especially as I have to allude to Platanus & shall see Heer[77] again. I will try to get him for his sake to visit Kew again. I am afraid you have not got his *Flora Tert. Helv.* at Kew or I sh.[d] have liked you to read, vol. 2, pp. 73–74, what Heer says of the Platanus occidentalis *acerifolia* being considered & given by Michaux as an

American tree, "while" (Heer adds) "every one acknowledges the American origin of Platanus occidentalis, Lin." But no doubt you have looked at Michaux.

Heer refers to its being often asserted that Pl. acerifolia & Pl. cuneata are [p. 95] found in Greece & Asia Minor. This he says may be, but if so it may be accounted for by the American Plane having, after its introduction into Europe, been spread far & wide so as to reach Asia Minor.

He considers his fossil (Pl. aceroides) to be nearest to the Pl. acerifolia & *therefore* to come nearest to the living American type or Pl. occidentalis of which he (Heer) thinks Pl. acerifolia a variety. [See Fig. 23.]

Fig. 23

a. variety
Pl. occidentalis Lin var. acerifolia

Platanus aceroides, Heer
Miocene fossil Switzerland

I have traced Heer's figures of the fossil & living. He says the living is [p. 96] more swollen at the top, but he also makes in the living *a* *b* a division *a·b* which he does not give in the fossils—& yet says nothing about its meaning which will puzzle the geologists for whose benefit I am copying the figures. Please tell me what this capsule means, I don't see it in your recent seeds.

Now in the first place your seeds of Pl. orientalis (you do not say whether it is var. aceroides or not) resemble the fossil P. aceroides both in size & form much more than they do the seeds you send of the American tree or P. occidentalis. Among the many vars. I can find some which agree with the fossil & ought [p. 97] to be figured by me as showing that if Heer gives up the attempt to distinguish the leaves he may also have to give up the seeds. Heer says the Pl. occidentalis seed is 3½ lines long, the fossil Pl. aceroides 2½. This would agree very well with the contract between the longer seeds of the Pl. occidentalis which you send & the shorter ones of the Pl. orientalis. Is there a difference between the seeds of Pl. orientalis & Pl. acerifolia? It seems to me that this question is of singular interest in reference to Heer's Atlantis theory & "creation by variation", for Pl. aceroides fossil abounds in the upper Miocene of Oeninghen in which there are so many American [p. 98] types of plants—& then it is found in the later Pliocene strata of Italy.

A leaf which Heer refers confidently to the genus [Platanus] & suspects by certain characters that it belongs to the species, Pl. aceroides, occurs with perfect leaves of Liriodendron & other Miocene types (Liriodendron is even Lower Miocene) in the *surturband* [78] of Iceland & is also fossil in Miocene of Mull.

This fossil Platanus, therefore, ranging over the old Atlantic continent, [shows] that land was most extensive in the Eocene & Lower Miocene period & sank first in the South & then [p. 99] more & more to northwards, this Miocene Plane tree, I say, left surviving varieties to the east & the west & it would be strange if they did not come to differ specifically in so many millions of years. And this may be the history of many representative species & may according to Darwin, admirably explain why those same representative species, whether they differ from each other, may each of them exactly resemble in some distinct characters their ancestor of the Miocene period.

I should like much to give figures of the typical form of [p. 100] the seed of Pl. acerifolia, var., if that came nearest to the fossil, or of Pl. orientalis L., if that approximates most nearly, & then figures of any variety in form of the Pl. orientalis L. (or of var. acerifolia) which may most closely resemble the fossil Pl. aceroides of which I have traced the outline from Heer.

Would it be too much to ask if you would set your artist to make those figures of nat. size & magnified as much as Heer's magnified figures, as given above & let me pay for his work, tho' I know that you would have to give [p. 101] more time to selecting specimens than he in drawing them. Perhaps you would like me to send back the specimens which you sent me. Some zoologists when they find fossil very large vars. of living mammalia, such as the marmot & others, give the Linnaean name & add *foss.* to imply that it is probably a variety of the olden time & so these other slight deviations occur. This is, I suppose, what you would do with all Heer's 30 or more instances of plants where the leaves are admitted to be indistinguishable from the living "homologues" but of which he anticipates that the fruits & seeds will turn out to [p. 102] be different where they are found. The reason given for christening all these leaves with new names when they occur in Miocene strata in Switzerland (& there are some 30 phanerogams in this category) is that in other cases, where the leaves were in like manner identical, the fruit, etc., when found, afforded sufficient characters to show a difference in species. Now this seems a most dangerous mode of proceeding. According to the analogy of the shells I admit that it is 4 or 5 to one against any given Miocene phaenogamous [p. 103] plant being of a recent species, but Heer at once makes *all* distinct even where the leaves are the same & nothing but the leaves known. He ought to speculate on ¼.th being of living species. But what is much stranger he makes the Woodwardia foss. different from the Woodwardia radicans tho' the fruit is same & only the leaves somewhat differently dentated!

There is no end to the embarrassment which Geologists have felt at all the plants being extinct when a 4.th or 5.th of the shells were still extant of these Miocene beds, & it is really high [p. 104] time to make the experiment & try whether the bucket of water with the fish does or does not weigh the same as the bucket without the fish.

All take Heer's determination as to the specific value of the differences as gospel, tho' he himself could not contend for the difference being greater than between the sessile & pedunculated oaks (&—Q[uercus]. robur.)

If the Platanus aceroides does not differ in its seeds, more than our var. of Pl. orientalis does from another, it seems to me that this ought to relieve us geologists of all trouble as to the supposed [p. 105] anomaly.

Nay more, it seems to me that one seed of the same individual Plat. orientalis (if all which you sent me came from one individual) differs as much from other seeds as does Heer's fossil from the average form of the living plane. If so, I ought to state it & tell Heer I mean to do so. I could so put it as not to disparage his work—or my great respect for him. Indeed geologists ought not to have reasoned from his (Heer's) determinations, without going into his data, his means for making species, & his avowed rules of philosophizing & nomenclature. That most of his work will stand I do not doubt. To make the illustration of the seeds complete there ought to be [p. 106] the two extreme varieties of Plat.^{s}. orientalis, the normal type & the one most like the fossil. The living ones which you sent seem more furrowed on the surface than Heer's fossil. How I wish you could give a month to the old tertiary flora & the Atlantis generation.

C. Lyell

[See Fig. 24.]

Fig. 24

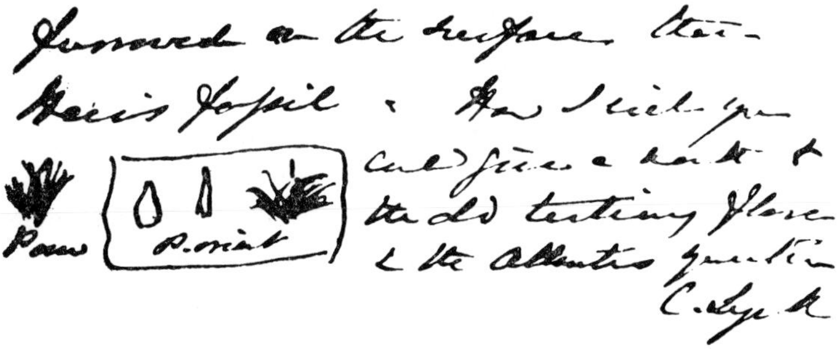

November 22, 1861

Lartet: Extract of Letter from C. Lyell to Lartet[79]

The difficulties which present themselves are almost entirely owing to your having visited Aurignac after there had been so much digging in the cave & removal of skeletons. Our antiquaries say that the instruments figured by [p. 107] you from Aurignac imply such a state of the arts as that which characterizes the age of stone in Denmark, or the Swiss lake-dwellings, & not the period of the flint hatchets of the valley of the Somme, which were certainly coeval with the mammoth & many extinct mammalia.

May not these instruments of bone have belonged to the people who used the Cave of Aurignac as a sepulchral vault *after* the epoch of Ursus spelaeus, Rhinoceros, etc.? May they not have found a cave in which there was the usual cavern-mud & breccia & bones of extinct animals? May they not have dug into the mud & bones, throwing out some of them from the interior of the cave, to make room [p. 108] for the skeletons? & may not this explain the mixture of cinders & bones of extinct mammalia on the outside of the cavern?

You have not, I think, stated what you consider the state of the bottom of the cavern to have been *before* it was used as a place of sepulture. Do you not imagine that there was mud & bones there before the introduction of the skeletons? If so, have we not two epochs, & perhaps a distinct fauna belonging to each?

The slab of rock could not have closed the whole of the opening? It surely was not large enough, if 2 metres 25 cm. high would it not have been too heavy to be removed? [p. 109] I mention this as one of the criticisms which I heard, but I do not think myself that it is one of much importance.

The following remark was made to me yesterday. "Your friends Lartet arrived too late. First the contents of the ossif.[s] cavern had been thrown into confusion over & over again by an ancient people who buried their dead there. 2.[ly] Rabbits & foxes & other burrowing animals caused a still greater mixture. 3.[d] M. Bonnemaison[80] rummaged the same loose materials. 4.th The Mayor of Aurignac & his myrmidons ransacked the already hopelessly disturbed mass & took most of it away, & then 8 years after all this, came for the first time a scientific observer."

I said in reply to this, that the [p. 110] reindeer's horns & the tusk of Ursus spelaeus had been made into utensils & must have been in a fresh state. They assert that this was the strongest part of your case, but they thought the reindeer may have continued to range to the south of France long after the mammoth was extinct. As to the tusk of Ursus spelaeus they suggested that it might have remained in a state fit for making instruments thousands of years after it had become extinct, like the fossil ivory of the mammoth in Siberia to this day.

I must say I am not satisfied with this last mentioned view

of the sceptics, unless we suppose the people [p. 111] who made the utensils were so ancient that the climate was colder then & had been so ever since the glacial epoch of the Pyrenees. I also pleaded Delesse's[81] analysis of the human & extinct bones. They said in reply that this argument was not good for much, because a bone embedded in a given matrix would go on losing animal matter up to a certain point, & then it would lose no more in that matrix, not even in tens of thousands of years. Another bone subsequently introduced into the same matrix might in 2000 or 3000 years lose the same quantity & then remain for ages in a stationary condition.

There may, I think, be some truth in this observation but it is nevertheless an hypothesis. We are not sure that there [p. 112] is really a maximum of decomposition for each kind of matrix. According to the hypothesis which I here put in their words, a difference in the quantity of nitrogen would imply a different degree of antiquity, but identity of condition would not mean equality of age.

I also cited you for the uniform manner in which the bones were broken in order to extract the marrow. They said this fact was only in favour of your theory if you could show that many long bones of extinct mammalia were so broken. There may have been some bones of a rhinoceros which the hyaenas had bitten through & broken open before the arrival of the men who used the cave as a burial place [p. 113] & had their repasts outside the cave. The hyaenas fed on a young rhinoceros & gnawed the bones. These were thrown out of the cave by the men who came to bury their dead (the dung of the hyaenas, coprolites, were also thrown out. Perhaps the people of the stone period may have taken some of the bones so cast out & cut them to see if they were in a fit state to be useful for making tools. They may also have amused themselves by splitting open a fossil mammoth's molar which had been thrown out when the graves were dug.

The criticisms which I send are all from geologists & antiquaries who believe that Man coexisted in Europe with the animals of extinct species found [p. 114] at Aurignac. They would not be surprised if you could prove that some human remains, contemporary with the hyaena, had been fossil in the cave before it became a place of sepulture. But they are not convinced by your paper that the people who buried their dead in the cave, or who left the cinders & ashes of their fires outside, were as old as the rhinoceros & cave bear. There are cases which I have seen in Belgium (Engiboul, etc.) in which the fauna of Aurignac including Man occurs fossil beneath a floor of stalagmite. Such may have been the state of things at Aurignac before the arrival of a more advanced tribe of the Stone period.

December 10, 1861

Extract of Letter from Mr. P.[82] to Sir William Hooker [p. 115]

I wish I were competent to describe Barren Island[83] & its formations geologically for they are most interesting. Of its general structure & forms, of course, I need not speak as every one knows that from Lyell, tho' he is mistaken wholly in one point for he says that the sea flows round the base of the cone *inside* the outer wall of the island. It is difficult to understand [p. 116] where his information came from, as not only it is not so, but the valley, filled with rugged & broken blocks of lava, is much elevated above the sea, I should say, generally 50 or 60 feet, and there are no appearances of any recent elevation of the island. We landed, a party of five early in the morning at the only practicable place, a breach in the summit of this submarine mountain out of which the lava bed flowed into the sea, (the steamer meanwhile steaming to & fro, as there is no [p. 117] anchorage.) It is only, then, opposite to this breach, as I call it, that you can get a sight of the inner cone as the general elevation of the island is about the same as that of the cone, namely about 800 feet. The cone is composed entirely of scoriae & loose cinders & is perfectly black & being excessively steep (about 45°) is, as you may imagine, very difficult to ascend. We ascended it however. Near the summit the loose material, of which the cone is composed, is cemented into a tolerable hard surface or crust [p. 118] from the deposits of sulphur & gypsum (I believe) which are precipitated from the vapours which are continually given out from it. There is a small crater on the top about 40 or 50 feet deep & about 100 feet in diameter. The bottom of it is quite smooth & firm & *cool,* i.e. was so at the time of our visit—it is along the *edge of this crater,* which is marked by long narrow fissures that the sulphureous vapors issue.

Speaking generally you may say that the cone is devoid of vegetation [p. 119] as there is not vegetation enough for it to be noticed at a distance or to alter in any degree its prevailing black colour—vegetation however does exist. Tufts of a species of Juncus are seen here & there, & a few miserably stunted ferns —viz. Nephrolepis hirsutula, Pteris longifolia & H. aurita & that curious little plant Psilotum. These were the only plants growing on the cone. Round the base & filling up the valley flows (or has flowed at some distant time) a stream of black lava, which has evidently emptied itself into the sea, at the breach in [p. 120] the side wall of the island before mentioned—I say a *stream,* because I think it is impossible to view it from the top of the cone without coming to the conclusion that it was once a stream flowing round the base & rushing out into the sea, but the surface of the lava is not smooth or even, but consists of a mass of loose blocks, some solid & crystalline in their texture & others (by far the greater number) porous & tufaceous, of every size & shape, thrown & heaped together in the greatest [p. 121] disorder as if (as I suppose must have been the case) thrown out from the crater on the top of the lava current as it began to cool. No vegetation grows on this bit, it is all as black as the cone & is the most painful stuff to walk over that can be imagined, from the sharp points presented to the feet, the looseness of the blocks of tuff & the horrible holes intervening.

The inner sides of the island facing the cone are also generally of the same material as the cone itself—i.e. of loose cinders & scoriae, black & steep, except [p. 122] here & there where the native rock projects & displays the stratification of the island. At the base of these slopes & encroaching here & there on the lava is tall rank grass—a low jungle of 3 or 4 low shrubs, one of which is a species of Mussaenda. I wished to have gathered specimens of everything that grew in the interior of the island, & left a man [illegible] [to do so] but he hurt his foot.

Looking to the interior of the island, it is well called Barren Island for it is truly a valley of desolation [p. 123] dark & gloomy —but as viewed from the sea it is extremely fertile, all the slopes seaward being clothed with thick vegetation tho' of what kind I had no opportunity of seeing. The opportunity of landings are very rare as there is but one spot & the water must be very still to make it practicable. I forgot to mention that the sea becomes hot as you approach this landing place till near the shore it becomes scalding hot. Some of our men, not expecting anything of the kind, jumped out of the boat as usual into the waters & began dancing [p. 124] about till they c.[d] either get in again or on shore. [See Fig. 25.]

Fig. 25

Barren island

Notes

1. In 1844, Richard Owen had identified a single fossil tooth, given him by Count Strzlecki as being from a cave in Wellington Valley, Australia, as that of a Mastodon and named it *Mastodon australis.* See Richard Owen, "Description of a Fossil Molar Tooth of a Mastodon discovered by Count Strzlecki in Australia," *Ann. & Mag. Nat. Hist.*, 1844, *14,* 268–71. The identification later proved erroneous.

2. Abraham Dee Bartlett (1813–97) was appointed Superintendent of the zoological gardens of the Zoological Society of London in 1859, a post he held until his death.

3. C. Darwin to C. Lyell, September 25, 1860. Darwin-Lyell mss.

4. Asa Gray (1810–88), botanist, was professor of natural history at Harvard University from 1842 until his death.

5. Heinrich Robert Göppert 1800–84, *Beobachtungen über das sogenannte Ueberwallen de Tannenstöcke . . . ,* Bonn, 1842, 26 pp.

6. Christian Erich Hermann von Meyer (1801–69), palaeontologist, of Frankfort-on-Main, Germany.

7. We have not been able to identify Albert Toilliez of Mons further.

8. The Maestrichien was a formation described by André Hubert Dumont (1809–57) in his geological map of Belgium. See André H. Dumont, *Carte Géologique de la Belgique indiquant les terrains qui se trouvent au-dessous du Limon Hesbayen et du Sable Campinien,* Brussels, 1855.

9. This memoir was not published until 1865. See Albert Toilliez, "Sur quelques faits géologiques pris pour le résultat du travail de l'homme," *Mem. Soc. Sci. Hainaut,* 1865, *10,* 9–28.

10. Charles Lyell, *Principles of Geology,* 9th ed., 3 vols., London, Murray, 1853, p. 700.

> However slowly a lake may be converted into a marsh, or a marsh into a meadow, it is evident that before the lacustrine plants can acquire the power of living in marshes, or the marsh-plants of

living in a less humid soil, other species, already existing in the region, and fitted for these several stations, will intrude and keep possession of the ground. So, if a tract of salt water becomes fresh by passing through every intermediate degree of brackishness, still the marine molluscs will never be permitted to be gradually metamorphosed into fluviatile species; because long before any such transformation can take place by slow and insensible degrees, other tribes, already formed to delight in brackish or fresh water, will avail themselves of the change in the fluid, and will, each in their turn, monopolize the space.

11. Robert Brown (1773–1858), botanist, was curator of the herbarium at the British Museum until his death.

12. Hugh Falconer (1808–1865), palaeontologist and botanist.

13. Thomas Henry Huxley (1825–95), zoologist and educator, was at this time professor of natural history at the Government School of Mines, Jermyn Street, London.

14. Joshua Trimmer (1795–1857), geologist.

15. Giuseppe Ponzi (1805–85), Italian geologist.

16. Winend Carel Hugo Staring (?–1877), Dutch geologist.

17. Heinrich Ernst Beyrich (1815–96), German geologist and palaeontologist.

18. Ernst Heinrich Carl von Dechen (1810–1869), German geologist, was a mining official at Bonn.

19. Ours Pierre Armand Petit Dufrénoy (1792–1857).

20. Pierre Henri Nyst (1813–80), Belgian conchologist.

21. Winend C. H. Staring, *De Bodem van Nederland. De zamenstellung en het ontstaan der gronden in Nederland ten behoeve van het algemeen beschreven,* 2 vols. Haarlem, 1856–60.

22. Lyell is here referring to a controversy that occurred in the pages of the *Athenaeum* in the late summer 1860. In the *Athenaeum* for August 25, 1860, Col. Sir Henry James, director general of the Ordnance Survey of Great Britain, published a letter (*2,* 256–57) in which he suggested that displacements in the earth's mass resulting from the elevation of mountain ranges may in the past have caused shifts in the earth's axis of rotation thereby producing "great and extraordinary changes in the climates of its different regions." His reasoning was based on the fact that the climate of northern Europe and other regions in high latitudes was

known to have been much warmer during the geological past, and on Newton's *Principia,* sec. 4, theor. 26, cor. 22, in which Newton shows that the addition of new matter to a rotating globe anywhere between its pole and equator will cause the pole to wander on its surface.

On September 8, Joseph Beete Jukes of the Irish branch of the Geological Survey published a letter in reply to James (*2,* 322–23), in which he pointed out that the mass of the largest mountain ranges and tablelands was extremely minute when compared with that of the equatorial bulge of the earth. He also questioned whether changes of climate had occurred in tropical regions, as would be required by Sir Henry's hypothesis.

Sir Henry James replied to Jukes on September 15 (*2,* 355–56), citing geological evidence in support of his view. However, in the September 22 issue, his conclusions were criticized in separate letters by Sir George Airy, Astronomer Royal, Mr. W. E. Hickson and Professor Henry Hennessy (*2,* 384–86). In the September 29 issue Sir Henry James replied again to his critics, and Professor Hennessy published a second letter to say that Sir Henry had misunderstood Newton (*2,* 415–16). On October 6, Jukes published a letter criticizing the catastrophic implications of Sir Henry's view (*2,* 451). James replied yet again on October 13 (*2,* 483) and on October 20, W. E. Hickson and Henry Hennessy published additional letters (*2,* 516–17). Hennessy pointed out that Newton was referring to a perfect globe or sphere not a spheroid like the earth. Lyell was frequently cited in the course of this controversy, but took no part himself.

23. J. S. Henslow, "Flints in the Drift," *Athenaeum,* 1860, *2,* 516. The Rev. John Stevens Henslow (1796–1861) was the English clergyman and botanist who exerted a guiding influence on the young Charles Darwin when the latter was a student at Cambridge.

24. "I found a comparison of the Coptic language with such roots and forms of the Old Egyptian as were then discovered, sufficient to remove from my mind all doubt as to the Asiatic origin of the Egyptians and their affinity with the Semitic or Aramaic stock." Christian K. J. Bunsen, *Egypt's Place in Universal History: an historical investigation in five books,* Charles H. Cottrell, trans., 4 vols., London, Longman, Brown, Green and Longmans, 1848–60, *1,* viii.

Baron Christian Karl Josias von Bunsen (1791–1860) was a Prussian diplomat, lay theologian, and scholar who from 1842 to 1854 served as Prussian ambassador to England. During this period, Lyell made his acquaintance.

25. Ibid., pp. ix–x:

> There my principal object was to find a universal formula for the relation which a colonial language (like the Icelandic) bears, on the one side, to the old tongue of the mother-country, and on the other to the modern idioms which there may have entirely superseded it . . . Now the Icelandic appeared to me to possess immense importance for the solution of the general problem, as being identical with the Old Norse, and as forming the point of departure for the Swedish and Danish, which in Scandinavia have succeeded that old idiom.

26. "The inference consequently was, that this period of 33 centuries between Menes and Alexander, in fixing which historical records and unimpeachable astronomical calculations combined, was neither indefinite nor undefinable, nor devoid of historical interest" (ibid., *4*, 14).

27. Ibid., p. 18:

> But as chronology commences with Egypt, as well as the possibility of analyzing the second epoch above referred to, or the modern history of the world, so the contemporary records of the Egyptian language go back to the fourth or fifth millennium B.C., an advantage possessed by no other nation for fixing the chronology not only of their own language, but of all the languages of the world.

28. Ibid., p. 21:

> But if we find almost 4,000 years before our era, a mighty empire, possessing organic members of very ancient type, a peculiar written character, and national art and science, we must admit that it required thousands of years to bring them to maturity in the retired valley of the Nile.

29. Bunsen wrote that the Arians in India had a tradition of the flood: "Neither the recollection of the great catastrophe [the flood] in the primeval country, nor that of the historical migration of their Arian fathers from their northern home, has been lost" (ibid., p. 431).

30. Bunsen dated the flood at 10,000 B.C. (ibid., p. 480).

31. James Smith (1782–1867) of Jordanhill, geologist and historian.

32. David Landsborough, "Description of a Newer Pliocene Deposit at Stevenston and Largs, in the County of Ayr," *Trans. Geol. Soc. London,* 1838–42, *3,* 444–45.

The Rev. David Landsborough (1779–1854), theologian and naturalist, was educated at the University of Edinburgh and, in 1811, was ordained a minister in the Church of Scotland. He especially studied cryptogamic botany.

33. Thomas F. Jamieson, "On the Drift and Rolled Gravel of the North of Scotland," *Quart. J. Geol. Soc.,* 1860, *16,* 347–71.

34. Robert Bald, "Notices regarding the Fossil Elephant of Scotland," *Mem. Wernerian Soc.,* 1821, *4,* 58–66.

35. John Scouler (1804–71), naturalist, studied medicine at the University of Glasgow and then went to Paris to study at the Jardin des Plantes. In 1824, he went as surgeon on the Hudson Bay Company's *William and Mary* to the Columbia River on the Pacific coast of North America, and, in 1829, he was appointed professor of geology, natural history, and mineralogy at Andersonian University, now part of the Glasgow and University of Scotland Technical College.

36. James Smith, "On the Occurrence of Marine Shells in the Stratified Beds below the Till. With a notice of the Occurrence of Marine Shells in the Till by J. C. Moore," *Quart. J. Geol. Soc.,* 1850, *6,* 386–89. Cf. Thomas F. Jamieson, "On the Drift and Rolled Gravel of the North of Scotland," 1860, p. 370.

37. On November 21, 1860, three days before Lyell wrote this letter, David Forbes had read a paper at a meeting of the Geological Society of London in which he said in part:

> That no very perceptible elevation has taken place in the immediate neighbourhood of the Morro of Arica . . . during the last 350 years, or since the Spanish conquest, appears from the numerous Indian tumuli found along the beach, for miles south of the Morro; many of these are not 20 feet, and some probably considerably less, above the present sea-level. That these tumuli have not been constructed since the Spanish invasion may be inferred from the ornaments of gold found in them, along with the mummies, one of which I was informed had been found by Mr. Evans, the Engineer of the Arica and Tacna Railroad, enveloped in a thin sheet of gold.

David Forbes, "On the Geology of Bolivia and Southern Peru," *Quart. J. Geol. Soc.,* 1861, *17,* 7–62, pp. 10–11.

38. Captain Thomas Spratt (1811–88), "Extract of a Letter from Captain Spratt R. N. on Crete," *J. Roy. Geogr. Soc.*, 1854, *24*, 238–39:

> I made an interesting discovery in the western part of the island [Crete], viz., that it had been subject to a series of elevations, amounting to the maximum of 24 feet 6 inches, which occurs near Pockilassos and Suia. In the middle of the island, at Messara, the Fair Havens, and Megalo Kastro, there is none. The eastern end of the island has dipped a little.

In two further letters, written directly to Lyell, Spratt had given additional information about the elevation and submergence of Crete. See Captain Thomas Spratt to Sir Charles Lyell. February 28, 1856 and February 29, 1856. Lyell mss.

39. Thomas F. Jamieson, "On the Drift and Rolled Gravel of the North of Scotland," 1860, pp. 368–69.

40. Darwin replied to this letter on December 4, 1860, Darwin-Lyell mss. His reply was printed, with some deletions, in Charles Darwin, *More Letters of Charles Darwin,* Francis Darwin, ed., 2 vols., New York, Appleton, 1903, *2,* 140–41.

41. The Rev. Samuel William King (1821–68), traveler and naturalist, was rector of Saxlingham, Nethergate, Norfolk. Educated at St. Catherine's College, Cambridge, he was an enthusiastic entomologist and geologist and helped Lyell with many investigations.

42. The Rev. Samuel W. King to Charles Lyell, October 3, 1860, Lyell mss.

43. The Rev. Samuel W. King to Charles Lyell, July 3, 1860, Lyell mss.

44. The letters from King to Lyell of June 9 and November 24 and 25, 1859, do not seem to have survived.

45. The Rev. Samuel W. King to Charles Lyell, January 11, 1860, Lyell mss.

46. The Rev. Samuel W. King to Charles Lyell, December 31, 1859, Lyell mss.

47. The Rev. Samuel W. King to Charles Lyell, November 16, 1860, Lyell mss.

48. The Rev. Samuel W. King to Charles Lyell, December 13, 1860, Lyell mss.

49. Charles Lyell, *A Manual of Elementary Geology,* 5th ed., London, Murray, 1855, p. 133.

50. Jean Joseph Leverrier (1811–77), "Sur le calcul des inégalités séculaires tel qu'il a été donné par M. de Pontécoulant dans le 3e volume du 'Système analytique du Monde' p. 387 à 401," *Comptes Rendus Acad. Sci.,* 1839, *9,* 550–52.

51. George Charles Wallich (1815–99), *Notes on the presence of Animal Life at vast depths in the sea,* London, 1860, 38 pp.

52. George E. Roberts, "High and low life," *Geol.,* 1861, *4,* 1–6.

53. J. W. Salter, "A Christmas lecture on 'Coal,'" *Geol.,* 1861, *4,* 6–13, 59–68, 100–02, 121–31, 177–83. Lyell is referring to the first installment of this lecture only.

54. S. J. Mackie, "The evidences of the geological age and human manufacture of the fossil flint implements (continued from vol. III, p. 404)," *Geol.,* 1861, *4,* 19–31.

55. Ebenezer Emmons (1799–1863), American geologist, had asserted the existence of a series of sedimentary strata lying beneath the Potsdam sandstone, the oldest and lowest formation of the New York series. To this supposed series Emmons gave the name the Taconic system. Other American geologists believed that the Taconic rocks constituted not a series, but a single formation much metamorphosed and disturbed. In 1856, Elkanah Billings, palaeontologist with the Geological Survey of Canada, found in a limestone at Point Levis, Quebec, trilobite fossils sufficiently different from those of the Trenton (New York) series to convince him that Emmons was right in postulating the existence of an older and lower series. In 1860, Joachim Barrande also adopted Emmons' conclusions in a paper that Lyell presumably had just read. See Joachim Barrande, "Documents, anciens et nouveaux, sur la Faune Primordiale et le Système Taconique en Amérique," *Bull. Soc. Geol. France,* 1860–61, *18,* 203–322.

56. Jules Marcou, "Additional notes to Joachim Barrande's paper on the primordial fauna and the taconic system," *Proc. Nat. Hist. Soc. Boston,* 1859–61, 7, 369–82.

57. Elkanah Billings (1820–76), palaeontologist with the Geological Survey of Canada.

58. See Elkanah Billings, "New species of Lower Silurian Fossils," *Canad. Geol. Surv. Repts.,* 1861–62.

59. Frank H. Bradley, "Description of a new Trilobite from the Potsdam Sandstone," *Canad. Natur. & Geol.* 1860, *5*, 420–25.

Frank H. Bradley (1838–79), American geologist, was at this time still a student at Yale College. He had collected the fossils he described in the above paper at High Bridge near Keesville, New York.

60. David Dale Owen (1807–60) was born in Scotland, but in 1827 came to America where his father had established a socialist community at New Harmony, Indiana. After studying medicine in the Medical College of Ohio at Cincinnati, he began, in 1836, to assist Gerard Troost with the geological survey of Tennessee. He later directed geological surveys of the territories that became the states of Iowa, Minnesota, Wisconsin, and Arkansas for the U.S. General Land Office.

61. George Busk F.R.S. (1807–86) surgeon and naturalist; Constantin Malaise, "Note sur quelques ossements humains fossiles et sur quelques silex taillés," *Bull. Acad. Sci. Bruxelles,* 1860, *10,* 538–46.

62. This was probably Henry William Bristow (1817–89), geologist on the staff of the Geological Survey of Great Britain.

63. George Robert Waterhouse (1810–88), naturalist, was keeper of mineralogy in the department of natural history, British Museum. Thomas Hawkins (1810–89), geologist, was a private collector of fossils.

64. Johann Friedrich Blumenbach (1752–1840), German zoologist and anthropologist, was professor of medicine at Göttingen from 1776.

65. Georges Cuvier (1769–1832), comparative anatomist and palaeontologist.

66. Isidore Geoffroy Saint-Hilaire (1805–61).

67. Wilhelm Heinrich Theodor von Plieninger (1795–1879), German palaeontologist.

68. Thomas Oldham (1816–78), geologist, a native of Dublin, was educated at Trinity College, Dublin, and the University of Edinburgh. In 1845, he was appointed professor of geology at Trinity College, Dublin, and, in 1850, superintendent of the Geological Survey in India. In 1859, as director of the survey, he began to issue parts of *Palaeontologica Indica, being figures and descrip-*

tions of the organic remains procured during the progress of the Survey. Lyell is referring to these published plates of fossils.

69. Jones showed that the little Triassic shell previously called Posidonia and Posidonomya minuta was not a mollusc, but a bivalved phyllopodous crustacean of the genus Estheria. See T. Rupert Jones, "Note on Estheria minuta," *Ann. & Mag. Nat. Hist.*, 1857, ser. 2, *19,* 104–06.

70. Pierre Louis Gratiolet (1815–65), French anatomist.

71. Thomas F. Jamieson (1829–1913) of Ellon, Aberdeenshire, geologist and farmer, was factor of the Ellon Castle Estate and tenant of the Mains of Waterton. He proved that the Parallel Roads of Glen Roy had formed the shores of a freshwater lake.

72. David Milne, "On the Parallel Roads of Lochaber, with remarks on the change of relative levels of sea and land in Scotland," *Edinburgh New Phil. J.*, 1847, *43,* 339–64. Cf. David Milne, "On the Parallel Roads of Lochaber, with remarks on the change of relative levels of sea and land in Scotland, and on the detrital deposits in that country," *Trans. Roy. Soc. Edinburgh,* 1849, *16,* 395–418.

73. Robert Chambers, "On the glacial phenomena in Scotland and some parts of England," *Edinburgh New Phil. J.*, 1853, *44,* 229–82, pp. 253–54.

74. Sir Thomas Lauder Dick, "On the Parallel Roads of Lochaber," *Trans. Roy. Soc. Edinburgh,* 1823, *9,* 1–64, p. 44:

> The numerous torrents that pour down the sides of the mountains of Loch Treig, have very much defaced the course of Shelf 4th around that lake. . . . It is to be observed at the southern extremity, and appears very visible on the western side, where it leads back and winds again into Glen Spean.

Sir Thomas Dick Lauder (1784–1848) of Fountainhall, Haddingtonshire, scientist and novelist.

75. Auguste Bravais [1811–63], "Sur les lignes d'ancien niveau de la mer dans le Finmark," Commission Scientifique du Nord de France, *Voyages . . . en Scandinavie, en Laponie, au Spitzberg et aux Faroe, pendant . . . 1838–1840 Sur la Recherche . . . ,* 16 vols. & 5 vols., atlas, Paris, 1842–55, vol. 1, part 1.

76. In both these instances, the problem was how boulders might be carried across a broad valley from the top of one range of

mountains to the ridges of another lower range. The Grampians are separated from the Sidlaw Hills by the broad shallow depression of the Vale of Strathmore and the Alps from the Jura by the deep valley in which lies Lake Geneva. The presence of boulders of Alpine granite on the slopes of the Jura had been a problem for geologists since their presence there had been described by Horace-Benedict de Saussure in his *Voyages dans les Alpes,* 1779–96.

77. Oswald Robert Heer (1809–83), palaeobotanist, was professor of natural history at the University of Zurich.

78. The surturband were beds of lignite of Miocene age in Iceland described by Oswald Heer.

79. Edouard Amand Isodore Hippolyte Lartet (1801–71) French geologist.

80. We have not been able to identify M. Bonnemaison.

81. Achille Ernest Oscar Joseph Delesse (1817–81), French geologist and mineralogist.

82. We have not been able to identify Mr. P.

83. Barren Island is a volcanic island in the Bay of Bengal, described by Lyell in the *Principles of Geology,* 11th ed., 2 vols., London, Murray, 1872, *2,* 74–75.

Index